Advanced Synthetic Materials in Detection Science

RSC Detection Science Series

Editor-in-Chief:
Professor Michael Thompson, *University of Toronto, Canada*

Series Editors:
Dr Subrayal Reddy, *University of Surrey, Guildford, UK*
Dr Damien Arrigan, *Curtin University, Perth, Australia*

Titles in the Series:
1: Sensor Technology in Neuroscience
2: Detection Challenges in Clinical Diagnostics
3: Advanced Synthetic Materials in Detection Science

How to obtain future titles on publication:
A standing order plan is available for this series. A standing order will bring delivery of each new volume immediately on publication.

For further information please contact:
Book Sales Department, Royal Society of Chemistry, Thomas Graham House, Science Park, Milton Road, Cambridge, CB4 0WF, UK
Telephone: +44 (0)1223 420066, Fax: +44 (0)1223 420247
Email: booksales@rsc.org

Visit our website at www.rsc.org/books

Advanced Synthetic Materials in Detection Science

Edited by

Subrayal Reddy
University of Surrey, Guildford, UK
Email: S.Reddy@surrey.ac.uk

RSC Detection Science Series No. 3

Print ISBN: 978-1-84973-593-3
PDF eISBN: 978-1-84973-707-4
ISSN: 2052-3068

A catalogue record for this book is available from the British Library

© The Royal Society of Chemistry 2014

Published by The Royal Society of Chemistry,
Thomas Graham House, Science Park, Milton Road,
Cambridge CB4 0WF, UK

Registered Charity Number 207890

For further information see our web site at www.rsc.org

Printed and bound by CPI Group (UK) Ltd, Croydon, CR04YY

Preface

Rapid advances are being made in the development of smart materials for detection science, from simple dry-strip colour-changing indicators to electronic biosensors and, more recently, using nanoparticle-based systems. Any reliable device needs to respond selectively to a target molecule or compound. Biology has helped in a big way here with enzymes and antibodies which show unrivalled specificity for molecular recognition. Bio-inspired materials are gaining popularity and capturing the imagination of scientific and lay audience alike, in the quest for synthetic materials to mimic biological function. The latter biomimics offer the promise of inexpensive and more stable alternatives to their biological counterparts. In some cases a fusion of materials chemistry and biochemistry is necessary to realise new and advanced functional materials. Consider, for example, that for proper function of antibodies or DNA in a biosensor detection strategy, molecular orientation is important and can be controlled with self-assembling chemistries.

The bedrock of most sensor strategies is getting the target molecule of interest to bind with a suitable complement, resulting in a measurable physical or chemical change. Of course, biological systems have had millennia to develop such strategies. Whereas materials chemistry is playing catch-up with biology, there have been significant strides in the past 20–30 years in developing advanced synthetic systems, such as complexing agents for the selective binding of small organic molecules and ions. Research in molecularly imprinted polymers has evolved from the development of materials for selective extraction of low molecular weight organics to the selective binding of large biomolecular systems such as proteins, DNA and viruses. The latter is heralding the era of functional plastic antibodies.

Nanomaterials have seen a surge in interest in the past 15–20 years, with research in carbon nanotubes, quantum dot semiconducting materials and

RSC Detection Science Series No. 3
Advanced Synthetic Materials in Detection Science
Edited by Subrayal Reddy
© The Royal Society of Chemistry 2014
Published by the Royal Society of Chemistry, www.rsc.org

gold nanoparticles for developing detection strategies. Their application to biosensing has met with variable successes and it is important to address the use of such materials in detection strategies and their fitness for purpose.

Where target molecule specificity is not achievable, or desirable for that matter, complex biological systems have developed natural intelligent algorithms to identify mixtures of compounds all in one sample, whether it be a liquid or gas. The ability of the human nose or tongue to discriminate between a variety of odours and flavours relies on receptors which have broad band selectivity. There have been significant efforts to mimic these systems in the laboratory with the development of electronic nose and electronic tongues.

The one overwhelming drawback in any strategy when interfacing the synthetic material world with the biological world is the lack of appreciation or, dare I say it, respect for the compatibility between the two. The issue of biocompatibility is important to address as there are many stumbling blocks *en route* to developing a reliable biosensor or detector, the most important being an understanding of the surface chemistry of the new material and being aware of the chemical composition and matrix effects therein of the interfacing biological sample. There are notable successes in improving the biocompatibility at the bio/sensor interface through rational design of polymeric materials that can withstand the harsh conditions presented by, for example, blood and thereby controlling bio-fouling, but at the same time the material must not elicit an adverse biological immune response. There is a lot we can learn from biology in this regard.

This book gives an introduction to bio-inspired materials in medicine in Chapter 1. Chapter 2 takes a more comprehensive and critical look at how biomimetic materials have been applied in various biosensor strategies. While the latter are important considerations for long-term exposure of a synthetic material to a bio-environment, there has been extensive research in developing extremely reliable 'single-use and dispose' devices. Chapter 3 focuses on the recent advances in molecularly imprinted polymers (MIPs) for imprinting of biological molecules (namely, proteins, viruses and DNA). Improvements in preparation of such antibody mimics could eventually see MIPs replacing biological antibodies in bioassays. We also review some established and new chemistries for the development of smart detectors finding applications in medicine, food and the environment. Chapter 4 gives a review of nanoparticle technologies that have been used in biosensor development and highlights some of the challenges of going down to the nanoscale. Chapter 5 explores some of the history of dry-strip colour-changing indicators and some of the more recent developments in smart indicators capable of responding to changes in their environment, with apparent applications in food packaging and biowarfare agent detection. Chapter 6 gives some of the physical chemistry and characterization techniques used to develop novel macrocyclic ligands based on calixpyrrole research for the selective binding of a range of ions. Chapter 7 gives a tutorial in pattern recognition and use of multivariate techniques and reviews the significant developments in electronic tongue research.

We have tried to capture a thread in this book that will hopefully weave a fabric of understanding and vision for potential new developments in advanced synthetic smart materials, endeavouring to bring the synthetic and biological worlds closer together to solve analysis and detection problems. Finally, I would like to thank the Royal Society of Chemistry publishing team for their support and chivvying along which has helped see this book come to completion.

Dr Subrayal Medapati Reddy
University of Surrey
Guildford, Surrey, UK

With deepest appreciation to Katharine, Jayesh, Anya, Arun and Leo for their friendship, patience and support.

Contents

RSC Detection Science Series No. 3
Advanced Synthetic Materials in Detection Science
Edited by Subrayal Reddy
© The Royal Society of Chemistry 2014
Published by the Royal Society of Chemistry, www.rsc.org

Biomimicry and Materials in Medicine

LARISA-EMILIA CHERAN,*[a] ALIN CHERAN[b] AND MICHAEL THOMPSON[a]

[a] Department of Chemistry, University of Toronto, 80 St. George Street, Toronto, Ontario M5S 3H6, Canada; [b] Ross University School of Medicine, 630 US Highway 1, North Brunswick, NJ 08902, USA
*Email: mikethom@chem.utoronto.ca

1.1　Biomimicry

Past generations have attempted to tame, harness and conquer nature, and bring her to her knees. Such efforts have transformed over the years into a wiser, enthusiastic pursuit to unravel her secrets and copy the "engineering" afforded by nature. Implicit in this is the recognition that literally billions of years of evolution have resulted in the refining of biological structures at the macro- and nano-scales.

A living cell that grows and multiplies looks much like a robot that has learned to build itself. We are still not sure exactly how this machine was first assembled and programmed. In comparison, strident announcements with respect to scientists "creating life" in the lab are only lame self-promoting marketing strategies. The reality is that there is a failure to acknowledge the fact that the inception of such efforts were initiated with existing living cells before any major manipulations were made inside the cells. There is no clear definition of life; we even struggle to ascertain the difference between a living cell and a dead one. All the chemicals and physical properties are essentially identical, with the most remarkable

RSC Detection Science Series No. 3
Advanced Synthetic Materials in Detection Science
Edited by Subrayal Reddy
© The Royal Society of Chemistry 2014
Published by the Royal Society of Chemistry, www.rsc.org

feature being the ineffable force that controls those 100 000 chemical reactions in each second, in each living cell. For that matter, we are required to recognize that we do not even have a real definition of reality, only hypotheses which contradict the Copenhagen interpretation because of the need of an observer, explicitly the multiverse theory,[1] Bohm's implicate order,[2] Everett's parallel universes,[3] and orchestrated quantum coherence.[4] Troublesome as it is, we do not have a definition of time either, since the vibrations of the cesium atoms and quartz crystals only measure something that does not exist when we dream, for example; something that expands exasperatedly when we are bored and contracts when we are in the flow.

Leaving such big questions under the rug of modern science, where they belong, we turn to the somewhat more uncomplicated quest of mimicking nature by creating synthetic materials capable of both detecting biochemical and biophysical changes at the molecular level that reflect sensory mechanisms. Additionally, synthetic materials can replace natural biological components for medical corrective procedures and/or healing. Such activities are often characterized as being incorporated in the relatively new field of "biomimicry", a word coined by Schmitt[5] some years ago.

1.2 Advanced Synthetic Materials

New materials which have features that can be modified under specific conditions are pushing forward the frontiers of scientific and technological capabilities. Synthetic materials designed to change their size or shape when exposed to heat, change from a liquid to a solid in magnetic fields, or dramatically change their volume, viscosity, conductivity, work function, *etc.*, will allow new and interesting applications to emerge, not only for sensing and detection purposes, but also for everyday use. Smart materials with piezoelectric, magneto-rheostatic, electro-rheostatic, and memory hysteresis are already used in cars, coffee pots, glasses, or in space missions. Relevant to this chapter, intelligent biomaterials that respond to biological signals show great promise in regenerative medicine, diagnostics, and drug delivery.

Common examples include piezoelectric materials which produce a charge under mechanical stress and, conversely, contact or expand when a voltage is applied.[6] In shape-memory alloys and polymers, large deformations can be induced and recovered by temperature changes or stress changes. This pseudoelasticity is the result of martensitic phase changes.[7] In magnetostrictive materials, changes in shape can be instigated by magnetic fields; on the other hand, these materials also exhibit changes in magnetization under mechanical stress.[8] Magnetic-shape memory alloys change their shape when under a magnetic field, and pH-sensitive polymers change in volume when exposed to alterations in the pH of the solution in which they are immersed. Halochromic materials change their color as a result of change in acidity; they are very suitable detection materials for applications such as corrosion detection.[9] Chromogenic materials change color as a response to electrical, thermal, or optical changes.[10] When an electric voltage is applied, electrochromic materials change their color or opacity, as in liquid crystal displays.[10] Thermochromic materials change in color depending on temperature changes and photochromic materials respond to light changes (used in light-sensitive sunglasses which darken when exposed to bright sunlight).[10] Photomechanical materials change shape when exposed to light. Dielectric elastomers produce large strains under external electrical fields.[11] Magnetocaloric materials change reversibly in temperature under a changing magnetic field.[12] Thermoelectric materials convert temperature differences into electricity.[13] Such active materials are ideal for sensor devices and for such applications as resorbable bioceramics, adaptive bioglasses, biomimetic polymers and gels, active nanoparticles, smart textiles, and active optical fibers. Manipulation of material properties at the atomic and molecular scale is leading to self-assembling materials, nanolithography, DNA-based technologies (such as DNA computing), nano- and micro-engineered devices for diagnostics, pharmaceuticals, therapies, drug delivery systems, biocompatible implants and prostheses, and bio-functional systems.

We now turn to the employment of new materials in a plethora of applications in biology and medicine.

1.2.1   Nitinol and Memory Metals

A shape memory transformation was first observed in 1932 in an alloy of gold and cadmium, and then later in brass in 1938.[14] The shape memory effect was seen in the gold–cadmium alloy in 1951, but this was of little use. Some 10 years later, in 1962, an equiatomic alloy of titanium and nickel was found to exhibit a significant shape memory effect and Nitinol (so named because it is made from *ni*ckel and *ti*tanium and its properties were discovered at the *N*aval *O*rdinance *L*aboratories) has become the most common shape memory alloy (SMA).[14] Other SMAs include those based on copper (in particular CuZnAl), NiAl, and FeMnSi, although it should be noted that the NiW alloy has by far the most superior properties.[14]

Metal alloys of nickel and titanium have the remarkable properties of remembering the shape by undergoing a phase change in which atoms are shifting their position in response to a specific stimulus, such as temperature or stress. The phase change temperature can be tuned by varying the ratio of nickel to titanium. The resulting structure has a highly symmetrical cube structure called austenite at temperatures above the phase change temperature and a much less symmetrical structure at temperatures below the phase change temperature, called the martensite state. In the latter state the material is very elastic, but the austenite state is rigid. The modification of the crystal lattice during the transformation can be carefully controlled. The shape change may exhibit itself as either an expansion or contraction. The transformation temperature can be tuned to within a couple of degrees by changing the alloy composition. Nitinol can be made with a transformation temperature anywhere between $-100\ ^\circ\text{C}$ and $+100\ ^\circ\text{C}$, which makes it very versatile.

The memory mechanism is based on the thermal energy acquired by the sample through heating, providing the energy necessary for the atoms to return to their original positions so the sample regains its original shape. Springs made of such alloys return to their original shape in warm water or in streams of hot air. Nitinol and superelastic materials made of shape memory alloys of gold–cadmium, copper–aluminum–nickel, copper–zinc–aluminum, and iron–manganese–silicon are used today in medical applications, aerospace, and the leisure industry, namely in vascular stents, for anchors attaching tendons to bones, medical guidewires, medical guidepins, root canal fillings, bendable surgical tools, cardiac catheters, orthodontic wires, flexible eyeglass frames, *etc.*[15]

Still many years away is the use of SMAs as artificial muscles, *i.e.* simulating the expansion and contraction of human muscles. This process will utilize a piece of SMA wire in place of a muscle on the finger of a robotic hand. When it is heated, by passing an electrical current through it, the material expands and extends the joint; on cooling, the wire contracts again, flexing the finger. In reality this is incredibly difficult to achieve since complex software and surrounding systems are also required. NASA has been researching the use of SMA muscles in robots which walk, fly, and swim.

SMA tubes can be used as couplings for connecting two tubes. The coupling diameter is made slightly smaller than the tubes it is to join. The coupling is deformed such that it slips over the tube ends and the temperature changed to activate the memory. The coupling tube shrinks to hold the two ends together but can never fully transform so it exerts a constant force on the joined tubes.

In addition to the shape memory effect, SMAs are also known to be very flexible or superelastic, which arises from the structure of the martensite. This property of SMAs has also been exploited in mobile phone aerials, spectacle frames, and underwires in bras. The kink resistance of the wires makes them useful in percutaneous angioplasty, a surgical procedure requiring catheters which need to remain straight as they are passed through the body. Nitinol can be bent significantly further than stainless steel without suffering permanent deformation.[16]

1.2.2 Smart Ceramics

Ceramics are more chemically stable and inert than metals. They can be classified into three distinct categories:

1. Oxides such alumina, beryllia, zirconia, and ceria
2. Non-oxides such as carbide, boride, nitride, and silicide
3. Composite materials such as particular reinforced fibers or combinations of oxides and non-oxides

Ceramics are used in numerous technical applications. Since ceramics are so much more biocompatible when used inside the human body, they play a major role as synthetic biomaterials that can be part of the systems which treat or replace a living tissue or lost function.[17] Depending on the particular application, the major requirements beside biocompatibility are resistance to abrasion and wear, fatigue, strength, durability, and resistance to corrosion, especially when they are used as an implant material. The most common materials are alumina, zirconia, bioglass, hydroxyapatite, and tricalcium phosphate. Not only they are inert, but they are also resorbable. They can dissolve and integrate actively in physiological processes such as bone healing. Hydroxyapatite can be actually found in the human body, in teeth and bones. The synthetic material is commonly used as a filler to replace amputated bone or as a coating to promote bone ingrowth into prosthetic implants. Natural coral skeletons can be transformed into hydroxyapatite at high temperatures. Their porous structure encourages the rapid ingrowths of bones. The high-temperature treatment destroys any organic molecules such as proteins, thus preventing immune rejection.

Zirconia is used on the artificial femoral heads employed in hip replacements.[17] It gives strength to the structure, so its dimensions can be smaller and less traumatic for the patient. It is also used in shoulder, knee, spinal implants, and phalangeal joints.[17] In dentistry, it is used with increased frequency for crowns, bridges, and implant abutments. Crystal zirconia is a modern dental ceramic replacement for the metal substructures used under porcelain crowns and bridges. It is translucent, thus giving the overlaid porcelain a brighter and more natural look. It is biocompatible and, unlike amalgams and metal alloys, does not generate adverse reactions or allergies. It is virtually unbreakable, so the dental work can last for a lifetime.[18]

Finally, a particularly interesting application of ceramic materials is their use in the treatment of cancer, through hyperthermia and radiotherapy.[19] In the quest to avoid the devastating effects of chemotherapy, glass microspheres are inserted into the tumor using a catheter and the radiation is focused on the tumor, similar to brachytherapy for prostate cancer. This causes minimal damage to the surrounding tissue. It is a simple treatment that can be performed on an outpatient basis.

1.2.3 Smart Polymers

Smart polymers are used for industrial purposes, in medicine, sports, and agriculture due to their inert or bioactive properties. Biodegradability is also a great advantage of such polymers. High-performance polyethylene is used in medicine for total or partial joint replacement of hip, knee, or intervertebral implants due to its high impact strength given by extremely long chains.[20] It is highly resistant to corrosion, has very low moisture absorption and a very low friction coefficient. It is self-lubricating and highly resistant to abrasion, more resistant than carbon steel.

Hydrogels, networks of hydrophilic polymer chains in colloidal gels, with water as the dispersion medium, are highly absorbent materials; they possess a high degree of flexibility, very similar to natural tissue, due to their significant water content.[21] Smart gels contain fluids, usually water in a matrix of large, complex polymers. These polymers are unique in that they respond to stimuli in an advanced way. Types of stimuli that affect smart gels are physical and chemical factors. Temperature, light, electric forces, magnetic forces, and mechanical forces are types of physical interactions on the gel that will precipitate a reaction. Chemical stimuli are usually pH changes or solvent exchanges. The reaction of the smart gel is always an expansion or contraction within milliseconds upon stimulation. When a gel swells, it absorbs additional fluid into it. When it deflates, it expels this fluid out of its membrane. The expansion and contraction are usually caused by a change in the polymer; the stimulus alters the polymer by making it more or less hydrophilic. For example, a significant pH decrease will neutralize ions in the gel, precipitating the polymers to be less hydrophilic and causing the gel to contract.

The effects of such synthetic gels are greatly aided by using nano-particles.[22] While microparticles usually allow the gel to function properly, smaller particles at the nanoscale increase intended effects dramatically. A great example of this is in the use of ferromagnetic particles. Ferromagnets are tiny particles that act as little bar magnets; applying a magnetic field on a smart gel encourages the ferromagnets to move.[22] This movement raises the temperature of the gel and consequently causes the gel to expand. While microparticles of iron still allow the gel to expand, nanoparticles make the gel more responsive to the magnetic field.

Gels are used as scaffolds in tissue engineering to support living cells for tissue repair, as coatings of wells for cell cultures. Smart hydrogels use their environmental sensitivity to detect changes in pH, temperature, or concentration.[23] They are also used in drug delivery systems, as biosensors, contact lenses (silicone hydrogels), EEG and ECG electrodes, and for dressings used in the healing of burns or hard-to-heal wounds.[24]

Applications of smart gels permeate into various fields, including both medical and industrial. The two main applications for smart gels are in artificial muscle fabrication and drug release.[24] In drug release, a smart gel containing the desired water-soluble drug is injected into the patient. After receiving a certain stimulus (usually temperature or pH), a hydrogel will expand by allowing the water and salt in the blood to enter the gel. The drug will be released from the gel in the desired environment. This concept can be used to release drugs to attack tumors or aid specific areas of the body (*i.e.* eye drops for the eye). This concept is beneficial because the area, duration, and speed of the release can be better controlled with a smart gel. Nano-particles will help this medical technology by increasing the effectiveness of the gel by increasing the surface area of its constituents.

Developments with respect to the electrical properties of smart gels could result in the future production of artificial muscles.[25] When an electric field is applied to certain types of gel, there is an asymmetric charge distribution within the gel. This asymmetry yields different rates of expansion through-out the gel layer. In fact, in some cases, one end might contract while the

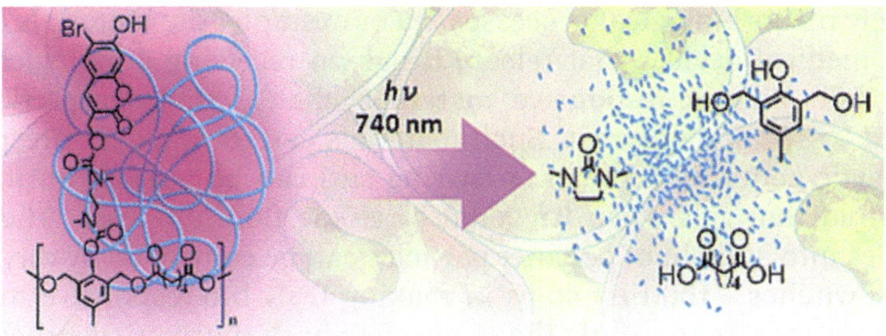

Figure 1.1 A polymeric material that is able to disassemble in response to biologically benign levels of NIR irradiation.

other expands. Asymmetry can also be created by producing heterogeneous gels with different rates of expansion throughout the gel. As a result of the electric field and asymmetry, the gel bends. This bending is significant because it mimics the role of muscles in the body which respond to electrical signals sent from the brain by creating mechanical energy. With more development, these gels could become prosthetic muscles for patients.

Poly(methyl methacrylate) is a transparent thermoplastic polymer often used as a shatter-resistant alternative to glass.[26] It has a good degree of biocompatibility and can be used for the replacement of intraocular lenses in the eye. In orthopedic surgery it is used as bone cement to affix implants or remodel lost bone. It is used for dentures and in dental fillings, and in cosmetic surgery in the form of suspended microspheres injected under the skin to permanently reduce scars.

Poly(glycolic acid) is a biodegradable, thermoplastic polymer with high initial tensile strength, smooth passage through tissue, easy to handle, excellent knot tying properties, and is commonly used for synthetic absorbable sutures, intracutaneous closures, implantable devices, tissue engineering scaffolds, bioabsorbable screws, and in abdominal and thoracic surgeries.[27] Scientists are also reporting the development and successful initial testing of the first practical "smart" material that may supply the missing link in efforts to use a form of light that can penetrate four inches into the human body. Near-infrared (NIR) light (which is just beyond what humans can see) penetrates through the skin and almost four inches into the body, with great potential for diagnosing and treating diseases. Low-power NIR does not damage body tissues as it passes. Figure 1.1 shows the new polymer, which has potential for use in diagnosing diseases and in engineering new human tissues in the laboratory.[28]

1.2.4 Enzyme-Responsive Materials

Intelligent biomaterials that can respond to a biological signal show revolutionary promise in regenerative medicine, diagnostics, and drug delivery. Enzyme-responsive materials have the potential to detect, respond to, and ultimately repair biological processes.[29] For example, the materials could be used in medical devices that release drugs on receiving a biological signal from a cell. Enzyme-responsive materials change their properties when triggered by specific enzymes. Such materials can form a gel in response to the catalytic action of a protease enzyme and can be used as an injectable cell-scaffold that gels when triggered by tissue fluid enzymes. The flow of molecules into and out of polymer particles can be controlled by very specific enzyme switches – the first steps in making truly bio-responsive materials. An essential goal is to mimic the *in vivo* feedback mechanisms that control secretory endocrine enzyme activity.

Smart advanced synthetic materials are combining the sensing activity with actuating activity, so they can be used in implementing systems that respond to different chemical changes, can learn to detect certain patterns by using neural networks and learning algorithms, then react accordingly.

1.3 Materials, Devices, and Sensing in Neuroscience

1.3.1 Material Brain

In order to understand the brain mechanisms and function, the last frontier of science and technology, subtle molecular processes must be detected and monitored using instruments of the same nanometric and atomic scale. The human brain has been called "a computer made of meat" by Marvin Minsky, an expert in artificial intelligence.[30] This computer apparently decides, chemically and genetically, how we feel, how we grow, live, and die. The only problem is that we are not sure how its unthinking atoms create our thoughts and experiences. Moreover, we are constantly reminded that it has quite a limited power, comparative to the sophisticated computers which can find solutions to astronomically complex mathematical problems at impressive processing speed. We cannot erase from our memory the embarrassment of Deep Blue, the chess-playing machine, beating Gary Kasparov, the reigning world chess champion, in that historic six-game match back in 1997. Deep Blue was making its decisions based on 200 million chess positions stored in its huge libraries of strategies, looking 20 moves ahead, while Kasparov could analyze just a few positions each second, relying more on intuition and experience than on processor power. The major difference in intelligence is that the brain can adapt to any changes in the rules of a game, while a computer is not able to adapt at all, unless it is programmed according to the new rules. Despite superhuman ability in chess, Deep Blue was not intelligent.

The brilliant mathematician Alan Turing, who delineated the foundation of computer science even before the first computer was built, defined the triumph of artificial intelligence as referring to the moment when an impartial judge will not be able to discern who the real person is and who the computer in a textual dialogue. Feeding a computer millions of bits of information and stuffing its huge memory space with enough rules to help it find a perfectly matching answer, a dialogue between a human and a machine now sounds like a normal "human" conversation. Then again, is a computer passing the Turing test really intelligent? There is no better test to define intelligence. Recent attempts in image recognition refer to the fact that, even if a computer can use a camera to detect an image, only an intelligent being can interpret what it represents, can analyze it, and reason about it. So far a computer can successfully make the difference between a cat and a dog, something that a toddler can do. No computer comes close to the richness of human perception.

Still, the "brain-as-computer" metaphor is an oversimplification at best. Conventional computers consist of transistors implementing a series of Boolean logical operations at very high speed. Brains function in a PARALLEL manner, doing thousands of operations simultaneously. The death of a neuron will not affect the brain, while if a single transistor is destroyed the whole operation of the computer is affected. A computer must be designed and programmed, while the brain comes "factory installed", capable of plasticity, regeneration and repair, of thinking and learning due to its dynamic thousand trillion synapses. Like a transistor, a neuron can send an excitation signal to thousands of other neurons. However, the architecture of the brain includes chemical modulation at synapses and inhibitory circuits that can change the nuances of the incoming signals in a way so complex and sophisticated that cannot be paralleled by any electronic devices.

Above all, computers cannot assign MEANING to their programmed processes, as the eminent mathematician and computer scientist John von Neumann famously remarked. On the other hand, human mind is all about meaning. It creates meaning even when none is present.

The big question is: will technology will ever be able to reproduce the human brain or is the brain something impossible to replicate? A brain does much more than applying specific algorithms to a set of data. It is surely the greatest mystery that science has yet to discover.

1.3.2 Quantum Dots

Quantum dots are small (5–8 nm) inorganic compounds made of semiconductor or metallic materials with well-defined quantum states and electronic structure.[31] They are composed of a metal core (cadmium, selenium, or cadmium telluride), a zinc sulfate shell, and an outer coating functionalized using bioactive molecules (Figure 1.2). Fluorescent quantum dots are used to visualizing molecular processes in neuron cells using fluorescent microscopy methods.[31] Small changes in the radius translate into distinct color changes so they can be used to replace bulky organic fluorophores, which interfere with the molecular structure of the object of investigation. The design of these nanostructures is based on the ability to control plasmonic behavior in metallic nanoparticles, quantum size effects in semiconductor heterostructures with designed asymmetries, and nanoparticles with implanted dopants possessing sharp emission spectra. Their

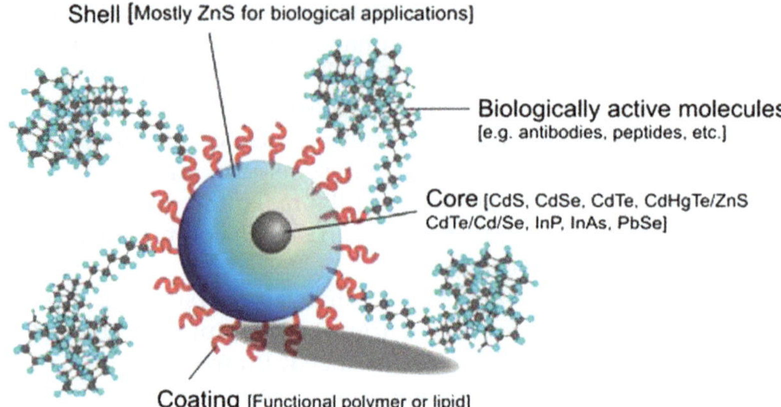

Figure 1.2   Structure of a semiconductor fluorescent quantum dot nanocrystal. The heavy metal core is responsible for the fluorescence properties of the quantum dot. The non-emissive shell stabilizes the core, whereas the coating layer provides anchor sites to organic and biological ligands such as antibodies, peptides, and other organic molecules. (Reproduced with permission from the Society for Neuroscience).

small size allows large and specific energy jumps between the energy band gaps of excited electrons or electron–hole pairs.

Inorganic nanoparticle optical probes can be tuned to match the photon energy requirements of the various excitation and detection systems. Unlike organic optical probes, they are photochemically robust during extended interrogation. For neuroscience studies, nanoparticles are combined with organic nanostructures for biofunctionalization, to attach them within neural cells configurations. The structure of semiconducting nanoparticles enables the generation of excitons, very sensitive to the external electric field. This sensitivity can turn these nanoparticles into reporters with externally modulated fluorescence intensity spectra. They may be combined with selective molecular binding moieties to confer sensitivity to changes in local neurotransmitter concentrations. Quantum dots can be used as local optical reporters for neuroscience, for visualizing dynamic molecular processes in neurons and glia on a large time scale, starting from seconds to many minutes, and on the small size scale of the synaptic cleft (20 nm) of neuron–neuron interactions or intracellular processes.[31] Owing to their intrinsic voltage sensitivity, they could be used directly as optical readouts of membrane potential. These reporters must be embedded into neural membranes (thickness ~ 2 nm) and react to local electric fields as well as local chemical environments. Functionalization with specific proteins make quantum dots capable of tracking receptors and functional responses in neurons (*e.g.* to glycine, nerve growth factor, glutamate, *etc.*).

Recent work using tools from atomic physics has shown that optically manipulated color centers in diamond provide exceptionally sensitive magnetic and electric field probes at sub-100 nm distances. Diamond is uniquely suited for studies of biological systems because it is chemically inert, cytocompatible, and ideal for coupling to biological molecules.

1.3.3 Nanotubes and Nanowires

Nanomaterials that can provide nanoscale topographical features have become popular materials, as culture substrates with nanoscale features have significantly different effects on neuronal adhesion and growth. Vertical nanowires were shown to selectively promote neuronal adhesion and guide neurite growth even without the use of a cell-adhesive coating.[31] Micropatterned islands of tangled carbon nanotubes also showed similar spontaneous adhesion and growth effects. Guided neuronal growth was reported on various nanotopographical substrates made of nanomesh carbon nanotubes, electrospun nanofibers, or patterned poly(urethane acrylate).[31]

One-dimensional structures such as nanotubes and nanowires may be used for highly local electrical measurements, for the delivery of photons to specific locations, and for the local release or collection of chemicals. These types of nanoparticles could be used alone, or combined with conventional organic chromophores, as they have been shown to greatly enhance optical signals, acting as an "antenna" for the light. Indeed, membrane-bound, antibody-linked gold nanoparticles have been already used for site-specific measurements of membrane potential. Traditional organic chromophores suffer from several drawbacks: they are large and can morphologically or chemically perturb the cellular environment. They can also bleach, that is, become ineffectual after exposure to light. Nanoparticles, by contrast, can be coated with a passivation layer or specialized shell that limits direct interaction with the surrounding media, which greatly minimizes bleaching and, inside the cell, the generation of reactive oxygen species. Present challenges include developing inorganic nanoparticles with enhanced voltage sensitivity, and orchestrating plasmonic enhancement of existing optical reporters. It should also be feasible to develop inorganic nanoparticles with voltage sensitivity, verify their biocompatibility, and identify and validate routes to targeted delivery. New organic nonlinear voltage probes and multifunctional nanoprobes that can locally report not only voltage but also chemical species (*e.g.* calcium, neurotransmitters, *etc.*), local temperature, or ionic environment are the focus for present research efforts.

Semiconductor nanowires can detect specific intracellular biomolecules, perform small-molecule drug screening, detect intracellular signaling, and also deliver drugs and genetic material into the cell.[31] This nanotube spearing necessitates an oscillating magnetic field to spear the nanotubes, followed by a static field to drive them inside the cells. Figure 1.3 shows how these nanowires are not fatal to the cell. The cell remains functional for a few days and can even differentiate from stem cells. The mechanism of cellular engulfment of the nanorods and subsequent normal function remains to be explored.

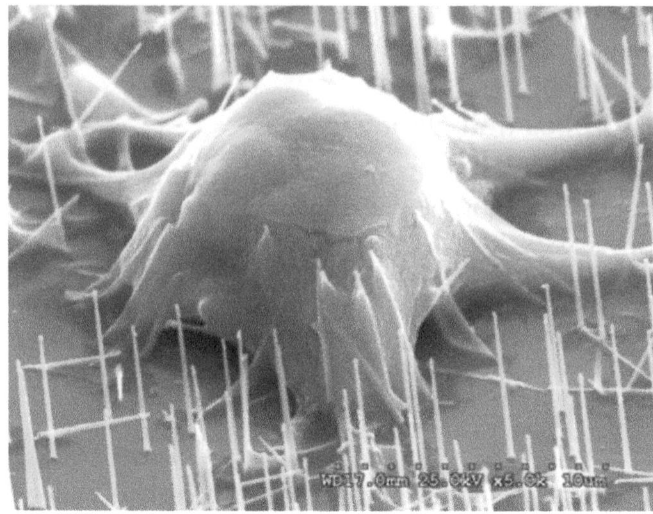

Figure 1.3 SEM image of mouse embryonic stem cells penetrated by Si nanowires on a Si substrate. The scale bar represents 10 microns. (Reproduced with permission from the American Chemical Society.)

Silicon nanowires have been implemented with either field-effect transistor-type active sensors or metal nanoelectrodes for *in vitro* neural sensors. The Lieber group at Harvard University has reported silicon nanowire field-effect transistor (NW-FET) arrays.[31] They showed that simultaneous recordings from the axon and dendrites of a single neuron were possible with NW-FET arrays. In addition, neural signals ranging from 0.3 to 3 mV were recorded from neural circuits in brain slices using a NW-FET array. NW-FET is a promising sensor that can provide sufficient sensitivity with unprecedented spatial selectivity.

Field-effect transistors (FETs) can also record electric potentials inside cells. Their performance does not depend on electrode impedance and they can be made much smaller than micropipettes and microelectrodes. FET arrays are better suited for multiplexed measurements. SiO_2 nanotubes synthetically integrated on top of a nanoscale FET penetrates the cell membrane, bringing the cell cytosol (Figure 1.4) into contact with the FET, which is then able to record the intracellular transmembrane potential.

A branched intracellular nanotube FET (BIT-FET) possess a bandwidth high enough to record fast action potentials even when the nanotube diameter is decreased to 3 nm, a length scale well below that accessible with other methods. Studies show that a stable and tight seal forms between the nanotube and cell membrane, not killing the cells because the diameter of the tubes is so small. Multiple BIT-FETs can record multiplexed intracellular signals from both single neurons and networks of neurons.

In the case of metal nanoelectrodes, the Park group at Harvard University developed a vertical silicon nanowire array with individual nanowires

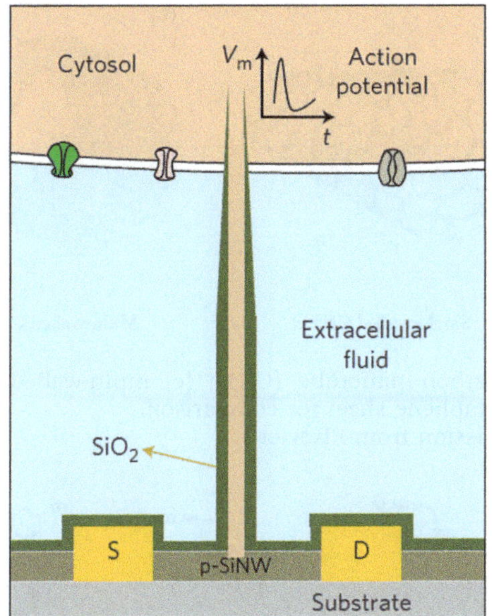

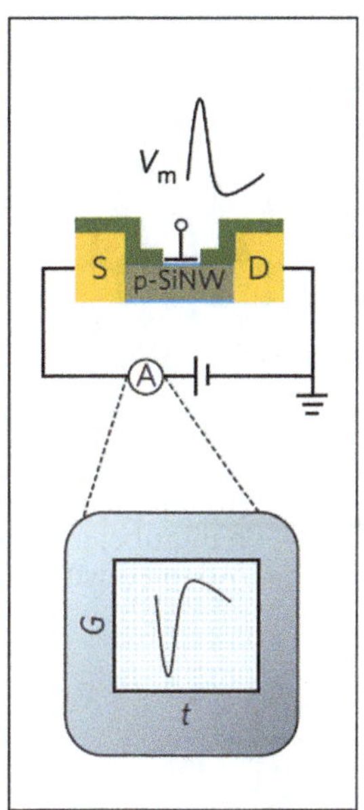

Figure 1.4   Schematic diagrams showing (*left*) a cell coupled to a BIT-FET and the variation in device conductance G (*right*) with time t during an action potential V_m. S and D indicate source and drain electrodes. The SiO_2 nanotube connects the cytosol (*orange*) to the p-type silicon nanowire FET and, together with the SiO_2 passivation (*green*), excludes the extracellular medium (*light blue*) from the active device channel. (Reproduced with permission from Nature Publishing).

150 nm thick and 3 μm high.[31] Several nanowires were grouped (2 μm spacing) to cover a single neuron, and an array of grouped nanowires was used to interrogate a small neural circuit. A high signal-to-noise ratio on the order of 100 was achieved, with the measured signal amplitude on the order of a few mV.

1.3.4   Carbon Nanostructures for Neuroregeneration

Carbon nanostructures are potential candidates to develop neural prostheses due to their similar nanoscale dimensions to neurites, as well as their unique electrical and mechanical properties.[11] When being used as a scaffold, they are able to repair injured nerves and even long gaps in severed nerves, by stimulating the healing of the severed ends in a nerve. Figure 1.5 shows the basic structures used as carbon scaffolds. These structures can be coated with thin layers of polymers in order to decrease the formation of glial scar tissue and to provide suitable sites for cell adhesion and proliferation.

Furthermore, the structures can be functionalized in order to improve biocompatibility by decreasing the toxicity. Figure 1.6 shows different

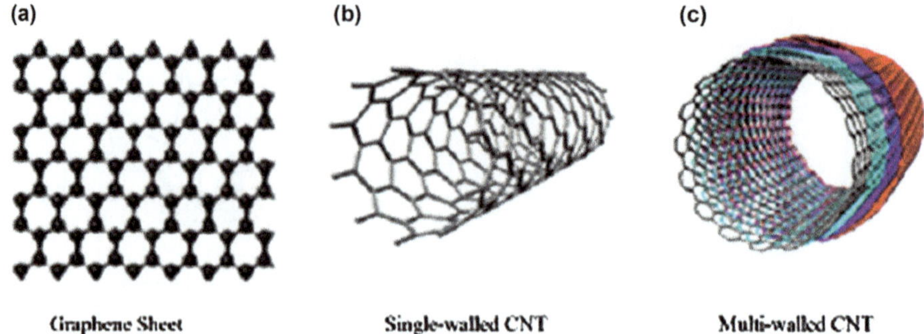

Figure 1.5   (b) Single-walled carbon nanotube (CNT); (c) multi-walled carbon nanotube; and (a) graphene sheet for comparison.
(Adapted with permission from Elsevier.)

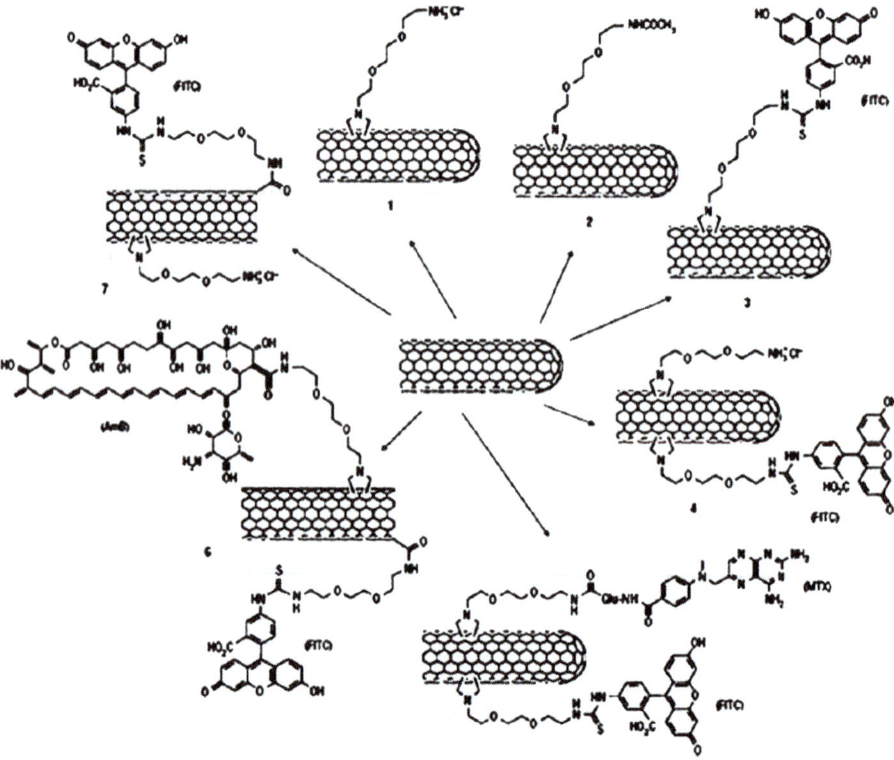

Figure 1.6   Molecular structures of CNTs functionalized covalently with different types of small molecules: (1) Ammonium-functionalized CNT; (2) acetamido-functionalized CNT; (3) CNT functionalized with fluorescein isothiocyanate (FITC); (4) CNT bifunctionalized with ammonium groups and FITC; (5) CNT bifunctionalized with methotrexate (MTX) and FITC; (6) shortened CNT bifunctionalized with amphotericin B (AmB) and FITC; (7) shortened CNT bifunctionalized with ammonium groups and FITC (through an amide linkage).
(Reproduced with permission from Elsevier.)

solutions implemented in order to explore the controversial aspect of carbon nanotube toxicity, which manifests itself by decreasing cell viability through blocking ion channels in the membrane, increase of cell oxidative stress, and reduction of cell adhesion or induction of apoptosis.

Recent improvements with respect to prosthetic medical devices have been associated with the growth of carbon nanotubes on biocompatible Pt used as a catalyst. The new material shows less cellular degeneration due to oxidative stress because the main product at a Pt catalyst from oxygen reduction is water. Guiding axon regeneration through tubes made of materials such as chitosan, a biocompatible and biodegradable natural material, can provide, as a gel sponge, a suitable scaffold for nerve regeneration.

1.3.5   Nanoribbons for Sensing Cellular Deformation

If the electrical response of neurons to applied voltages has been studied extensively, the mechanical response has been largely ignored; such research could advance knowledge of cellular function and physiology, especially in the area of axon elongation and dendrite formation. Using piezoelectric nanoribbons made of PbZr or $Ti_{1-x}O_3$ it was found that the cells deflect by 1 nm when 120 mV is applied to the cell membrane.[11] Such depolarization induces changes in the membrane tension so it is accommodating the stimulus by equalizing the overall pressure across the membrane through a process resembling converse flexoelectricity.

Figure 1.7 shows piezoelectric nanoribbons suspended over a trench as nanobeams to maximize deflection. The use of an underlying substrate of transparent MgO as well as transparent indium tin oxide (ITO) electrodes facilitates backside chip visualization during electrophysiology measurements. The electrodes are electrically isolated by a coating of SiN_x to ensure no cross-signal response when the chip is placed into solution. PC12 cells, a rat pheochromocytoma cell line that acquires many of the characteristics of sympathetic neurons when treated with nerve growth factor (NGF), were used. PC12 cells were cultured on the piezoelectric chip, and those cells located on the nanobeam arrays were patch-clamped with a standard glass electrode for membrane voltage stimulation.

In general terms, it is not clear how neurons interact with nanostructures, why they continue to function impaled by nanospears from all directions, and how they heal when confined in nanoscaffolds, in closing the gap between severed nerves ends. The reports on the toxicity of nanomaterials are extremely controversial. However, if any such undesirable effect can be minimized and these new materials can contribute to brain regeneration and repair, the outcome might be a positive one for the future. Like the miniaturized submarine crew in the 1966 movie Fantastic Voyage, smart nanodevices might one day be sent beyond the blood/brain barrier (BBB) to perform lifesaving surgery inside the brain tissue and destroy cancerous brain tumors with extreme precision and in a minimally invasive mode, by

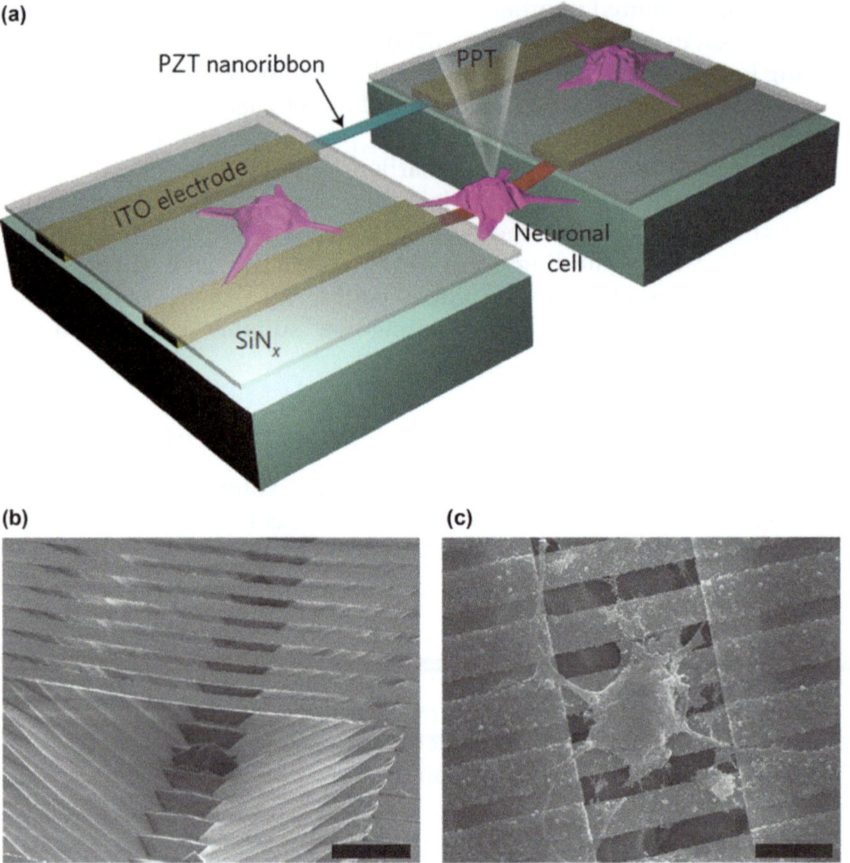

Figure 1.7 Interfacing of PZT nanoribbons with cultured neuronal cells. (a) Schematic of the piezoelectric nanoribbon device with cultured neuronal cells. The suspended nanoribbons record cellular mechanical deflections while the glass pipette (PPT) applies and records membrane potentials. (b) SEM image of suspended PZT nanoribbons (scale bar, 5 mm). (c) SEM image of a single PC12 cell directly interfaced with suspended PZT nanoribbons (scale bar, 15 mm). (Reproduced with permission of Nature Publishing.)

delivering the necessary drug, then leaving the body in a harmless way. It would represent a less expensive and far more effective treatment because only the target cells will be affected, without the side effects of today's chemotherapies. Such an approach to medicine has the potential to transform health care for everybody.

1.4   Biosensor Devices Based on Graphene

Ever since the first isolation of free-standing graphene sheets were described in 2004, this two-dimensional (2D) carbon crystal has been highly anticipated to provide unique and new opportunities for sensor applications.[31] Significant potential for application of the material in various novel sensors has been demonstrated. Useful features include the exceptional electrical properties of graphene (extremely high carrier mobility and capacity), electrochemical properties (high electron transfer rate), optical properties (excellent ability to quench fluorescence), structural properties (one-atom thickness and extremely high surface-to-volume ratio), and its mechanical properties (outstanding robustness and flexibility).

Graphene nanostructures exhibiting such excellent properties are very suitable for the fabrication of channel structures in field-effect transistors (FETs), which are typically used as electronic sensor devices to detect biomolecules and neural cell activity. This involves the incorporation of graphene into FETs, *via* insertion of a material with superior sensing properties in a structure of high sensitivity, simple device configuration, low cost, high miniaturization, and capable of real-time detection. A typical FET consists of a semiconducting channel between two metal electrodes, the drain and source electrodes, through which the current is injected and collected. Varying the gate potential through a thin dielectric layer, typically 300 nm SiO_2, can capacitively modulate the conductance of the channel. In a typical p-type metal oxide semiconductor field-effect transistor (MOSFET), the negative gate potential leads to the accumulation of holes (majority charge carries), resulting in an increase of the channel conductance, while the positive gate potential leads to the depletion of holes and hence a decrease of the conductance. In the case of the electronic sensor, the adsorption of molecules on the surface of the semiconducting channel either changes its local surface potential or directly dopes the channel, resulting in change of the FET conductance. This makes the FET a promising sensing device with easily adaptable configuration, high sensitivity, and real-time capability, providing, again, that the nonspecific absorption problem is solved using smart chemistry to prevent the fouling of the surface when the device is exposed to complex media such as human serum or blood.

In some cases, the gate electrode is removed, in order to simplify the device structure, to form a chemoresistor. Such a configuration is suitable for the fabrication of graphene-based sensors on polymer substrates for flexible electronic applications.[31] Despite the lack of modulation by the gate potential, the working principle of the chemoresistor is the same as a normal FET sensor. Figure 1.8a shows a typical gas sensing system, where the channel is directly exposed to a target gas. The adsorption of gas molecules results in the doping of the semiconducting channel, leading to a conductance change of the FET device. The charge transfer from the adsorbed gas molecules to the semiconducting channel is the dominant mechanism for the current response, which is similar to carbon nanotube-based gas sensors.

In order to detect biospecies, the graphene FETs should operate in an aqueous environment. As shown in Figure 1.8b, the graphene channel is usually immersed in a flow or sensing chamber, which is used to confine the solution. The drain and source electrodes are electrically insulated to

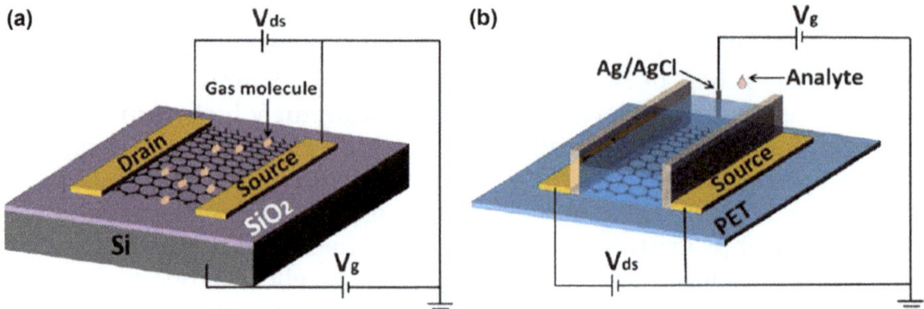

Figure 1.8   (a) Typical backgate graphene FET on a Si/SiO$_2$ substrate used as a gas sensor. (b) Solution-gate graphene FET on flexible poly(ethylene terephthalate) (PET) substrate used as a chemical and biological sensor in aqueous solution.

prevent current leakage from ionic conduction. Different insulators, including poly(dimethylsiloxane) (PDMS)/silicone rubber, SiO$_2$ thin film, SU8 passivation, and silicone rubber, are used in different device structures. The gate electrode, usually Ag/AgCl or Pt, is immersed in the solution. The gate potential is applied through the thin electric double layer capacitance formed at the channel/solution interface. The double-layer thickness (or Debye length) is determined by the ionic strength in the solution, typically within 1 nm. Normally, the solution-gate FET is over two orders of magnitude more sensitive than the typical backgate FET.

Two major sensing mechanisms have been proposed for graphene-based biosensors in solution, *i.e.* the electrostatic gating effect and the doping effect. The gating effect suggests that the charged molecules adsorbed on graphene act as an additional gating capacitance, which alters the conductance of the graphene channel. On the other hand, the doping effect suggests a direct charge transfer between the adsorbed molecules and the grapheme channel, similar to gas sensing. In a real case, the actual sensing mechanism might be a combination of both mechanisms, or involve more complicated mechanisms.

The detection of living cells is more challenging as the interaction between the graphene channel and living cell membranes is much more complicated. Graphene offers a improved opportunity to study the cell/nanomaterial interface since its 2D structure provides a homogeneous contact with the 2D cell membrane. A solution-gate FET to investigate living cell behavior is shown in Figure 1.9. The devices were fabricated based on large-scale micropatterned reduced graphene oxide (rGO) thin films with a thickness of 1–3 nm. Living neuron cells (PC12) were directly cultured on the rGO channel to obtain an intimate contact. The rGO FET was able to detect the adsorption of hormonal catecholamine molecules and those secreted from the living PC12 cells with a high signal-to-noise ratio. Moreover, the rGO FET can be fabricated on the flexible PET substrate and functions well during bending, which might be useful in complicated *in vivo* biosensing.

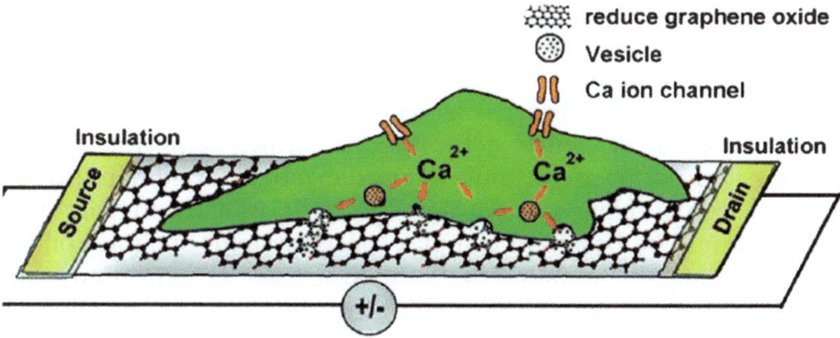

Figure 1.9 Schematic illustration of the interface between a PC12 cell and a reduced graphene oxide FET. The detection was realized by the real-time monitoring of I_{ds} during the dynamic secretion of catecholamines.

1.5   Bio-inspired Computation

Collections of individual animals, birds, and fish sometimes act in precisely coordinated ways, as if moved by a group mind in perfectly choreographed movements. Herds of animals, flocks of birds, or schools of fish seem to exhibit intelligent group behavior. This "swarm intelligence" is used in the new "swarm theory" with interesting practical applications in artificial intelligence, to optimize the distribution of aircraft to the gates, move masses of passengers in airports as efficiently as possible, as well as route trucks in a smooth operating transportation system.

To explain bird flocking, fish schooling, beehives, or ant and locust colonies, scientists observed that in such large groups, neighbor-to-neighbor changes take less than 15 ms. Originated by single individuals somewhere in the group, such changes of direction propagate as a rapid wave throughout the group. When trying to mimic such behavior using artificially simulated graphic units, three main conditions were incorporated in the computer program:

1. Do not crowd nearby units
2. Fly in the average direction of nearby units
3. Stay close to nearby units

After running the program, the graphic result was a striking life-like movement. The conclusion: there is no coordinating command that has to propagate through the entire flock of birds or school of fish; the emergent behavior is the result of individual optical cues received by the members of the group.

Bio-inspired computation can be used in modeling cancer and bacterial growth, for testing dedicated algorithms for global optimization. Further applications mimicking insect swarms are the swarms of robots performing dangerous tasks such as mine sweeping, risky search-and-rescue missions, or even exploring the surface of Mars.[32]

"Living" machines based on advanced materials and sophisticated designs would have new characteristics such as fault tolerance, self-repair, self-replication, reproduction, evolution, adaptation, and learning. Such artificial life would be based on evolutionary computation algorithms and artificial or living neural networks, the former being interfaced with computing devices, for use in applications dangerous for humans (to control unmanned fighter planes, for example). Their interaction with the environment would be based on complex sensor systems such as computing tissues mimicking the human skin. Computing tissues can be used to design novel man/computer interfaces, intelligent and adaptive prostheses, intelligent doors, floors, walls, black boards, displays, *etc.* Artificial systems capable to grow, adapt, and reproduce in hardware are imagined and designed by scientists working in artificial intelligence. Figure 1.10 shows the fundamental element of a sensing computing tissue. It consists of a transparent touch-sensitive element, a LED color display, and a reconfigurable chip, a new fine-grained field-programmable gate array (FPGA). This element of the reconfigurable computing tissue consists of an input, an output, and the FPGA computing unit, organized in three hierarchical levels as shown in Figure 1.10. The inputs might include temperature sensors, force sensors, cameras, *etc.*, and the outputs might include microphones, motors, speakers, displays, *etc.*

Parallel systems made of high numbers of such miniature elements would be able to compute in parallel. Such configurations, based on smart materials already employed in display technology and organic electronics, confirm the new trend towards intelligent interactive systems.[33]

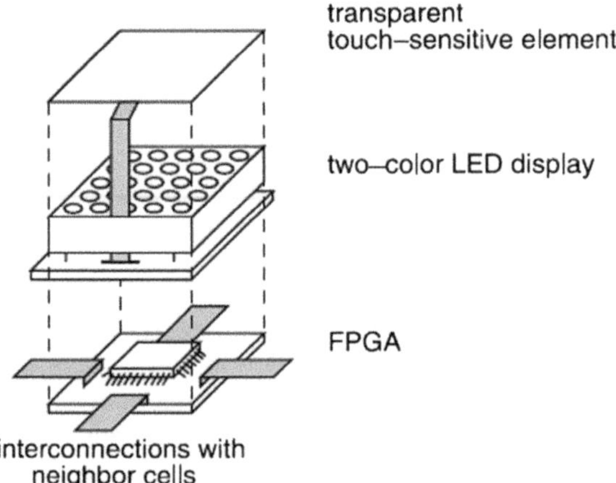

Figure 1.10 A fundamental element of a sensing computing tissue. It consists of a transparent touch-sensitive element, a LED color display, and a reconfigurable chip, a new fine-grained field-programmable gate array (FPGA).
(Reproduced with permission of Elsevier.)

1.6 Surface-Modified Materials to Enhance Biocompatibility

The nature of the interaction of material surfaces with biological fluids such as urine, blood, plasma, and serum is crucial in terms of both the biocompatibility of medical implants and biosensor devices, whether the latter are implanted or used in the clinical laboratory. Serious medical problems can arise from deleterious effects of materials on blood proteins and cells. For example, formation of thrombotic emboli upon contact of blood with materials employed in bypass circuitry during extracorporeal circulation (*e.g.* in cardiopulmonary bypass and renal dialysis procedures) is a well-recognized clinical issue.[34] Indeed, it is now known that micro-clots produced in such apparatus can result in cognitive disability in later years for patients subjected to dialysis and bypass surgery.[35,36] With regard to implantable biosensor technology, the lack of a closed-loop system with incorporation of a glucose device for operation with an artificial pancreas system has become a legendary research problem. Biosensors do not even figure prominently in the standard clinical biochemistry laboratory, largely because of the biological fluid/surface interaction problem.[37]

A key aspect of negative effects caused by biologically foreign materials is the spontaneous adsorption of blood-based proteins on substrate surfaces.[38] This process is often referred to as "fouling" by those concerned with implant biocompatibility, whereas the analogous effect with regard to sensor technology is called "nonspecific adsorption" (NSA). The distinction between these two terms, however, is somewhat artificial since there is a great deal of common ground in their physical chemistry. Not surprisingly, tremendous research effort has been devoted over many years towards the development of protein-resistant surfaces in order to ameliorate the effects of deleterious interfacial interactions. The many strategies employed for this purpose, involving both coatings and covalently-bound layers, have been reviewed comprehensively by Blaszykowski *et al.*[39] Of the vast array of surface-modifying agents used in this field, peptides, poly(ethylene oxide)s/glycols, and zwitterionic sulfo- and carboxybetaines have been prominent. Despite the massive attention apparently paid to attempts to avoid fouling, it is fair to say that by far the majority of the effort has been on surface interactions with solutions containing single protein species. Far less research has been conducted on important biological matrices such as blood and serum. Here, we briefly describe a strategy originating from our own laboratory that does function in such complex media.

As alluded to above, the devices employed for hospital renal dialysis and coronary bypass surgery, often termed bypass circuitry, are generally composed of polymers such as polycarbonate (PC), polymethacrylate, and polyesters. In our own research we have been concerned with the covalent surface modification of PC using silanization chemistry. We report a new mono(ethylene glycol) (MEG-OH) silane adlayer coating for PC (Figure 1.11)

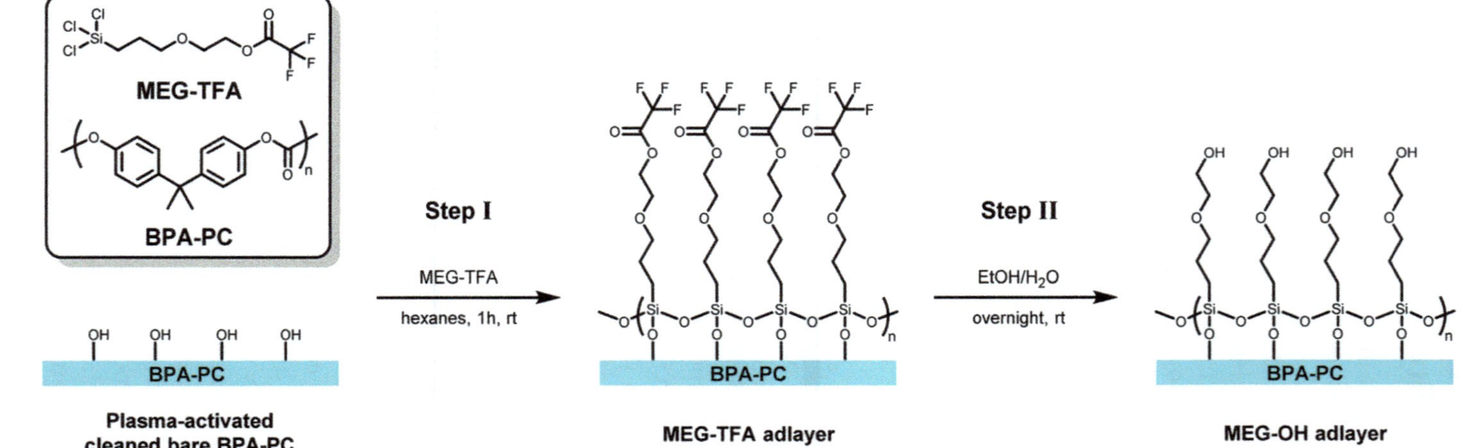

Figure 1.11   Surface modification of bisphenol A polycarbonate (BPA-PC) polymer with a MEG-OH silane adlayer. The MEG-TFA molecule is 2-[3-(trichlorosilyl)propyloxy]ethyl trifluoroacetate. (Reproduced by permission of the American Chemical Society.)

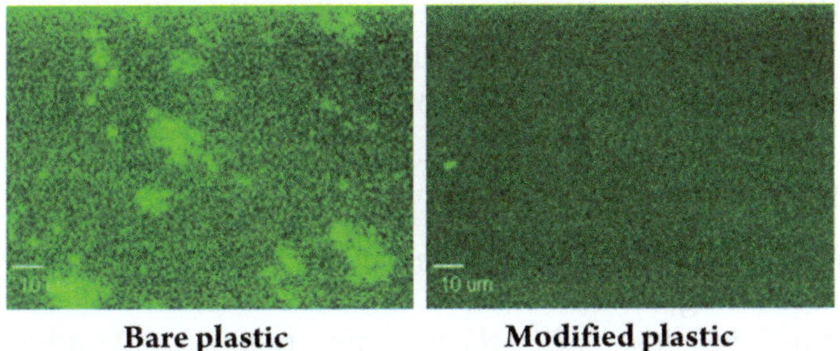

Figure 1.12   Representative video frames showing platelet adhesion, aggregation, and thrombus formation (or lack of) on bare (*left*) and MEG-OH modified (*right*) BPA-PC substrates ($32\times$ magnification) after 5 min exposure to whole human blood at a shear rate of 1000 s^{-1}.
(Reproduced by permission of the American Chemical Society.)

that displays excellent anti-thrombogenicity, far exceeding that of the bare substrate.[40]

Following thorough surface characterization by X-ray photoelectron spectroscopy, the treated PC surfaces were tested for anti-thrombogenicity in a real-time fashion with whole human blood labeled with fluorescent dye. Thrombus formation was visualized with an inverted fluorescence microscope. The MEG-OH surface displayed a 97% reduction in thrombus events when compared to the response of the bare polymer (Figure 1.12).

The mechanism responsible for this sort of dramatic result is still under debate, but it appears from neutron reflectometry experiments that interstitial water in the adlayer is at least partly involved.[41] An important aspect of this chemistry is not simply that a water "barrier" prevents protein adsorption at surfaces, but that the adlayer-instigated structure of the water plays a key role. Circling back to the beginning refrain of this chapter, it is highly noteworthy that structured water is also considered to be a crucial component of the behavior of the head-group zone at the surface of biological membranes.[42]

1.7   A Final Remark

The emerging field of biomimicry tells us that the most sophisticated technologies which involve future smart materials and sensing devices will be ones which emulate biological systems. There are already many examples of technological advances being proposed and developed that provide evidence for this. In the world of medicine, multifunctional materials will have the capability to detect, select, and execute specific "intelligent" functions in response to changes in the local environment, which is obviously crucial in terms of maintaining homeostasis. Their ability to recognize, analyze, and discriminate will be based on superior properties of self-repair, self-multiplication, self-degradation, self-learning, and artificial homeostasis. Although much remains to be accomplished, not least obviation of the serious biocompatibility issue referred to above, it is clear that smart materials science offers huge potential for future treatment of a plethora of disease and injury conditions. Maybe one day we will see a true "bionic man"!

References

1. S. Weinberg, in *Universe or Multiverse?*, ed. B. Carr, Cambridge University Press, Cambridge, 2005.
2. D. Bohm, *Wholeness and Implicate Order*, Routledge, London, 1980.
3. H. Everett, *Rev. Mod. Phys.*, 1957, **29**, 454–462.
4. S. Hameroff and R. Penrose, *Phys. Life Rev.*, 2014, **11**, 39–78.
5. J. F. V. Vincent, O. A. Bogatyreva, N. R. Bogatyrev, A. Bowyer and A.-K. Pahl, *J. R. Soc. Interface*, 2006, **3**, 471–482.
6. S. Trolier-McKinstry, in *Piezoelectric and Acoustic Materials for Transducer Applications*, ed. A. Safari and E. K. Akdogan, Springer, New York, 2008.
7. H. K. D. Bhadeshia, *Geometry of Crystals*, Institute of Materials, *London*, 2001.
8. J. P. Joule, *London, Edinburgh Dublin Philos. Mag. J. Sci.*, 1847, **30**, 76–87, 225–241.
9. J. B. Conant, L. F. Small and B. S. Taylor, *J. Am. Chem. Soc.*, 1925, **47**, 1959–1974.
10. C.-G. Granqvist, *Handbook of Inorganic Electrochromic Materials*, Elsevier, Amsterdam, 1995.
11. T. McKay, B. O'Brien, E. Calius and I. Anderson, *Smart Mater. Struct.*, 2010, **19**, 055025; http://dx.doi.org/10.1088/0964-1726/19/5/055025.
12. A. M. Tishin and Y. I. Spichkin, *The Magnetocaloric Effect and its Applications*, Institute of Physics, Bristol, 2003.
13. W. Thomson, *Proc. R. Soc. Edinburgh*, 1851, 91–98.
14. J. M. Jani, M. Leary, A. Subic and M. A. Gibson, *Mater. Des.*, 2014, **56**, 1078.
15. A. Pelton, S. Russell and J. DiCello, *JOM (J. Miner. Met. Mater. Soc.)*, 2003, **55**(5), 33–37.
16. *Shape Memory Alloys: Manufacture, Properties and Applications*, ed. H. R. Chen, Nova Science, New York, 2010.
17. *Bioceramics: Applications of Ceramic and Glass Materials in Medicine*, ed. J. F. Shackelford, Trans Tech Publications, Durnten-Zurich, Switzerland, 1999.
18. www.CrystalZirconia.com.
19. D. Shi, *Adv. Mater.*, 2013; DOI: 10.1002/adma.201301376.
20. S. M. Kurtz, O. K. Muratoglu, M. Evans and A. A. Edidin, *Biomaterials*, 1999, **20**, 1659–1688.
21. C. R. Nuttelman, M. C. Tripodi and K. S. Anseth, *Matrix Biol.*, 2005, **24**, 208–218.
22. G. Pasparakis and M. Vamvakaki, *Polym. Chem.*, 2011, **2**, 1234–1248.
23. C. L. Chang, *Sensors, 2010, IEEE*, 2010, 1617–1621.
24. Y. Qiu and K. Park, *Adv. Drug Deliv. Rev.*, 2001, **53**, 321–339.

25. M. Shahinpoor, *Smart Mater. Struct. Int. J.*, 1997, **3**, 367–372.
26. R. Q. Frazer, R. T. Byron, P. B. Osborne and K. P. West, *J. Long Term Eff. Med. Implants*, 2005, **15**, 629–639.
27. D. K. Gilding and A. M. Reed, *Polymer*, 1979, **20**, 1459–1464.
28. N. Fomina, *Macromolecules*, 2011, **44**, 8590–8597.
29. D. Roy, J. N. Cambre and B. S. Sumerlin, *Prog. Polym. Sci.*, 2010, **35**, 278–301.
30. M. L. Minsky, *The Emotion Machine: Commonsense Thinking, Artificial Intelligence, and the Future of the Human Mind*, Simon & Schuster, New York, 2007.
31. M. Thompson, L.-E. Cheran and S. Sadeghi, *Sensor Technology in Neuroscience*, RSC, Cambridge, 2013.
32. *Swarm Intelligence and Bio-Inspired Computation – Theory and Application*, ed. X.-S. Yang, Z. Cui, R. Xiao, A. H. Gandomi and M. Karamanoglu, Elsevier, Amsterdam, 2013.
33. C. Teuscher, D. Mange, A. Stauffer and G. Tempesti, *Biosystems*, 2003, **68**, 235–244.
34. B. Nilsson, K. N. Ekdahl, T. E. Mollnes and J. D. Lambris, *Mol. Immunol.*, 2007, **44**, 82.
35. A. M. Murray, *Adv. Chronic Kidney Dis.*, 2008, **15**, 123–132.
36. N. Stroobant, G. Van Nooten, Y. Van Belleghem and G. Vingerhoets, *Chest*, 2005, **127**, 1967–1976.
37. M. Thompson, S. Sheikh, C. Blaszykowski and A. Romaschin, in *Detection Challenges in Clinical Diagnostics*, ed. P. Vadgama and S. Peteu, RSC, Cambridge, 2013, p. 1.
38. S. Franz, S. Rammelt, D. Scharnweber and J. C. Simon, *Biomaterials*, 2011, **32**, 6692.
39. C. Blaszykowski, S. Sheikh and M. Thompson, *Chem. Soc. Rev.*, 2012, **41**, 5599–5612.
40. K. Federov, C. Blaszykowski, S. Sheikh, A. Reheman, A. Romaschin, H. Ni and M. Thompson, *Langmuir*, 2014, **30**, 3217–3222.
41. N. M. Pawlowska, H. Fritszche, C. Blaszykowski, S. Sheikh, M. Vezvaie and M. Thompson, *Langmuir*, 2014, **30**, 1199–1203.
42. M. J. Higgins, M. Polcik, T. Fukuma, J. E. Sader, Y. Nakayama and S. P. Jarvis, *Biophys. J.*, 2006, **91**, 2532–2542.

Biomimetic Materials and Surfaces in Detection

RYAN D. BOEHM AND ROGER J. NARAYAN*

Joint Department of Biomedical Engineering, University of North Carolina
and North Carolina State University, 911 Oval Drive, Raleigh,
NC 27695-7115, USA
*Email: roger_narayan@msn.com

2.1 Introduction

Interest in the use of biomimicry in biosensing applications has been
growing in the past couple of decades, as advances in fabrication techniques
have allowed for the creation of materials that mimic those found in nature
and couple them to transduction mechanisms that produce useable information in sensing applications. Creation of useful biosensors has relied
upon the interface between the transduction platform and the sensing
mechanism, which traditionally has fallen under the umbrella of a few
categories: bioaffinity, catalysis using enzymes, or whole cell sensing.[1] In
this chapter, we touch on a few of the ways in which biomimetic materials
may accomplish successful biosensing by utilizing or mimicking one of
these aforementioned sensing mechanisms. The categories that are covered
with respect to the biomimetic biosensors include:

- Lipid membranes which mimic cellular membranes, allowing for
  examination of membrane properties and/or immobilization of relevant
  biomolecules (*e.g.*, enzymes, peptides)

RSC Detection Science Series No. 3
Advanced Synthetic Materials in Detection Science
Edited by Subrayal Reddy

Published by the Royal Society of Chemistry, www.rsc.org

- Biomolecule immobilization strategies, featuring the biomolecules as recognition or catalytic elements, as well as constructs that mimic such activities
- Whole cells affixed upon transducers as sensing entities and/or mimics of tissue constructs

While the depth and breadth of these topics in the literature is extensive, some of the main concepts are covered in the following sections. These topics are highlighted by short review summaries as examples of the means by which they may be utilized in biosensing applications.

2.2 Lipid Membranes

2.2.1 Lipids Introduction

Biosensors that utilize protein capture layers on a sensing pad or transducer have traditionally involved either spontaneous adsorption of proteins onto the sensing surface or covalent immobilization of the proteins onto the sensing surface after modification to the protein and/or surface.[2] However, problems are associated with capture layer formation *via* both protein adsorption (*e.g.*, surface dependent adsorption rates; difficulty in predicting protein–surface binding energies; instability of proteins in their natural form and the potential for denaturation following adsorption; desorption and replacement of the protein with an analyte molecule) and covalent immobilization (*e.g.*, lack of specificity of the orientation by which the protein is immobilized; lack of protein functionality resulting from the immobilization chemistry; irreversible protein alterations; inability to immobilize a single monolayer; changes to protein functionality as a result of local pH changes during the immobilization process).[2] An alternative approach to protein immobilization is to coat the transducer surface with a lipid bilayer (*e.g.*, *via* the Langmuir–Blodgett technique on a planar hydrophilic surface or vesicle deposition on non-planar geometries), as a biomimetic means of immobilizing proteins at the lipid bilayer surface. The proteins are then configured in a more natural state (preferably with the hydrophobic tail anchored within the membrane). This method has the advantages of limiting non-specific binding, likelihood of denaturation, or desorption from the membrane, in addition to avoiding the pitfalls of covalent immobilization.[2] Furthermore, this biomimetic approach takes advantage of lipid membrane fluidity, allowing the proteins to orient themselves in an optimal functional fashion.

Transmembrane sensors utilize a biomimetic membrane, meant to be similar to a cellular membrane, immobilizing a biological recognition unit within the membrane, which is deposited on the sensor's transducer[1] (*e.g.*, Na^+/D-glucose transporter protein immobilized in a bilayer lipid membrane to monitor Na^+ ion transit across the membrane, as initiated by D-glucose[3]). Supported lipid bilayers (SLBs) have been studied extensively over the course

of several decades, allowing researchers to investigate biomolecule interactions with these model cell membranes, helping to elucidate biological
phenomena at the nanoscale. More recent advances in this field have been
extensively reviewed, pointing to keys to the use of SLBs in biomimetic
research, including:[4]

- The use of these membranes for interactions with cells, peptides, and
  other biomolecules
- How SLBs may be tailored to specific applications on numerous types of
  substrates (*e.g.*, metals, oxides, SAM-treated substrates, polymers,
  porous substrates, nanoparticles, *etc.*)
- The wide range of SLB patterning techniques, such as stamping, robotic
  deposition, microfluidic deposition, photopatterning, and polymer
  lift-off techniques

The properties inherent to these SLB constructs, as influenced by such
factors as their composition, tethering technique, biomolecule functionalization, or substrate characteristics, lend them to a variety of biosensing
applications. While there is extensive literature covering the many aspects of
lipid membranes as cellular membrane mimics, the following subsections
will look at how some of these supported lipid bilayers, or tethered bilayer
lipid membranes (tBLMs), are utilized in biosensing applications. We will
briefly look at a few examples of how membranes can be analyzed for use in
biosensing, as well as their immobilization on planar surfaces with the
incorporation of pore-forming or catalytic moieties. Lastly, a few examples
are discussed on the use of these membranes on porous substrates and/or
nanostructured substrates, as well as on discrete particles.

2.2.2 Lipids Tethered to Planar Surface

2.2.2.1 *Membrane Characteristics*

The properties of the bilayer membranes that are used in biosensing
applications are of interest as they are needed for determining how they
influence the biosensor design. For instance, determination of the electrochemical properties of the membrane is useful to know, as incorporation of
proteins or enzymes for analyte detection may influence and change those
properties. In terms of electrochemical responses, these factors may result
in a change in the membrane resistance, conductance, capacitance, or
impedance values, which can be utilized for analyte detection.

For instance, Chilcott *et al.* examined the electrochemical properties of
a biomimetic hybrid bilayer membrane.[5] In this study, deposition of
hybrid bilayer membranes (hBLMs) consisted of an octadecane monolayer,
Si–C linked to a silicon substrate, with either lecithin : cholesterol, phosphatidylcholine : cholesterol (both referred to as PC : cholesterol), or POPC :
cholesterol deposited as the upper leaflets of the hBLMs. The conductance

and capacitive values of these hBLMs were calculated from the total impedance '$Z(\omega)$' data obtained from the electrical impedance spectroscopy (EIS) using Maxwell–Wagner fits by applying an equation as a model for sandwiches of multiple electrically homogenous planar layers (*e.g.*, hBLMs immobilized on silicon wafers) (see Eqn 2.1):[5]

$$Z(\omega) = \frac{1}{G} + \sum_{i=1}^{N} \frac{1}{G_i + j\omega C_i} \qquad \text{where } j \equiv \sqrt{-1} = e^{\frac{j\pi}{2}} \qquad (2.1)$$

In these equations, ω is the angular frequency of the ac current, G is the resistance component of conductance for the electrolyte, G_i is the conductance of the ith layer of the hBLM sandwich, and C_i is the capacitance of the ith layer of the hBLM sandwich.[5]

The calculated capacitance values for the octadecane monolayer from these calculations were seen to be 18 ± 2 mF m^{-2}; the data are in good agreement with previously reported values.[6,7] The data for a PC:cholesterol anchored on the alkanethiol monolayer gave a capacitance value of 10 mF m^{-2}, which is lower than the value that was reported for freestanding PC:cholesterol BLMs (13.2 ± 0.4 mF m^{-2});[8] this lower capacitance is as expected, as a thicker leaflet should be formed on the densely packed alkanethiol monolayer compared to a free-standing BLM.

Furthermore, one of the benefits of immobilizing biomolecules within a lipid bilayer membrane is the fluidity which the membrane imparts, allowing for some movement within the bilayer. At the same time, adhesion of the lipid film to the bilayer is of benefit, preventing breakdown of the membrane and unwanted transit of analytes to the sensing surface. Tethering of the lipid membrane may be accomplished through self-assembled monolayers, peptide linkages, membrane spanning molecules, polymer links, or other methods. One such example of a polymer linkage and its effect on membrane properties was presented by Lehmann and Rühe.[9] As mentioned in their work, the use of a polymer membrane monolayer can serve as a support layer for lipid membranes attached to solid surfaces (*e.g.*, gold, silicon oxide), allowing for greater membrane fluidity and consequently more functionality of embedded proteins (*e.g.*, receptors in protein–receptor couples), while balancing sufficiently strong attachment of the membrane to the surface of interest. In this study, the authors investigated chemisorptions of polyethyloxazoline monolayers by grafting to solid substrates or cationic chain-opening polymerization growth from the substrate surface. The method of grafting the polymer onto the substrate provided limited control of film thickness to a few nanometers, due to multiple binding sites of the polymer chain to the substrate; polymer chains of the monolayer grown on the substrate by cationic chain opening polymerization by a tosylate initiator facilitated control over the film growth to greater thicknesses (20–30 nm). Additionally, the authors investigated the swelling of polymer in humid environments and the non-specific adsorption of fibrinogen. Exposure of the monolayers to humid environments resulted in the polymer layer swelling to a thickness up to 50%; the swelling is important to understand for the stability of the polymer, as most biological assays are aqueous. The non-specific protein adsorption investigation indicated that when polymer repeat units were above $n = 150$, the non-specific fibrinogen adsorption layers were 0–0.3 nm thick, indicating negligible non-specific binding. Limiting non-specific protein binding is critical to allow protein–receptor specific interactions for biosensing applications.

The conformations of biomolecules influence their efficacy, so maintenance of their functionality while immobilized within a lipid membrane is important for their use as analyte recognition elements. Presentation of the appropriate catalytic or receptor/ligand binding moieties may be influenced by membrane fluidity, as referenced previously, but membrane charge and hydrophobicity or hydrophilicity may also play a role in appropriate presentation. This concept was investigated, as an example, by Li *et al.*, for functionality of glucose oxidase.[10] The conformational changes of glucose oxidase (GOx), embedded within a lipid membrane, were shown to be strongly influenced by the charge of the lipid membrane. Langmuir–Blodgett films of eight glycolipids (alkyl 2-amino-2-deoxy-β-D-glucopyranoside and 1,2-*o*-dialkyl-3-*o*-β-D-glycosylglycerols) of varying chain lengths were embedded with GOx and changes in the conformation of the protein were observed using circular dichroism. For the neutrally charged lipids (1,2-*o*-dialkyl-3-*o*-β-D-glycosylglycerols), the strength of GOx immobilization within the lipid was governed by hydrophobic chain length, showing stronger interactions with longer chains. However, in the case of the positively charged lipids the chain lengths showed no effect on immobilization, suggesting that electrostatic forces between the positively charged lipid and the negatively charged GOx dominate the interaction. Additionally, the hydrophobic forces playing a role in the interaction between the neutral lipid and GOx allow the protein to penetrate more easily into the membrane, maintaining an α-helical formation. Conversely, the electrostatic forces influencing GOx interaction with the positive lipid result in unfolding of the protein into a β-sheet configuration. Circular dichroic spectral analysis of the enzymatic activity of GOx with indigo carmine indicated that GOx retains more active enzymatic activity when immobilized within the lipid membranes in the α-helical conformation; this information is important to understand for future lipid selections for GOx in biomimetic sensor applications.

Another example of appropriate biomolecule presentation was illustrated by Hilbig *et al.* in their investigation of antibody detection.[11] In this study, multilamellar vesicles of cholesterol/phosphatidylcholine/cardiolipin (25%/65%/10% by mass) were deposited as membranes onto transducers (glass with 10 nm Ta_2O_5 and 330 nm SiO_2) that were silanized and coated with DCPEG/PEG (70%/30% by mass). β_2GP-I molecules were immobilized within the membranes as recognition units for the anti-β_2GP-I antibodies, which are critical markers for the diagnosis of antiphospholipid syndrome (APS). Accurate detection of antibodies for APS has suffered from a lack of sensitivity for the key epitopes of β_2GP-I; immobilization of β_2GP-I within a negatively charged biomimetic lipid membrane allows for the presentation of the key epitopes of the molecules that are needed for appropriate antibody binding. The sensor chip created in this study was designed for use in a flow cell for reflectometric interference spectroscopy (RIfS) for rapid, point-of-care diagnosis. The biosensor chip was able to successfully detect anti-β_2GP-I in a concentration-dependent manner, could be regenerated rapidly for additional assay testing of new samples by washing away the immobilized antigens and antibodies with a mild solution of tris(2-carboxyethyl)phosphine, and was not prone to false antibody recognition when tested with another polyclonal IgG antibody (anti-bovine serum albumin).

2.2.2.2 Enzymatic

Enzymes may be incorporated into these lipid membranes and measurements of the reaction cascade resulting from analyte interaction with the enzymes may be measured. This function may be utilized to detect a target analyte in solution if it effectively interacts with an enzyme. As such, detection of the target analyte can be accomplished as a result of the signal that is produced from the reaction. Often, these enzymatic interactions are measured electrochemically. For instance, Naumann *et al.* prepared tBLMs from the rupture and deposition of large unilamellar vesicles (LUVs) of phosphatidylcholine from egg yolk, anchored to a gold substrate by thiopeptides.[12] A transmembrane protein, cytochrome *c* oxidase (COX), was incorporated into the lipid membranes as a means of characterizing proton transport across the membrane to the electrode surface in the presence of cytochrome *c*. COX is capable of transporting protons against an electrochemical gradient, utilizing the reduction of dioxygen in cytochrome *c* as an electron source. Values of the membrane thickness were determined from surface plasmon fluorescence spectroscopy (SPFS) as well as surface plasmon spectroscopy (SPS); the optical thickness of the lipid bilayer membrane atop the thiopeptide spacer, with incorporated COX, measured at 12.2 nm thick. Changes in thickness could be monitored in real-time *via* SPS as the lipid membrane was forming and the COX enzyme was being incorporated. Activity of the COX with the addition of cytochrome *c*, as analyzed by square wave voltammetry (SWV) and chronoamperometry, indicated cathodic and anodic current peaks at -0.7 V and $+0.3$ V (*vs.* Ag/AgCl), respectively. There was a notable decrease in resistance of the membrane in a dose-dependent manner with respect to cytochrome *c* concentration. This observation could be reversed in the presence of cyanide, which acted as a COX inhibitor by binding to the enzyme, preventing its activity as a proton pump. However, while analysis of the electrochemical impedance data was able to show active transport mediated by COX for the first time in a tBLM, the data indicated that the tBLM covered only 70% of the electrode (consequently 30% of the coverage was defective). Improvements are still needed, but the ability to utilize a combination of optical and electrochemical characterization techniques using this model holds promise for its applicability to biosensing technology.

Another example of enzyme incorporation into a lipid membrane was described by Nakaminami *et al.*[13] A gold electrode was prepared with a self-assembled monolayer (SAM) of *n*-octanethiolate, followed by deposition of a redox agent, 1-methoxy-5-methylphenazinium (MMP), as an electron mediator and a biomimetic lipid bilayer of L-α-phosphatidylcholine β-oleoyl-γ-palmitoyl (PCOP) embedded with uricase (urate oxidase, UOx). The coupling of UOx with MMP allowed for more selective detection of uric acid, limiting interference from competing analytes (*e.g.*, ascorbic acid), as MMP possesses a lower redox potential than the naturally occurring O_2/H_2O_2 mediator and UOx provides selectivity for uric acid over other blood compounds. Cyclic voltammetric characterization of the biosensor confirmed that MMP was successfully sequestered between the gold electrode and the lipid bilayer, as part of the oxidation mechanism of uric acid with UOx. Amperometric detection of uric acid solutions was carried out at 0 V *vs.* saturated calomel electrode (this potential is positive enough to oxidize the MMP) and detection was successful within the range typically seen within blood samples of healthy humans (0.15–0.4 mmol dm^{-3}), with a detection time of 20 s. Furthermore, detection of uric acid was successful without interference from the competing analyte, ascorbic acid. However, the authors noted that they observed changes in sensitivity over successive measurements with an electrode, potentially due to leakage of the UOx molecules from the lipid membrane. There are plans for investigation into the effect of bonding together the lipid molecules on the electrode as a means of more securely anchoring the UOx molecules within the matrix.

Godoy *et al.* utilized an electrochemiluminescence detection mechanism for an enzymatic recognition of acetylcholine.[14] This study utilized the immobilization of acetylcholinesterase (AChE) within a lipid bilayer membrane, coupled to a non-inhibiting IgG monoclonal antibody that holds the enzyme in a functional orientation, deposited upon a cross-linked thin film of poly(vinyl alcohol) with styrylpyridinium groups (PVA-SbQ) encapsulating choline oxidase (ChOD) to facilitate an electrochemiluminescent (ECL) reaction which can be measured to detect the concentration of acetylcholine in the sample of interest. The PVA-SbQ layer with ChOD is photopolymerized onto a screen-printed graphite working electrode, providing the electrical connection for electrochemiluminescence. The series of steps involving the breakdown of acetylcholine to hydrogen peroxide, which can then interact with luminol in solution to emit light,[14] is shown in Eqns (2.2)–(2.4):

$$\text{Acetylcholine} + H_2O \xrightarrow{\text{AChE}} \text{Acetate} + \text{Choline} \tag{2.2}$$

$$\text{Choline} + H_2O + 2O_2 \xrightarrow{\text{ChOD}} \text{Betaine} + 2H_2O_2 \tag{2.3}$$

$$\text{Luminol} + H_2O_2 \xrightarrow{\text{graphite} + 0.45\,\text{V}\,vs.\,\text{Ag/AgCl}} \text{3-Aminophthalate} + 2N_2 + 3H_2O$$
$$+ h\nu\,(\lambda_{max} = 425\,\text{nm})$$

$$\tag{2.4}$$

10-Undecyloxymethyl-3,6,9,12-tetraoxatricosyl 2-acetamido-2-deoxy-β-D-gluco-pyranoside (a glycolipid) with IgG incorporation was deposited onto the electrode *via* a modified Langmuir–Blodgett and vesicle fusion process, and the bilayer was subsequently functionalized with AChE by incubation in AChE solution. The electrode was then placed into a glass cuvette and a calibration curve for the ECL reaction with choline was created as a background comparison for acetylcholine recognition. The linear range for choline detection is 4×10^{-7} M to $\sim 10^{-4}$ M, with a sensitivity of ~ 40 arbitrary units (a.u.) μM^{-1}. The authors note that the lipid bilayer does not serve as a diffusion barrier for either choline or luminol, likely due to the ~ 17 nm thickness of the bilayer. The linear range for acetylcholine detection is 4×10^{-7} M to 7×10^{-5} M, a sensitivity of ~ 40 a.u. μM^{-1}, and light intensity range of 40–2000 a.u. Furthermore, the device is able to maintain functionality over an extended period of time. At days 12, 15, and 26, the choline/acetylcholine detection sensitivity ratios are 0.81, 0.85, and 0.90, respectively; the theoretical values of these calculated ratios must not exceed the stoichiometric value of 1.0, as a single mole of acetylcholine can only produce a single mole of choline. The devices experience a decrease in the absolute values of their detection capacity, but the ratio remains effective as a means of detecting acetylcholine. This device also offers a platform for the integration of other antibodies and recognition molecules.

A later study by the same group built upon this ECL concept.[15] The innovation in their design was the use of a reagentless sensor which integrates the luminol directly into the lipid bilayer membrane. An amphiphilic luminol derivative named TF46 [*N*-(1,4-dioxo-1,2,3,4-tetrahydrophthalazin-5-yl)-13-undecyloxymethyl-3,6,9,12,15-pentaoxahexacosanamide], functionalized with IgG and choline oxidase (ChOD), was applied to the graphite working electrode and confirmed by AFM. Two types of reagentless lipid coatings were tested with ECL. The first type utilized a bilayer of TF46 with anti-ChOD IgG and bound ChOD [(TF46-IgG)-ChOD]. The second type utilized a bilayer of TF46 that was subsequently coated with a bilayer of GC11 glycolipid that was functionalized with anti-ChOD IgG and bound ChOD. These two types of sensors displayed sensitivities of 20 a.u. M^{-1} and 50 a.u. M^{-1}, respectively, with response times of ~ 30 s. However, the authors note that additional work is needed to improve the sensitivity of the devices.

A more purely electrochemical sensor was devised for ATP detection with an ATPase by Bücking *et al.*[16] An adenosine triphosphate (ATP) enzyme, EF_0F_1-H^+-ATPase, was immobilized within a lipid bilayer membrane, covalently coupled to a gold electrode by 16-mercaptohexadecanoic acid, and examined electrochemically with cyclic voltammetry (CV). Following the deposition of the self-assembled monolayer of 16-mercaptohexadecanoic acid, a lipid monolayer of dimyristoylphosphatidylethanolamine (DMPE) was applied. Proteolysosomes of DMPE with incorporated EF_0F_1-H^+-ATPase were immobilized onto the lipid monolayer surface to create a bilayer membrane. The capacitance of the SAM, monolayer, and bilayers were characterized using the following formula[16] (Eqn 2.5):

$$C = \Delta I / 2Av \tag{2.5}$$

In this case, C is the capacitance, ΔI is the difference in value between cathodic and anodic currents, A is the electrode surface area, and v represents the scan rate used in the CV.[17] Using this formula, the capacitance values for the SAM, monolayer, and bilayer were 60 nF cm^{-2}, 2.8 μF cm^{-2}, and 500 nF cm^{-2}, respectively. The authors note the increase in capacitance with the addition of the monolayer to the SAM, followed by a decrease in capacitance when the bilayer is deposited is likely due to the hydrophobic nature of the monolayer, while the SAM and bilayer are hydrophilic and polar.

Next, the ATPase activity was examined, as active proton transport occurs through the membrane to the electrode surface as ATP is hydrolyzed, according to the formula (Eqn 2.6):[16]

$$2H^+ + 2e^- \leftrightarrow H_2 \tag{2.6}$$

The authors observed an increase in reductive current with increasing ATP concentrations (2.5–60 mM), covering the physiologically relevant range of 2.5–10 mM.[16,18] However, the authors unexpectedly observed that the peak reduction current occurred at a potential of $E = -1.1$ V $vs.$ Ag/AgCl, when the theoretically calculated value ($E = -0.059$ V × pH) for the pH-dependent reaction should have been $E = -0.1$ V; the buffer used in the system was 100 mM NH$_4$Cl at pH 5.1. The authors attributed this observation to a localized pH decrease at the electrode surface, where the rate of proton reduction was exceeding the rate by which ATPase could transport protons through the membrane. However, despite this unexpected result, the ATPase activity was selective, as there were much lower current densities observed in the presence of ATP for membranes without ATPase incorporation. Additionally, the sensors were highly sensitive, with a calculated sensitivity for ATP of 50 μA mM^{-1} cm^{-2}.

In a final enzymatic example, Choi *et al.* utilized a supported lipid bilayer (SLB) of 1,2-dipalmitoyl-*sn*-glycero-3-phosphocholine (DPPC) on a SAM of 11-mercaptoundecanoic acid (MUA), analyzed by AFM and surface plasmon resonance spectroscopy (SPRS), as a means of examining the effects of copper/zinc superoxide dismutase (SOD1) aggregates on the biomimetic membrane's integrity.[19] Human SOD1 variants are associated with the disease amyotrophic lateral sclerosis (ALS)[20] and the aggregates of SOD1 are believed to be a potential cause.[21] Deposition of the DPPC SLB by unilamellar vesicles was confirmed by both AFM and SPRS. Furthermore, upon introduction of the SOD1 aggregates, a decrease in the SPRS incident angle was observed over time, indicating a lower refractive index due to a loss of surface coverage of the SLB. This loss of surface coverage is attributed to dissociation of DPPC molecules from the SLB as a result of interaction with the SOD1. AFM confirmed that the SLB had areas that were multiple layers thick as a result of the lipid dissociation from the SLB and deposition onto intact areas of the SLB. Furthermore, SPRS reflectance measured a decrease in reflectance of the SLB after exposure to SOD1 aggregates, indicating defect formation in the SLB. The authors suggest that these results indicate the feasibility of this device for *in vitro* screening of protein aggregates associated with Alzheimer's disease, Huntington's disease, Parkinson's disease, and other neurodegenerative diseases.[22]

2.2.2.3 Pore Formation and Ionic Transit

In addition to enzymatic reactions, another means by which lipid membranes can be functionalized and examined is with pore forming complexes or ionic transporters. The pore forming and ionic transit complexes are composed of biomolecules incorporated into the membrane, facilitating movement between the analyte solution and the surface of the transducer. Some of the pore forming complexes include gramicidin, alamethicin, α-hemolysin, and classes of porins, which are obtained from various microbes and may be immobilized within the lipid membranes to examine their influences over the membrane integrity and ionic transit. Furthermore, valinomycin, a potassium ion-selective ionic transporter, can be incorporated into membranes to move potassium ions against an electrochemical gradient. One benefit of deliberately incorporating porosity into a membrane or ionic transporters is that the effects of drug compounds meant to target those transit elements can be examined. For instance, many of these pore forming or ionic transporting complexes have origins in microbes, which bind to cellular membranes, disrupting the cellular integrity. Pharmaceutical compounds that can target the proteins and peptides which cause pore formation are of interest and can be studied on these model cellular membranes, typically when immobilized on an electrode for sensing changes to the electrochemical properties of the model membrane.

One such example of the concept of pore formation was illustrated by Woodhouse *et al.* to create an ion channel biosensor.[23] In this work, the authors described a biosensor platform that utilized tBLMs with a combination of diphytanylphosphatidylcholine (DPEPC) and glycerodiphytanyl (DPG) lipids linked to a gold membrane electrode *via* mercaptoacetic acid disulfide. Both mobile gramicidin A-coupled (gA) and tethered membrane spanning lipid-coupled (tMSL) receptors were utilized to facilitate ion transport through the membrane. The alignment of the mobile gA in the upper tBLM leaflet and the tethered gA in the lower tBLM leaflet resulted in a gA dimer that allowed ion transport across the membrane. The device utilized either a sandwich assay or a competitive assay for analyte recognition. In the sandwich assay, receptors are coupled to both mobile gramicidin A ion channels in the outer leaflet of the tBLM and to membrane spanning lipid (MSL) complexes that are tethered to the gold electrode. The membrane has a higher percentage of receptors bound to MSL complexes, compared to the mobile phase receptors. As the analytes bind to the receptors, the analyte has a higher chance of binding to the receptors on the bound MSL complexes, but the mobile phase receptors are able to migrate laterally within the bilayer and bind to the analytes in close proximity to these immobilized locations. Consequently, the mobile phase receptors are pulled away from the tethered gA channels in the lower leaflet of the tBLM in order to bind with the analyte, reducing the ability for ions to pass through the tBLM, and thus reducing the conductivity of the membrane. In the competitive binding assay, antibody complexes corresponding to the analyte of interest are multiply-bound to both immobilized and mobile receptor complexes in the membrane. Upon introduction of the analyte, the receptors can dissociate from the antibody complex and bind the analyte. During this process, the mobile receptor units are able to move laterally within the membrane upper leaflet, aligning with tethered gA channels in the lower membrane leaflet. The alignment of the upper and lower leaflet gA channels due to the analyte binding (*e.g.*, digoxin) creates the gA dimers, facilitating transport of ions across the membrane, and consequently increasing the conductance of the membrane (see Figure 2.1b below). Electrochemical impedance measurements and mathematical models were applied to these tBLM biosensors for detection of *Listeria monocytogenes* DNA (using a DNA 19-base probe), *Escherichia coli* (bacterium), thyroid stimulating hormone (protein), and digoxin (small molecule cardiac drug). There was good agreement between the modeling and the detection for these targets, with clear conductance changes of the membrane upon target molecule presentation.

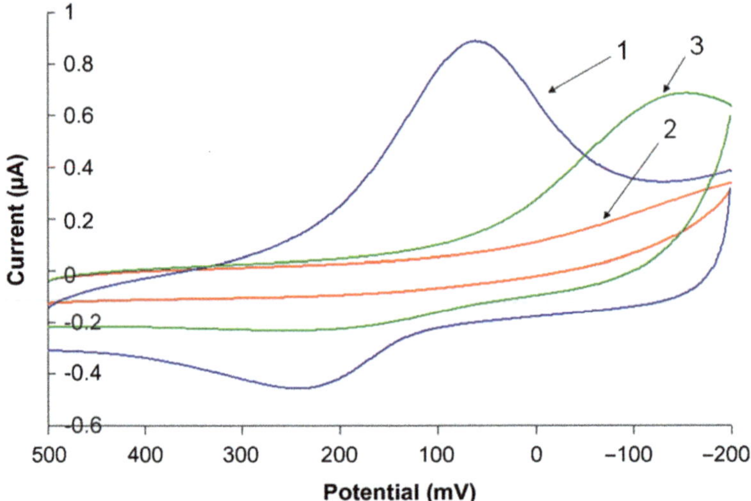

Figure 2.1 CV curves for a bare gold electrode coated with a tethering lipid monolayer (1), a tBLM formed by deposition and rupture of DOPC liposomes on top of the tethering lipid monolayer (2), and the tBLM after PorB incorporation (3). The redox peaks were absent after bilayer formation but were observed after PorB incorporation, suggesting PorB-mediated transport of ferricyanide ions through the tBLM.
(Reproduced from Jadhav *et al.*,[31] with permission from Elsevier.)

Gufler *et al.* looked into pore formation in lipids using alamethicin and potassium transit with valinomycin.[24] Langmuir–Blodgett (LB) bilayer membranes of a phospholipid (DPhyPC) and monolayer membranes of bolaamphiphilic main tetraether phospholipid (MPL) were formed on a re-crystallized bacterial surface layer (S-layer) protein, SbpA, which was obtained from *Bacillus sphaericus*; the S-layers were recrystallized upon gold substrates. The S-layer supported membranes formed on gold were compared against lipid membranes formed on S-layer functionalized porous ultrafiltration membranes (SUMs) and microfiltration membranes (MFMs). The SUMs consist of polyamide MFMs with deposited fragments of S-layer from *B. sphaericus* CCM 2120, with/without cross-linking by gluteraldehyde. These S-layer type proteins, which have origins from bacteria or archaea, typically possess a molecular weight between 40 and 200 kDa, forming lattices of 5–10 nm in thickness, and porosity varying in size (2–8 nm) and prevalence (30–70% porosity).[25,26] As summarized by Schuster *et al.*, proteins from these prokaryotic microorganisms may be recrystallized to create supports onto a variety of substrates, allowing for semi-fluid, solid-supported, free-standing, or porous substrate-supported lipid membranes.[23] These S-layer supported lipid membranes can subsequently be functionalized with proteins, peptides, or other relevant biomolecules for use in detection studies.[25,26]

The peptides alamethicin and valinomycin, which serve as a voltage-dependent pore-forming peptide and as a potassium-selective ion carrier, respectively, were incorporated into the membranes. Incorporation of alamethicin into the DPhyPC membranes showed a shift in the membrane toward resistive behavior in the middle of the frequency range of the EIS: the resistance lowered from ~ 80 MΩ cm^2 to ~ 950 Ω cm^2; no observable change in capacitance was noted. This shift in behavior occurred without an applied dc voltage and is indicative of pore formation by the alamethicin in the membrane. Additionally, the pores formed by the alamethicin could be blocked by the addition of the sodium channel blocking compound 5-(*N*-ethyl-*N*-isopropyl)amiloride, resulting in a dose-dependent partial return of resistance to the membrane. Upon the addition of 10 μM and 20 μM amiloride, the resistance values of the DPhyPC membranes containing SbpA increased to 1120 and 1310 Ω cm^2, respectively. A similar effect was seen for the MPL membranes supported by a SbpA lattice, but could be enhanced by applying an electric field. Ion channel formation was optimized at a dc bias voltage of -5 mV. Furthermore, the pores could be partially blocked by the addition of the channel inhibitor amiloride–HCl–H$_2$O, as indicated by an increase in the membrane resistance.

Furthermore, valinomycin incorporation as a transmembrane carrier for potassium ions (K$^+$) was tested in the MPL membranes on the SbpA supports with gluteraldehyde cross-linking. The addition of the valinomycin showed a drop from ~ 30 to ~ 17.4 MΩ cm^2 in the presence of the buffer containing 10 mM Tris and 50 mM NaCl (pH 7.5) that was used throughout the experiments. However, as valinomycin is more selective for transport for

K^+ compared to Na^+, the inclusion of 50 mM of KCl lowered the membrane resistance to $R_m = 42.5$ kΩ cm^2. The increase in conductivity indicated that valinomycin-mediated K^+ transport was occurring and imparted a K^+/Na^+ ion transport ratio of 410, showing similar values to those seen in the literature for black lipid membranes[27,28] and tethered lipid membranes.[29] The ability to form membranes with comparable impedance measurements to the literature, incorporate peptides to facilitate transmembrane transport of ions, and subsequently block the peptide-mediated ion transport with pharmaceutically relevant compounds lends this technology to biosensing applications.

In a similar capacity to gramicidin and alamethicin, the protein α-hemolysin can cause pores in lipid membranes and may be incorporated to provide porosity, as confirmed by electrochemical changes in membrane behavior. Bao *et al.* were able to prepare lipid bilayer membranes on gold by vesicle deposition of HDMC [*N*-hexadecyl-3,6-di(*p*-mercaptophenylacetylene)-carbazole].[30] Vesicle deposition of HDMC on gold results in a disorganized self-assembled monolayer; the disorganization is due to the molecular defects that form during the self-assembly, allowing for loose packing of the molecules, which have a large and rigid head group while the alkyl-chain tail is small and flexible. The resulting SAM that forms is much more permeable to hydrophobic species than those that are hydrated, though experimentally both hydrophilic and hydrophobic redox molecules could permeate the monolayer. Furthermore, a hybrid bilayer membrane (HBM) was formed by vesicle deposition of didecanoyl-L-R-phosphatidylcholine (DPPC) onto the SAM. Integration of the enzyme horseradish peroxidase (HRP) as well as the pore-forming protein α-hemolysin (αHL) was investigated on the HBM. The integrated proteins in the HBMs demonstrated the feasibility of the device for biomimetic membrane research and examination of the membranes *via* CV and impedance spectroscopy. CV and impedance measurements of the αHL-incorporated HBMs indicated that they exhibited an increase in reductive current and a decrease in impedance, respectively, for ferricyanide ions; these results are indicative of pore formation in the HBMs and an increase in membrane permeability. Furthermore, HRP incorporation into the HBMs was confirmed *via* redox peaks for phosphate buffered saline during CV, indicating direct electron transfer by HRP to the gold electrode surface.

PorB class II porin, a protein from *Neisseria meningitidis* that can create pores in host-cell membranes during *Neisserial* infections, was used to functionalize a tBLM on a gold electrode for EIS and cyclic voltammetric investigation, in a study by Jadhav *et al.*[31] The gold electrode was exposed to the phospholipid 1,2-dipalmitoyl-*sn*-glycero-3-phosphothioethanol (DPPTE) to create a lipid monolayer, tethered by its thiol group, and exposed to liposomes of 1,2-dioleoyl-*sn*-glycero-3-phosphocholine (DOPC) to create the upper leaflet of the tBLM. Introduction of PorB into the membrane permeabilized it to ions (*e.g.*, Na^+ and Ca^{2+}). EIS analysis confirmed the formation of the tBLM as capacitance and resistance values for the membranes in NaCl solutions were 0.709 μF cm^{-2} and 2.50 MΩ cm^2, respectively; in CaCl$_2$ solutions, the tBLM had a capacitance of 0.81 μF cm^{-2} and a resistance of 24 MΩ cm^2. Furthermore, PorB incorporation was shown to drop the resistance of the membranes to 604 kΩ cm^2 in NaCl solution and 1.93 MΩ cm^2 in CaCl$_2$ solution. These results indicate the integration of the pore-forming protein within the membrane, facilitating the transit of the Na$^+$ and Ca^{2+} ions across the tBLM. Additionally, the authors confirmed that cyclic voltammetry (CV) could be performed with solutions of ferricyanide, on tBLM with PorB incorporation (Figure 2.1). Prior to PorB introduction, no redox peaks for ferricyanide were observed during CV scans of the tBLMs. However, following the formation of pores with the PorB, a reduction peak for ferricyanide was observed. These results indicate that this technology may be lent to screening of pharmaceuticals that target PorB class II porin pore formation or as a model to study ion channels.

2.2.3 Lipids on Porous or Nanostructured Substrates

While the previous sections examined the immobilization of lipid membranes on planar surfaces, this section looks at efforts to deposit lipid membranes on porous or nanostructured substrates. Incorporating porosity into the substrate can allow access to both sides of the deposited membranes, limit reagent volumes, isolate membranes into arrays for screening processes (*e.g.*, pharmaceutical testing), allow for imaging on more than just the exterior surface, or create microcavities to limit diffusion distances during reactions. Furthermore, incorporation of nanomaterials or nanostructured substrates may provide enhancement to lipid membrane stability and/or functionality of incorporated proteins and peptides, by more naturally mimicking their material interactions at the nanoscale.

Costello *et al.* described the creation of a sandwiched agarose gel biosensor, incorporating either a dioleoylphosphatidylcholine (DOPC) or polysiloxane cross-linked phosphatidylcholine (PSPC) lipid bilayer-coated channel to regulate the passage of electrolyte.[32] Mica substrates were used as the outer layers of the sandwiched biosensors, with the bottom layer prepared with an Ag/AgCl electrode for ion detection. A thin piece of PTFE was perforated with a 50–100 μm pore and placed onto the bottom slab of agarose. A thin layer of lipid solution (DOPC, DOPC with valinomycin, or PSPC) was spread over the PTFE membrane to form a lipid bilayer over the pore, and a second slab of agarose was placed on top (covered by a mica layer with a pore for drop-casting electrolyte samples onto the agarose). Conductance values for the analytes showed values <25 and <2 S m^{-2} for DOPC and PSPC, respectively; furthermore, DOPC that was exposed to gramicidin (a bilayer pore-forming agent) showed an order of magnitude increase in conductance, indicative of pore formation in the bilayer which allows transport of ions. Capacitance measures of the lipid layers upheld expected values (8–9.5 mF m^{-2}) for a bilayer, as opposed to stacks of bilayers. Additionally, lipid layers prepared with valinomycin for potassium ion transport showed conductance values as expected when exposed to KCl solution. These results indicate the potential use of the sandwich design as a lipid-bilayer, gel-supported biosensor.

Evans and Fritsch described the development of a microcavity electrode with a suspended polymer and phospholipid membrane for the electro-chemical detection of analytes.[33] Their microcavity was fabricated by reactive ion etching of two gold-coated polyimide layers (4 μm thick) atop a gold-coated silicon wafer. The etching resulted in a microcavity with diameter of 10 μm, depth of 8 μm, and geometric volume of 0.6 pL. A tubular nanoband (TNB) of gold around the middle of the well formed from the etching of the first gold-coated polyimide layer and gold on the top surface/microcavity rim of the macrostructure from the etching of the second of the gold-coated polyimide layers served as the self-contained electrodes (working and auxiliary/pseudo-reference, respectively) for the electrochemical character-ization of the device. The gold surfaces were functionalized with a self-assembled monolayer (SAM) of 11-mercaptoundecanoic acid (MUA) as anchor points for the polymer and lipid layer; 10,12-pentacosadiynoic acid (PDA) was the polymer and dipalmitoylphosphatidylcholine (DPPC) was the lipid used as the membrane suspended over the microcavity. The SAMs were electrochemically removed from the gold surface on the bottom of the microcavity and the TNB to avoid cross-linking of PDA within the well. The PDA layer was applied by Langmuir–Blodgett application of a thin film and UV-polymerized in place. Subsequently, a Langmuir–Blodgett application of the DPPC was applied to fill any gaps in the membrane formed by the PDA while imparting additional fluidity as well as amphiphilicity to the mem-brane. The formation of the membrane over the microcavity was confirmed by Langmuir–Blodgett isotherms and polarization-modulation Fourier transform infrared reflectance absorption spectroscopy (PM-FTIR). Electro-chemical characterization using CV was performed with a subphase of 10 mM $K_4Fe(CN)_6$ in 0.1 M KCl or 1 mM $CaCl_2$; the electrochemical charac-terization with the ferrocyanide solution demonstrated that the electrolyte was enclosed within the microcavity and could be analyzed amperome-trically. This type of microcavity with self-contained electrodes and a bio-mimetic membrane can be utilized for biosensor fabrication, allowing for amperometric and voltammetric analysis of analytes that would otherwise be difficult to achieve with electrode configurations where the working, aux-iliary, and reference electrodes are separated by greater distances.

Thin silicon membranes with micron and submicron pores serving as small chambers over which lipid bilayers may be suspended by Langmuir–Blodgett deposition have also been investigated.[34] These silicon membranes (thickness 300 nm) contained arrays of square pores with sizes of 300, 650, and 1000 nm, which were created by a combination of UV lithography, chemical vapor deposition, and wet etching. The goal of having a 300 nm-thick porous membrane for lipid suspension is to provide access to both surfaces of the lipid membranes for tasks such as analyte exchange, mem-brane functionalization (*e.g.*, ion channel introduction), fluorescence detection, and characterization of membrane protein organization upon ligand binding. The resulting pores were examined using atomic force microscopy (AFM) and scanning electron microscopy (SEM). These studies revealed that the pores were not perfectly squared and that the 300 nm pores were closer to 390 nm in size due to over-etching, but overall the etching was successful. Furthermore, the lipid suspension of dimyristoyl-L-α-phosphatidic acid (DMPA) and dimyristoyl-L-α-phosphatidylcholine (DMPC) (2:1 molar ratio) suspended over the pores by Langmuir–Blodgett deposition were confirmed by AFM. For instance, the indentation depth by a AFM tip under a load of 0.5 nN entered 144 nm into the bare 300 nm pores, while the AFM tip only descended 12 nm in the presence of the suspended lipid bilayers. Additionally, it was observed that lipid bilayers suspended over the smaller pores were more stable over time and could withstand higher AFM force loads than lipids suspended over larger pores. Overall, the fabrication was successful and lipid bilayer depositions were confirmed.

Another pore-array structure was examined by Hansen *et al.*[35] This study examined the fabrication of arrays of black lipid membranes (BLMs) suspended over pores (diameter 300 ± 5 μm, center-to-center spacing 400 μm) ablated into 50.8 μm-thick films of ethylene tetrafluoroethylene (ETFE) by a CO_2 laser for optical (fluorescence) and electrochemical (voltage clamp) characterization of the BLMs; the arrays were created as 8×8 and 24×24 rectangles, as well as 24×27 hexagonal arrays. The BLMs were composed of 1,2-diphytanoyl-*sn*-glycero-3-phosphocholine (DPhPC) (a lipid) that was doped with 1 mol% of 1-oleoyl-2-{6-[(7-nitro-2-1,3-benzoxadiazol-4-yl)-amino]hexanoyl}-*sn*-glycero-3-phosphocholine (NBD-PC) (a fluorescent lipid) and were deposited by a modified lipid painting process. Brightfield and fluorescence imaging of the NBD-PC fluorescence at the Plateau–Gibbs border facilitated confirmation of the lipid painting process (Figure 2.2). Furthermore, the authors were able to functionalize the 8×8 array membranes with the peptides valinomycin and gramicidin A, as well as the proteins α-HL and the porin FomA. By utilizing these peptides and proteins, ionic transport was facilitated across the membrane, which was verified by the voltage-clamp current and capacitance measurements. The 24×24 and 24×27 arrays were also successfully functionalized with α-HL and confirmed by electrochemical measurements. Additionally, it was noted that the larger arrays had longer functional lifetimes (~ 16 h) compared to the 8×8 arrays. The scalability of these arrays may lend them to pharmaceutical screening processes or other biosensing applications.

Lastly are a couple of examples of nanomaterial integration with these lipid films. Largueze *et al.* investigated the tethering of lipid membranes into the nanoscopic channels of alumina membranes.[36] Tethered bilayer lipid membranes (tBLMs) of a egg yolk L-α-phosphatidylcholine type XVI-E (EggPC), 1,2-dioleoyl-*sn*-glycero-3-phosphatidylethanolamine (DOPE), 1,2-distearoyl-*sn*-glycero-3-phosphoethanolamine-*N*-carboxy-poly(ethylene glycol) 2000) (DSPE-PEG$_{2000}$COOH), and ubiquinone (co-enzyme Q$_{10}$, UQ) were deposited from lipid vesicle fusion within a nanoporous alumina membrane

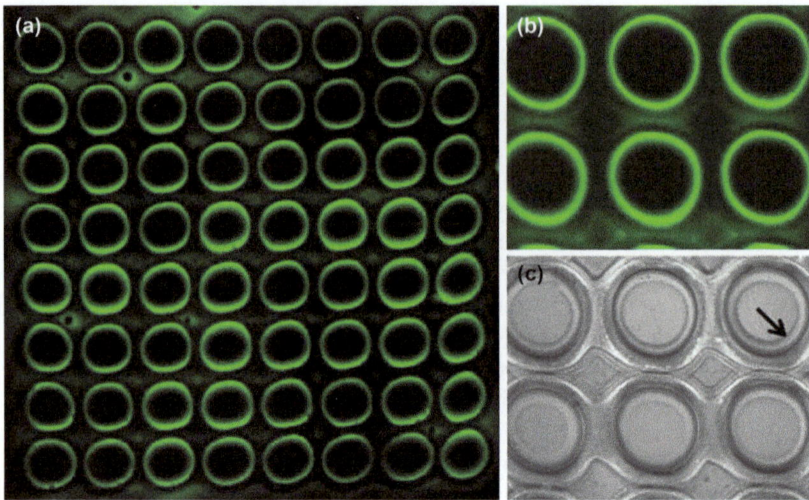

Figure 2.2   Fluorescent and brightfield images of established horizontal 8×8 bilayer arrays. (a) Fluorescent image of established 8×8 bilayer array with 25 mg mL^{-1} DPhPC in *n*-decane doped with NBD-PC (1 mol%). Image was acquired with a 2.5× air-corrected objective. (b) Fluorescent image acquired with a 10× air-corrected objective, and (c) the corresponding brightfield image. Indicated is the visual appearance of the Plateau–Gibbs border (*black arrow*).
(Reproduced from Hansen *et al.*,[35] with permission from Springer.)

(11.6 ± 0.5 µm average thickness; 184 ± 9 nm average pore diameter). The alumina was coated on one face with a gold–chromium layer (800 nm and 3 nm thick, respectively) as an electrical connection for electrochemical characterization of the tBLMs. The walls of the alumina pores were functionalized with (3-aminopropyl)triethoxysilane (APTES), the gold was functionalized with undecanethiol, and vesicle fusion (vesicles were EggPC/DOPE/DSPE-PEG$_{2000}$COOH/UQ, 67:30:1:2 mol) to the functionalized surfaces was triggered by incubation with *N*-(3-dimethylaminopropyl)-*N*′-ethylcarbodiimide (EDC) and PEG$_{8000}$. The breakdown of the tBLMs due to exposure to Triton X-100 (a non-ionic detergent), and the subsequent reformation of the membranes, was successfully monitored *via* CV of UQ in solution in the pores of the alumina membranes. In this manner, membrane integrity following exposure to detergents or lipases, or *in situ* diffusion of analytes across the membrane, may be monitored.

A fibrous mineral clay, sepiolite, has been utilized in a "bionanohybrid" support structure for the attachment of supported lipid bilayers (BL) of phosphatidylcholine (PC) with immobilization of the enzyme urease for biosensing applications.[37] The sepiolite-BL-PC (S-BL-PC) hybrid materials were immobilized onto gold electrodes that were altered by 5,5′-dithiobis(2-nitrobenzoic acid) (DTNB) to allow for a pH dependent redox couple of NHOH/NO for use as a reagentless pH sensor during CV characterizations.

The sensing of urea could then be carried out due to the observed shift toward a negative potential during CV, as urea oxidation is catalyzed by the immobilized urease in the S-BL-PC sensing platform, causing an increase in pH. Subsequent testing involved the incorporation of the S-BL-PC substrates into a film of poly(vinyl alcohol) on the DNTB gold electrodes for characterization. A nearly instantaneous peak shift was detectable with the biosensor in the presence of urea, with sensitivity of 30.8 ± 0.7 V M^{-1}. Additionally, the use of the PVA was shown to provide a benefit, forming a partially selective barrier layer against the electrochemically interfering agent ascorbic acid during testing of biologically relevant blood serum levels.

2.2.4 Lipids on Discrete Particles

In this last subsection, the concept of attaching lipid membranes to discrete particles is discussed. Discrete particles, when dispersed in an analyte solution, may be utilized for target detection through mechanisms such as fluorescence detection, fluorescence self-quenching, or fluorescence resonance energy transfer (FRET). Depending on the particle composition, these discrete particles could be used in solution samples, *in vitro* testing, or potentially *in vivo* diagnostics and treatment, when the lipids are functionalized with appropriate targeting functionality (*e.g.*, toxin detection, tumor cell targeting specificity).

Song and Swanson investigated the development of biosensors that utilized fluorescence self-quenching and resonant energy transfer for optical detection of cholera toxin (CT) by protein–receptor recognition with ganglioside GM1.[38] The fluorescently tagged ganglioside B$_{581/591}$-GM1 receptors were embedded within lipid membranes of POPC vesicles and POPC bilayers on silica microspheres for both the fluorescence resonant energy transfer as well as self-quenching experiments. The polyvalent nature of the CT (it has five binding sites for GM1) allow the binding of the CT to bring the GM1 receptors into close proximity to one another, triggering either the self-quenching or resonant energy transfer mechanism of detection. The addition of CT to the suspension of POPC-coated glass beads resulted in a significant decrease in fluorescence due to binding of the toxin at the GM1 receptor, and subsequently the self-quenching of the fluorescence. For the experiments with distance-dependent resonant energy transfer, the GM1 receptors were labeled with B$_{TMR}$ (B$_{TMR}$-GM1 as the fluorescence donor) and B$_{TR}$ (B$_{TR}$-GM1 as the fluorescence acceptor) as an energy transfer pair. Binding by the CT to the GM1 brings the energy transfer pair into close proximity of one another, allowing the acceptor to fluoresce strongly while the donor fluoresces only weakly. This study shows promise for the development of biomimetic sensor with a quick response time (<5 min) and a detection limit <0.05 nM CT. This platform allows for the modification of the technique to incorporate other protein–receptor pairs, as well as those requiring either multiple receptors or co-receptors.

Building upon these concepts for detection of CT, a later study examined GM1 incorporation into lipid bilayer membranes in free vesicle form or as immobilized layers on silica microbeads.[39] Three schemes were investigated for the optical detection of CT when interacting with the GM1 receptor within the lipid bilayer: (1) fluorescence self-quenching resulting from the polyvalent binding of the fluorophore-tagged GM1 receptor; (2) resonant energy transfer, utilizing a donor–acceptor transfer couple for a fluorescence shift that results from the binding of CT to GM1 receptors; (3) a combination of the self-quenching and energy transfer mechanisms, resulting in a signal amplification. POPC lipid bilayers with fluorophore-tagged donor/acceptor pairs (B_{TMR}-GM1/B_{TR}-GM1; B_{FL}-GM1/$B_{558/568}$-GM1; B_{R6G}-GM1/$B_{576/589}$-GM1) were integrated with the receptors for the self-quenching and energy transfer schemes. Following optimization of surface density of the B-GM1 complexes, detection of CT below 50 pM was achieved. For the scheme combining the self-quenching and energy transfer mechanisms, fluorescent-probe derivatized lipids [1,2-bis(1-pyrenedecanoyl)-*sn*-glycero-3-phosphocholine (BP-PC) and 1-hexadecanoyl-2-(1-pyrenehexanoyl)-*sn*-glycero-3-phosphocholine (P-PC)] were mixed with POPC to provide fluorescence directly within the lipid bilayer. For instance, BP-PC within the POPC bilayer will serve as fluorescence donors, while the fluorophore labeled GM1 will serve as the fluorescence acceptor. During binding of B-GM1 receptors to the multiple binding sites of CT there is a conformational change in the lipid bilayer, leading to aggregation of GM1 receptors. The aggregation of the B-GM1 receptors results in decreased fluorescence intensity of the acceptors as a consequence of self-quenching by the acceptors and a decrease in the quenching efficiency of the donors. Furthermore, the fluorescence intensity of the donors is subsequently increased, resulting from the decrease in distance-dependent resonant energy transfer efficiency. However, in this scheme the authors note that non-specific response to albumin was seen at high concentrations, likely due to conformational changes in the lipid matrix that result from albumin binding; they noted that the ratio of the CT/albumin fluorescence change was 30. Overall, this technology is a promising platform for use in optical sensor development based on polyvalent binding approaches.

In another study of lipid-coated silica beads, Freeman *et al.* also examined incorporation of fluorescence into lipid coatings.[40] Microresonators, composed of ∼5 μm silica microbeads, were coated with lipid bilayers of DPPC:DOPC:cholesterol (2:2:1 mass ratio) by vesicle fusion. The bilayer nature of the lipid coatings on the microresonators was confirmed by integrating Texas Red-modified 1,2-dipalmitoyl-*sn*-glycero-3-phosphoethanolamine (TR-DMPE) into the bilayer and using fluorescence quenching to examine the lipid coating. A fluorescence quencher (QSY-7 amine) resulted in a 51% fluorescence decrease, indicating that only the outer membrane leaflet was exposed to the quencher, and there were only two leaflets coating the microresonator. Fluorescence bleaching experiments confirmed membrane fluidity, as a bleached spot on the membrane was reduced to 53%

fluorescence, and subsequently returned to 97% fluorescence after 20 min. Additionally, the membrane was functionalized with a Cy5-labeled DMPE lipid (Cy5-DMPE), demonstrating the ability to conjugate a fluorophore onto the membrane that can be used for protein labeling. Laser excitation (200 µW, $\lambda = 633$ nm) successfully excited the Cy5 with a peak emission wavelength of 670.11 nm; this value is in good agreement with the reported emission peak wavelength of 670 nm.[41]

Multi-component polymer nanotube particles have also been fabricated *via* template growth in anodic alumina filters out of 3,4,9,10-perylenetetracarboxylic dianhydride (PTCDA) and polyethyleneimine (PEI), creating (PTCDA/PEI)$_6$ nanoparticles.[42] The incorporation of the PTCDA into the nanotubes imparts an optoelectronic response allowing for fluorescence imaging of the composite material. The (PEI/PTCDA)$_6$ nanotubes were 50–60 µm in length, with outer diameter of ~ 260 nm, inner diameter of $\sim 90 \pm 6$ nm, and a wall thickness of 84 ± 6 nm. The nanotubes were further functionalized with a supported lipid bilayer of egg L-α-phosphatidylcholine (egg PC), by vesicle fusion and rupture, creating a biomimetic surface that could be functionalized with proteins, enzymes, or other biologically relevant molecules. The lipid vesicles (~ 150–180 nm diameter) were prepared *via* extrusion out of the zwitterionic egg PC, which is negatively charged at pH 7.2,[43] and fused onto the hydrophilic and positively charged PEI surface,[44,45] resulting in a lipid bilayer that was $\sim 14 \pm 2$ nm in thickness. The electrostatic forces repelling the lipid vesicles from one another, while promoting their attachment to the positively charged PEI, facilitated formation of a single supported lipid bilayer as opposed to multiple layers.[46,47] Texas red was incorporated into the lipid vesicles, allowing for fluorescence recovery after photobleaching (FRAP) to examine the fluidity of the membrane; FRAP experiments indicated a mobile fraction of the lipid membrane of 77%, with an estimated diffusion coefficient of 0.01 μm^2 s^{-1}. These lipid-coated nanotubes offer promise for use in a variety of biomedical applications, including incorporation into biosensing platforms.

A benefit to utilizing a lipid-coated particle approach to analyte detection is that the discrete particle format can be analyzed with some common bench-top equipment. For instance, coupling of this type of lipid-coated particle technology with a flow cytometer could help to facilitate rapid detection of target analytes in a well-accepted and commonly used piece of equipment. This concept is highlighted in a study by Song and Swanson,[48] which built on their previous work with lipid-coated particles.[38,39] Flow cytometry was utilized for CT detection by functionalized lipid bilayer membranes supported by silica microbeads.[48] The lipid coating on the particles had FRET fluorophore-tagged GM1 receptors for CT, utilizing B$_{Fl}$/B$_{558/568}$-GM1 as the energy transfer pair. A low-cost flow cytometer featuring an argon ion laser excitation source ($\lambda = 488$ nm) was used in the study. Owing to the ability of flow cytometers to measure the fluorescence of individual particles (*e.g.*, lipid bilayer functionalized microbeads), as opposed to that of the bulk solution, the flow cytometric method was able to detect CT down to 10 pM concentrations by comparing the ratio of the mean acceptor/ donor fluorescence intensity. This detection limit is an improvement over the 100 pM CT detection limit of standard fluorometry for the microbeads and data collection can be achieved within 30 min. Furthermore, the authors showed that by utilizing the apparent binding rates calculated from the slope of early mean acceptor/donor fluorescence intensity, the dynamic detection range for CT can be extended out to 2.5–5.0 nM; this determination is important, as detection of CT *via* mean acceptor/donor fluorescence plateaus at 500 pM. As is the case with the authors' previous studies, the control of surface density of the B-GM1 receptors is critical to achieving optimization of detection and needs additional work. Nonetheless, this study provides an example of the feasibility in coupling discrete particles with a readily available detection platform.

2.3   Biomolecule Stabilization and Mimicry

2.3.1   Biomolecule Stabilization and Mimicry Introduction

While the previous section covered the immobilization of biomolecules within lipid membranes as a means of stabilizing and maintaining functionality of the biomolecules, there are other instances where they are immobilized directly onto a transducer, or in combination with polymers or surface coatings for stabilization, allowing them to function as sensing units in signal transduction. From a biomimetic standpoint, various stabilization procedures may be employed for maintenance of the functionality of the biomolecules, or the biomolecules themselves are utilized to mimic an environment that favors cellular binding.

The use of short peptides as sensing units has been reviewed previously by Pavan and Berti,[49] highlighting how self-organized synthetic peptides, peptides harvested from phage display, and those used in receptor–ligand binding pairs can be used as sensing elements. An extensive list of biomolecules can be detected with peptide sensors,[49] as well as targets such as metallic ions,[50–55] chemicals and pollutants,[56–58] and whole cells.[59,60]

Proteins and other biomolecules (*e.g.*, amino acids, nucleotides) may be utilized in a similar capacity to the sensing peptides, serving as biosensing units. A few examples include the following: olfactory receptor proteins that have been isolated from bullfrog olfactory epithelium and immobilized onto quartz crystal microbalances (QCMs) for odorant molecule detection;[61] hemoglobin that has been entrapped within thin films for direct electron transfer to a glassy carbon electrode during electrocatalysis of nitric oxide;[62] glucose levels in solution may be monitored *via* glucose oxidase on a graphite rod electrode;[63] and insulin antibodies have been deposited onto polymer-coated electrode surfaces for insulin detection from blood samples *via* a bioaffinity interaction.[64]

Furthermore, there has been a push toward creation of synthetic molecular constructs to mimic biomolecules, which aim to avoid the pitfalls of denaturation that may lead to inhibition of functionality. As such, enzymes or molecular recognition sites may be mimicked in order to facilitate sensing of target analytes.

This section covers in more detail additional examples of the use of peptides, proteins, and other biomolecules incorporated into sensing platforms, harnessing their functions as receptors, ligands, or catalysts. There is also a growing push in sensing technology to move toward the use of nanomaterials, harnessing their beneficial material properties (*e.g.*, high surface-to-volume ratios for enhanced signal transduction/amplification). Examples of the biomolecule incorporation with nanomaterials as a means of enhancing detection are discussed. Furthermore, while biomolecules can function as sensing units, there are concerns about their long-term stability in sensing platforms. As such, there are a few sensing platforms in the literature that employ molecular constructs to mimic the activities of biomolecules through artificial means. In this respect, these biomimetic constructs take on functional conformations to mimic the activity of a biomolecule.

2.3.2 Biomolecule Immobilization for Analyte Detection

As mentioned previously, the use of biomolecules as sensing mechanisms is well documented in the literature. These constructs serve as sensing units through mechanisms such as ligand–receptor binding, catalysis reactions, or protein–metal ion interactions, and when coupled to a transduction element can lead to a useful signal in detection. This transduction is often through optical detection (*e.g.*, fluorescence), mass-based changes to the sensor surface (*e.g.*, changes in QCM resonance), or electrochemical reactions (*e.g.*, oxidative peaks in CV). By mimicking their natural affinities for target analytes, these biomolecule-functionalized transducers can detect the presence of a target in solution. Furthermore, biomolecule immobilization may harness the molecular properties of the material to merely function as a direct electron transfer moiety or molecular wire to the transducer surface.

For instance, in a study by Yin *et al.* a Langmuir–Blodgett (LB) monolayer of hemoglobin (Hb)–linoleic acid (LA) was spread onto a gold electrode for use as a means of studying the direct electron transfer behavior of Hb.[65] The linoleic fatty acid served to protect the hemoglobin from denaturation and preserve its electrocatalytic activity in a way similar to the lipid membranes discussed previously. Coupling of a catalytic protein, such as hemoglobin, to an electrode surface can facilitate electron transit in biosensing designs. The cyclic voltammetric measurements on these constructs with phosphate buffer (pH 5.0) with 0.5 M KCl, using a 200 mV s^{-1} scan rate, showed reversible redox peaks; the anodic and cathodic peak potentials were -170 mV and -250 mV (*vs.* Ag/AgCl), respectively, with a formal potential (average of the anodic and cathodic peaks) of $E_0' = -210$ mV (*vs.* Ag/AgCl) and a peak separation of -80 mV (indicative of a fast electron transfer response). Furthermore, an increase in the solution pH between 4.0 to 8.0 resulted in a linear negative shift of the cathodic potentials and anodic potentials, with slope values of the reduction potential, oxidation potential, and formal potential of -49.5, -45.8, and -47.65 mV pH^{-1}, respectively; these values were in good agreement with a reversible electron-coupled single electron transfer value of -55.7 mV pH^{-1}.[66] Additionally, the electrodes retained 82% of initial sensitivity after 7 days of storage in 0.01 M phosphate buffer (pH 7.0, temperature 4 °C).

In another study, metallic ion detection was accomplished with metalloproteins geared toward specific ion recognition. Lee *et al.* fabricated a biomimetic sensor with the purpose of detecting Cu(II) ions, which can be toxic in the human body.[67] Gold-coated glass with a SAM of 11-mercaptoundecanoic acid (MUA) was functionalized with Cu-demetallated metalloproteins called Cu/Zn-superoxide dismutase (E/Zn-SOD1) that were re-metalized by Zn(II) ions; these proteins were expressed by *E. coli* with transfected human SOD1 genes. Solutions containing the divalent metal ions Mg(II), Cu(II), Fe(II), Ni(II), and Zn(II) were exposed to the sensor, which was examined by surface plasmon resonance. Insignificant changes in SPR reflectance measures were observed for all but the Cu(II) ion solutions, indicating specificity of the E/Zn-SOD1 proteins for these ions. Detection was linear in a narrow range of 10–100 μM Cu(II) in solution.

Moreover, as has been mentioned before, enzymes may be immobilized onto sensing platforms for catalytic detection schemes, such as acetylcholine esterase for acetylcholine detection. Polyacrylonitrile (PAN) membranes can be modified by the addition of either physically or chemically bound chitosan and examined for enzyme functionality upon immobilization of acetylcholinesterase (AChE).[68] Physical coatings were applied by filtering high M_w chitosan (400 kDa) solution through the PAN–NaOH + ethylenediamine (EDA) membranes. Chemically bound membranes were prepared by tethering high and low M_w chitosan (400 kDa and 10 kDa, respectively) in the presence of gluteraldehyde. The membranes were subsequently functionalized with AChE. Physical and chemical immobilization of AChE on the membranes shifted the optimum pH stability of the AChE compared to free AChE to slightly lower pH values. The chemically tethered membranes showed improved resistance to elevated temperature (60 °C) levels compared to free AChE and membranes with physically immobilized AChE. While activity of free AChE with a substrate of ATChI and the chromogen of 5,5′-dithiobis(2-nitrobenzoic acid) was shown to be nearly 10 times higher than the immobilized enzymes on the dual-layer membranes, the activity of the free enzyme was reduced to only 3% of its original activity following 60 day storage. Contrastingly, the low M_w chitosan treated membranes retained 75% of their activity after storage. Furthermore, this set of membranes was reusable up to 10 times, while maintaining greater than 50% of its activity, showing reusability and lending it to potential industrial use (*e.g.*, biotechnology applications).

Tian *et al.* also created enzymatic biosensors, in this case for detection of adenosine.[69] A microbiosensor was created on a platinum microelectrode (50 μm diameter, 0.5 mm length). The microelectrode was prepared for adenosine detection by electropolymerizing a layer of poly(pyrrolepropanoic acid), which was cross-linked by poly-L-lysine [aided by *N*-hydroxysulfosuccinimide sodium salt (NHSS) and EDC], prior to a biomimetic silicification step. The silicification was carried out with a mixture of hydrolyzed silanes [tetramethyl orthosilicate (TMOS), 1,2-bis(trimethoxysilyl)ethane (BISTMOS), and (3-glycidoxypropyl)(dimethoxy)methylsilane (GOPTMOS)] and the enzymes adenosine deaminase (AD, EC 3.5.4.4), xanthine oxidase (XO, EC 1.1.3.22), and nucleoside phosphorylase (NP, EC 2.4.2.1) in Tris buffer (pH 7.2), with the silicification catalyzed by the poly-L-lysine. The long chains ($\sim$400 repeat units) of the positively charged poly-L-lysine allowed mobility for the negative enzymes and silane ions to create the functionalized silica film, composed of numerous interconnected silica particles. Adenosine detection was performed by the conversion of adenosine to inosine by AD, inosine to hypoxanthine by NP, and hypoxanthine to hydrogen peroxide and urate, where the electrochemical signal of hydrogen peroxide was measured by the electrode (+600 mV *vs.* Ag/AgCl, in saturated KCl). The sensor was tested on solution of 6–20 µM, allowing for calibration of a linear range of 0.2–50 µM, with a limit of detection (LOD) of 40 nM and a sensitivity to adenosine of 153.0 ± 2.4 µA mM^{-1} cm^{-2}. The sensor achieved 10–90% response in approximately 25 ± 2 s ($n = 4$). Furthermore, utilizing the electropolymerization of diaminobenzene monomers through the silica layer onto the sensor surface, a protective layer was formed that gave the microbiosensor a lack of susceptibility to the interfering electroactive species dopamine, ascorbic acid, uric acid, serotonin hydrochloride, and catechol (10 µM concentrations); a lack of interference was observed for ascorbic acid, uric acid, and 5-HT even at 100 µM concentrations. These biosensors were stable in dry storage for up to one month, retaining 90% original activity. Furthermore, L-glutamate and lactate biosensors were fashioned in a similar manner, by substituting either L-glutamate oxidase or lactate oxidase, respectively, into the microbiosensors silica film. These microbiosensors displayed high sensitivities to glutamate (135.6 ± 7.1 µA mM^{-1} cm^{-2}) and lactate (365.2 ± 9.0 µA mM^{-1} cm^{-2}).

In addition to enzymatic reactions, ligand–receptor binding is a means by which detection of an analyte may take place. However, non-specific binding of proteins is problematic and can lead to errors in detection. To address such issues, surface properties may be tailored to facilitate specific receptor or ligand immobilization, while preventing non-specific adhesion of biomolecules. For instance, surface plasmon resonance biosensors, composed of the zwitterionic polymer poly(carboxybetaine acrylamide) (polyCBAA), were fabricated as ultralow fouling devices for screening of a potential cancer biomarker, activated leukocyte adhesion molecule (ALCAM).[70] The polyCBAA, due to its zwitterionic nature, can be chemically activated in a facile manner for protein functionalization and subsequently deactivated against non-specific binding. In this case, the polyCBAA was activated with EDC and *N*-hydroxysuccinimide (NHS), followed by sodium acetate buffer at pH 5.0, to facilitate amino coupling during protein immobilization. Subsequently, anti-ALCAM antibody solution in sodium borate buffer (pH 8.5–9.0) was used to functionalize the activated sites of the polyCBAA. The antibody application added ALCAM recognition specificity to the polyCBAA and deactivated (*via* hydrolysis) the residual activated groups. Approximately 250 ng cm^{-2} of the anti-ALCAM antibodies were applied to the polyCBAA surface. When testing the fouling properties, there was binding by less than 3 ng cm^{-2} of non-specific protein from human blood plasma samples (surface functionalization with anti-*Salmonella* sp. antibodies). The sensors were able to achieve a detection limit ~10 ng mL^{-1} in ALCAM-spiked human blood plasma. The polyCBAA sensors were compared to a set of sensors featuring a commonly prepared anti-fouling surface chemistry of oligo(ethylene glycol) (COOH/OH OEG) groups. The polyCBAA sensors outperformed the COOH/OH OEG sensors in ease of surface deactivation, anti-fouling levels, and target protein recognition sensitivity.

Similarly, in a study by Brault *et al.*, SiO$_2$-coated glass substrates were prepared with a zwitterionic polymer coating of carboxybetaine methacrylate (CBMA) that had been grafted onto the SiO$_2$ coating *via* linkage to the adhesive compound 3,4-dihydroxy-L-phenylalanine (DOPA).[71] The DOPA$_2$–pCBMA$_2$ was further functionalized *via* the addition of anti-ALCAM and anti-Salm to examine the specificity of protein binding of ALCAM in 100% human serum. SPR measurements were unable to detect an indication of non-specific binding of the ALCAM onto anti-Salm units. However, ALCAM-specific binding onto the anti-ALCAM antibodies was successful, with a detection limit of 64 ng mL^{-1} in 100% human serum. The grafting technique was successful in this study to create the biomimetic biosensor, which should lend the technique to substrates other than those typically used in biosensor designs (*e.g.*, gold).

2.3.3 Biomolecule Immobilization for Cellular Detection

Cellular function and adhesion may also be monitored using biomolecule immobilization, with modifications to the sensor surface that mimic their natural environments. These modifications may promote attachment of specific cell types. Once immobilized, the viability and growth characteristics can be monitored for the cells. This would be beneficial to monitoring, for instance, when exposing the cells to pharmaceutical agents.

Quartz crystal microbalances (QCMs) as biosensors have been utilized to this end.[72] QCMs were treated with an electropolymerized layer of tyrosineamide or tyronsineamide with the peptide Arg-Gly-Asp-Tyr (RGDY) in molar ratios between 1:20 and 1:3 (RGDY:tyrosineamide). The tyrosine (Y) group is covalently linked within the tyrosineamide film, leaving an exposed RGD sequence, which is a peptide found in fibronectin and vitronectin that can be recognized by endothelial cells. Growth of endothelial cells could then be analyzed by changes in frequency and motional resistance of the QCM. These QCM parameters can be used to monitor changes in cellular growth as a biosensing mechanism during altered growth conditions (*e.g.*, cell exposure to drug regimens). Furthermore, endothelial cells were grown in electric cell–substrate impedance sensing (ECIS) chambers on the tyrosineamide and RGDY-incorporated tyrosineamide layers, stained with Coomassie blue, and examined for cell count and morphology. Cells grown on tyrosineamide layers with higher RGDY concentrations showed greater numbers, and morphology resembling those seen in standard tissue culture plasticware; cells were more resistant to attachment without the RGDY-incorporated tyrosineamide and had uncharacteristic morphologies. The incorporation of the RGDY component to the QCM device aided in forming a biomimetic surface, promoting cell adhesion and growth, lending itself to functionality as a biosensor.

Similarly, poly(ethylene glycol) (PEG)-treated QCM surfaces with RGD functionalization has been utilized for cellular immobilization.[73] PEG has the benefit of preventing non-specific protein binding, while simultaneously facilitating tethering of biomolecules that promote specific protein or cellular adsorption.[74–76] In this study, a dynamic application of a QCM sensor was investigated by examining cellular adhesion and growth medium composition. A flow cell was coupled to a QCM crystal, allowing dynamic changes to the medium, QCM surface modification, and medium composition to be monitored in real time. A PEG $(NH_2PEG_{2000}C_{11}S)_2$ self-assembled monolayer (SAM) on the gold QCM surface was shown to prevent adhesion of murine fibroblast 3T3-L1 cells and Sprague–Dawley rat marrow stromal cells (rMSCs) that were otherwise able to adhere to the unmodified gold surfaces of the QCMs; adhesion of 3T3-L1 cells and rMSC cells resulted in frequency shifts of -2.0 and -6.7 mHz cell^{-1}, respectively. Subsequent modification of the PEG SAM with the peptide sequence Gly-Arg-Gly-Asp-Ser (GRGDS) presents an adhesion motif, RGD, which promotes cellular binding by cells expressing the integrins $\alpha_v\beta_3$, $\alpha_5\beta_1$, and $\alpha_{IIb}\beta_3$.[77] The RGD-modified QCMs promoted adhesion of the rMSCs in a measurable manner by the QCM. Subsequent addition of either soluble GRGDS or the enzyme trypsin allowed for rapid detachment of the cells from the RGD-modified PEG SAM. Additionally, medium modification with Mn^{2+} could be detected by the QCM, as addition of Mn^{2+} promoted interaction between the cells and the RGD surface, leading to greater cellular adhesion and a shift in the frequency of the QCM resonance (-6.5 mHz cell^{-1} for rMSCs). This design displays a sensitive and dynamic means of biosensing, with the ability to examine cellular adhesion characteristics *via* either surface or medium modification within the flow-cell QCM setup.

With a similar goal to that displayed in the previous paper by Knerr *et al.*, another study described the use of dually-functional polymer surface coatings of poly(carboxybetaine methacrylate) (polyCBMA), which are resistant to non-specific protein binding and cellular adhesion.[78] Simultaneously, these coatings possess the functional groups that allow attachment of proteins or peptides which can facilitate specific ligand binding for proteins or cells. Gold surfaces, modified with polyCBMA, were resistant to adsorption

(<0.3 ng cm^{-2}) by three different proteins [human fibrinogen, human chorionic gonadotropin (hCG), and lysozyme] when analyzed by surface plasmon resonance. However, upon functionalization of the polyCBMA with anti-hCG antibodies, hCG adsorption was promoted (~ 24 ng cm^{-2}). Furthermore, a cross-linked hydrogel of the polyCBMA was examined for cellular adhesion. Unmodified and fibronectin-modified hydrogels were tested for cellular adhesion of bovine aortic endothelial cells (BAECs). The unmodified hydrogels were resistant to cellular adhesion, while the fibronectin-modified hydrogels promoted cellular adhesion.

Using modifications in the binding moieties, bacterial cells may be similarly detected. Integration of peptides, for instance, which are specific for interactions with microbes, can facilitate their binding to a sensor surface. A study by Mannoor *et al.* utilized an electronic biosensor, composed of arrays of gold microelectrodes, functionalized with an antimicrobial peptide (AMP), magainin I, which can bind selectively to Gram-negative bacterial cells such as pathogenic *E. coli* O157:H7.[79] These bacteria can be harmful to humans *via* contaminated food.[80] Changes in impedance of the AMP-functionalized electrodes were measured in the range of 10 Hz to 100 kHz upon exposure to heat-killed bacteria (*Listeria monocytogenes*, *Salmonella typhimurium*, pathogenic *E. coli* O157:H7, and non-pathogenic *E. coli*). The microcapacitive devices displayed greatest sensitivities at 10 Hz, which was consequently chosen for impedance detection of bacteria. The results indicated that binding by all bacteria showed an increase in impedance over the blank control. The AMP showed a preferential binding to the pathogenic bacteria (*Salmonella* and *E. coli* 0157:H7) over non-pathogenic bacteria, with particular affinity for the pathogenic *E. coli*; additionally, the sensor was more selective for Gram-negative bacteria (*Salmonella* and both strains of *E. coli*) compared to the Gram-positive *Listeria*. Real-time monitoring of bacteria-spiked PBS (10^4–10^7 CFU mL^{-1}) under continuous flow indicated the ability of the sensor to mimic water sampling detection, featuring a measurable response within 5 min and a saturation in response after ~ 20 min; however, the authors noted that static incubation produced greater response, likely due to the lack of shear on the binding moieties caused by fluid flow. Under static conditions, the sensitivity approached a detection limit of 1 bacterium μL^{-1} for pathogenic bacteria. This design featured a label- and antibody-free detection system, with portability and real-time monitoring capabilities that may translate well to diagnostics in the field.

Lastly, an enzyme-sensor integration example is provided by a study by Luo *et al.*, examining the detection of reactive oxygen species by superoxide dismutase (SOD).[81] A sensor was developed for the detection of these reactive oxygen species (ROS), which mainly consist of O_2^-, and normally undergo the process of dismutation by the SOD enzyme into H_2O_2 and O_2.[82] However, in cases of various disease states and altered physiological conditions, O_2^- may be present in greater concentrations.[83–91] In this device, the synthetic compound Mn-TPAA (manganese coupled to tris{2-[*N*-(2-pyridylmethyl)amino]ethyl}amine) was immobilized onto TiO$_2$ nanoneedles (which was spin-coated onto ITO-coated glass), acting to mimic SOD, allowing for electrochemical detection of ROS. Electrochemical investigations indicated that both anodic peaks and cathodic peaks of dismutation in the potential range of ~ 0.4–0.7 V (*vs.* Ag/AgCl) allowed for amperometric detection of O_2^{-}. Detection of O_2^{-} was set at a potential of 0 mV (*vs.* Ag/AgCl) to avoid interfering current responses from physiologic analytes such as ascorbic acid, dopamine, and L-cysteine; the resulting current responses for these competing analytes at 0 mV (*vs.* Ag/AgCl) were less than 2%. Furthermore, the electrodes were stable out to three months, when tested three times daily, with less than a 3% change *versus* the initial current response. When HeLa cells were incubated on the sensors and monitored for Ang II-mediated O_2^{-} release, a cathodic current could be detected within ~ 25 s and linked to changes in intracellular release of Ca^{2+}, as confirmed by fluorescence confocal imaging of Ca^{2+} concentration changes by the cells. These results suggested that an increase of intracellular concentration of Ca^{2+} corresponds to an increase in extracellular O_2^{-} levels from HeLa cells, when stimulated by Ang II.

2.3.4 Nanostructures Functionalized with Biomolecules

As has been discussed with regard to the lipid membrane section, the integration of nanomaterials with sensory platforms is likely to increase in future investigations. This holds true with biomolecular immobilizations. Nanostructured materials may, for instance, facilitate biomolecule arrangements, transducer patterning, and/or signal transduction.

Müller *et al.* demonstrated such a concept in a study featuring immobilization of peptides into specific orientations to aid in the development of biomimetic surfaces for applications in biosensing.[92] The arrangement of stiff polypeptides into polyelectrolyte multilayers (PEMs) were studied on silicon substrates with both smooth and texturized Si surfaces. Texturized surfaces on the Si were composed of nanometer scale grooves of ~ 50–70 nm widths and 5–8 nm depths. Using pH-controlled conditions, the oppositely charged polyelectrolytes (PELs) pair of poly(L-lysine hydrobromide) (PLL) with poly(vinyl sulfate) (PVS) were ordered in the nano-grooves of the Si substrate with high-order parameter values. Furthermore, oriented PEMs of poly(L-glutamic acid) (PLG) with poly(diallyldimethylammonium chloride) (PDADMAC) were achieved. The coils were influenced into ordered α-helical orientations by several factors, including molecular weight, as larger PLL and PLG molecules with longer contour lengths were forced into orientations of greater alignment within the nano-grooves. Additionally, factors such as the number of layers within the PEM, the concentration of the PEM components, the molecular weights of the PVS and PDADMAC, and whether or not the sample was wet or dry when the orientation was measured *via* ATR–FTIR–dichroism, influenced the orientation of the PELs within the PEMs. Control over these parameters can be utilized to provide surface exposure of PELs in oriented α-helical conformations.

In another nanomaterial application, an amperometric biosensor with a demonstrated ability to detect hydrogen peroxide in solution was created by modifying the surface of a glassy carbon electrode (GCE) through a biomimetic co-precipitation of nanoparticles consisting of calcium phosphate compounds (CaPs) and the protein hemoglobin.[93] The GCE, when incubated with a solution containing calcium chloride and hemoglobin [in addition to NaCl, $K_2PO_4 \cdot H_2O$, tris(hydroxymethyl)aminomethane, and HCl], underwent precipitation of the hemoglobin,[94] which subsequently resulted in surface nucleation of the calcium phosphate compounds. The nucleation sites were then built upon by the adsorption of the calcium ions, phosphate ions, and hemoglobin, resulting in a composite network of hemoglobin and CaP crystals.[95–104] The measured formal potential $(E^{\circ\prime})$ for the amperometric detection of H_2O_2 was -0.365 V, with a linear detection range of 0.1 µM–2.6 mM, a 0.1 µM limit of detection, and a Michaelis–Menten constant of 8.08 µM. The device was able to reach 95% of steady-state current within 10 s and displayed a relative standard deviation of 4.6% between devices $(n = 8)$. The pH of the PBS played an important role in detection, with a negative shift in both the anodic and cathodic peaks with increasing pH, and a linear relationship for the formal potential between pH 3.0–9.0; consequently, a pH of 7.0 was chosen for H_2O_2 detection.

Tognalli *et al.* provided an example of molecular "wiring" with a redox sensor design, incorporating glucose oxidase as part of the redox mechanism linkage to a substrate.[105] A nanobiosensor was fabricated for surface enhanced plasmon resonance (SERS) detection of the redox changes to glucose oxidase (GOx) in the presence of the target molecule, glucose. Layer-by-layer assembly of the complex between poly(allylamine) and osmium pyridinebipyridine (PAH-Os) with GOx was completed on gold nanoparticles to form core-shell nanoparticles. The PAH-Os is immobilized in its oxidized state, providing no Raman scattering, but can provide a SERS signal upon its reduction from Os(III) to Os(II) by a charge transfer from the GOx, when GOx converts glucose to gluconic acid; SERS enhancement is provided by the gold nanoparticles, amplifying the SERS signal. The SERS signal was further enhanced by dropcasting the core-shell nanoparticles in glucose solution onto a nano-patterned substrate of silver, presenting cavities measuring 700 nm in diameter; the nanocavities of the silver substrate provided an improvement in signal-to-noise and 500× enhancement of the signal. An estimate of less than 1000 core-shell nanoparticles were utilized (60.5 pM suspension) during measurements of the various glucose concentrations, up to a saturation point of 7–8 mM glucose; only 1 µL of the nanoparticle solution with varying glucose concentrations was needed for each test. The molecular "wiring" of the redox molecules with a biomimetic recognition unit of glucose oxidase was demonstrated as an example for the use of redox molecule SERS detection biosensors.

Another interesting nanomaterial application was examined by Hnilova *et al.*[106] A patterned silica and gold substrate was prepared *via* nanosphere lithography,[107] where 1.5 µm and 5.0 µm microsphere suspensions were dropcast onto silica coverslips. Following drying, a 25 nm layer of gold was deposited by ion beam sputtering to create islands of gold upon the silica substrate. The patterned substrates were then treated with peptides whose sequences had been selected to target either silica or gold as a binding substrate. The gold binding peptides (AuBP2), which were biotinylated (AuBP2-bio), were coupled to streptavidin-functionalized QDots605 (AuBP2-bio/SA-QD605) for fluorescence imaging upon binding to the gold islands. The SiO_2-specific peptides (QBP1) were FITC-modified (QBP1-FITC), as illustrated in Figure 2.3a. AFM was able to confirm the formation of the gold islands, as well as the peptide immobilization on the surface of the substrates. Confocal microscopy confirmed the co-immobilization of the QDots- and FITC-modified peptides, with substrate-specific affinities (Figure 2.3b and 2.3c). The ability to co-assemble the fluorescently modified peptide

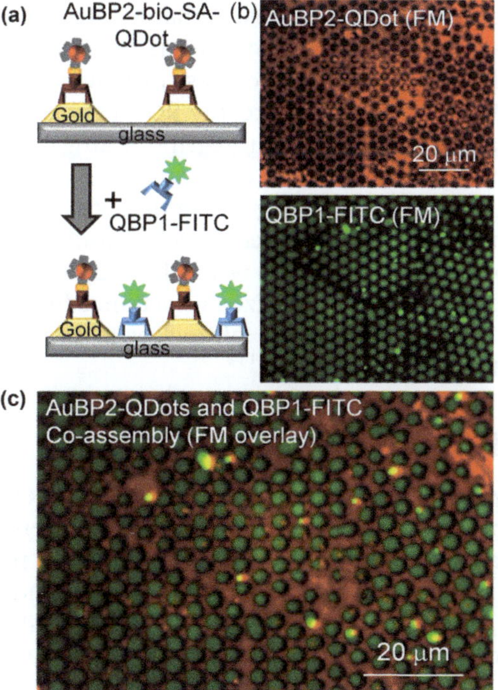

Figure 2.3 Simultaneous peptide-directed co-assembly of QDots and FITC fluorophore on the NSL substrate. (a) Schematic illustration of the simultaneous immobilization process using two peptide constructs. (b) Fluorescence images of the SA-QDot immobilized *via* gold-binding peptide (AuBP2-bio) onto gold and the QBP1-FITC construct immobilized onto silica glass. (c) The overlay of the fluorescence images is shown at the bottom, revealing that the FITC molecules (attached to QBP1) are immobilized on the circular regions of the silica and the QDots are immobilized onto the gold triangular island nanostructures *via* AuBP2-bio.[106]

sequences into specific regions of a nanopatterned substrate can lead to future biosensing applications.

As a final nanomaterial example, an electrochemical biosensor was created by immobilization of multiwall carbon nanotubes (MWNTs) on the surface of a GCE, which were functionalized with fibronectin (Fn) to promote cellular adhesion, and utilized as a cytosensor for the detection of cellular attachment (SMMC 7721 cancer line) to the sensor surface.[108] A custom heating element of copper, with direct current for heat production, was examined as a means of improving detection, as local heating of the electrode surface promoted local convection and increased diffusion. The MWNTs facilitated functionalization by Fn after acid treatment (*e.g.*, attachment of the $-NH_2$ of the Fn to the $-COOH$ groups on the MWNTs), created a rough and porous surface that promoted cellular attachment,[109] increased the electrode surface area by 4.2 times, and facilitated the transfer of electrons to the electrode surface from guanine molecules in the cytoplasm of adhered cells. It is proposed that electrochemical detection of adhered cells can occur through an irreversible oxidation of guanine to 8-oxoguanine, with the guanine molecules traversing the membrane of the cell to the Fn-MWNTs.[110] Furthermore, the RGD peptide sequence within Fn promoted cellular attachment *via* integrin-binding, as MWNTs without functionalization were resistant to cellular adhesion. Cell detection was conducted at 42 °C, after optimizing the temperature for a combination of improved detection and maintenance of cell viability. At 42 °C, a cellular detection limit of 5.0×10^3 cell mL^{-1} was achieved, illustrating an improvement over other cytosensors that utilized impedance for K562A cell detection $(LOD = 7.1 \times 10^3$ cells $mL^{-1})$[111] and piezoelectric immunosensing of *Salmonella* $(LOD = 1.0 \times 10^4$ cells $mL^{-1})$.[112]

2.3.5 Biomolecular Mimics

While immobilization of biomolecules can facilitate sensing applications, there have also been developments in molecular mimicry to serve the same purpose as natural biomolecules. These strategies have produced a handful of sensors which successfully detect target analytes and have shown wide detection ranges, low detection limits, and reasonable shelf-lives that could lend the technologies to commercial uses. A few of these sensing techniques are highlighted in the following subsection, with their analyte detection ranges and limits listed in Table 2.1.

In an example of protein binding site mimicry, a silicon substrate was patterned with hydrophobic and hydrophilic areas in a method to imitate the binding sites for the different protein conformers of calmodulin (CaM).[113] A hydrophobic coating was applied *via* octyltrimethoxysilane (OTMS) incubation, followed by a hydrophilic coating of bovine serum albumin (BSA). Notches were carved out of the BSA layer using AFM, creating trenches that were 10 ± 1 nm with depths of -0.5 ± 0.1 nm, that were then functionalized with Ca^{2+}-CaM since the trenches mimic the natural binding

Table 2.1 Biomolecule mimic detection.

Analyte	Biomimic[a]	Detection Range	Limit of Detection	Ref.
Catechol	$Fe^{III}T4MpyP$	0.6–6.0 µM	0.35 µM	114
Hydrogen peroxide	$[Cu(2pymehist)Cl]ClO_4$	1–30 mM	4×10^{-6} M	118
Paracetamol (acetaminophen)	FeTPyPz	1.0×10^{-5} to 5.0×10^{-2} M	1.0 mM	122
Dopamine	CuPc	10–140 µM	0.6 µM	123
O^{2-}	Mn^{2+}-ZSM/PDDA	5.0×10^{-7} to 1.2×10^{-3} M	0.37 µM	124

[a]$Fe^{III}T4MpyP$ = iron(III) tetra(*N*-methyl-4-pyridyl)porphyrin; $[Cu(2pymehist)Cl]ClO_4$ = $[CuClN_4C_{12}H_{14}]ClO_4(H_2O)$; FeTPyPz = iron(III) tetrapyridinoporphyrazine; CuPc = copper(II) phthalocyanine; ZSM = zeolite; PDDA = poly(diallyldimethylammonium chloride).

sites of Ca^{2+}-CaM. Incubation with anti-CaM antibodies showed an increase in the height of the trenches to 4.5 ± 0.8 nm, indicating specific binding of the Ca^{2+}-CaM within the trenches. Furthermore, incubation of the conformer apo-CaM did not provide an indication of binding, leading to the conclusion that the hydrophobic/hydrophilic patterning was specific to mimicking the Ca^{2+}-CaM conformer binding site. The Ca^{2+}-CaM could be reversibly removed with a Ca^{2+}-chelating agent, EGTA.

There have also been a number of studies which utilize molecular constructs to mimic the behavior of enzymes, catalyzing a reaction for detection by an electrochemical transducer. In the amperometric biosensor described by Damos *et al.*, iron(III) tetra(*N*-methyl-4-pyridyl)porphyrin ($Fe^{III}T4MpyP$), which mimics the action of horseradish peroxidase (HRP) as an electron mediator for phenolic compound detection, was immobilized onto the surface of a GCE.[114] In addition to porphyrin, molecules of the amino acid histidine (His) were embedded within a Nafion® matrix on the surface of the GCE. The combination of the $Fe^{III}T4MpyP$ and His [optimized 1:2 ratio (w/w)] mimic the active site of HRP. Biosensors based on the immobilization of peroxidases such as HRP have been developed previously,[115–117] but the authors note that they suffer a lack of sensitivity due to high currents in the presence of peroxide from the direct electron exchange between the peroxidase and the electrode. By embedding the biomimetic complexes within the Nafion® membrane in an enzymeless biosensor, excess current can be reduced, improving sensitivity, and allowing for more accurate detection of phenolic compounds (*e.g.*, catechol, dopamine). The results of this study indicated that this technique was successful in creating a biomimetic biosensor capable of repeatable detection of catechol. When using optimized conditions, detection of a 1.2 µmol L^{-1} catechol solution during seven consecutive measurements yielded a relative standard deviation of only 4%. A 25% reduction in sensitivity was observed after 10 measurements, maintaining stability through 100 measurements, and the sensor displayed a response of 65% of the initial sensitivity after 200 measurements. Furthermore, these detections were obtained within 0.5 s after catechol introduction, indicating a fast and reliable response in a linear range of 0.6–6.0 µmol L^{-1}, with a detection limit of 0.35 µmol L^{-1}. Additional compounds that were successfully detected at a 1.2 µmol L^{-1} concentration in solution were dopamine, *o*-phenylenediamine, 2-amino-4-chlorophenol, 4-aminophenol, hydroquinone, 3-aminophenol, *p*-phenylenediamine dihydrochloride, L-DOPA, and phenol. The sensor was most sensitive for the detection of dopamine, exhibiting a relative response of 273% compared to catechol, allowing for detection of this compound in nmol L^{-1} concentrations.

Similarly, Alves *et al.* created an electrochemical sensor with an enzyme mimic for hydrogen peroxide detection.[118] Their study described the fabrication of an inorganic–organic polymer network of a linear chain chloride-bridged imine-copper(II), covalently linked to poly(*N*-vinylimidazole) (PVI), and coupled to a GCE for the amperometric detection of hydrogen peroxide. While the sensor was enzyme-free, it sought to mimic the active site of the enzyme catalase, which serves as a cellular protectant against damage caused by hydrogen peroxide; hydrogen peroxide in the body may result through the events of aging or disease states such as diabetes and cancer.[119–121] For this biosensor, the $[Cu(2pymehist)Cl](ClO_4)$ is cross-linked to the PVI with poly-(ethylene glycol) diglycidyl ether (PEGDGE) to form a hydrogel on the GCE. The copper complex mimics the catalase activity while remaining enzyme-free, forming Cu^{I}–µ-$O_2{}^{2-}$–Cu^{I} sites, where hydrogen peroxide is reduced to water upon oxidation of the copper ions. The reduction current at the electrode as a result of this process was measured amperometrically, displaying a linear range (concentration range 1–30 mmol L^{-1}) for H_2O_2 in PBS (pH 7.0), with a sensitivity of 268 mA mol^{-1} L cm^{-2} (−0.4 V *vs.* SCE). One benefit of enzyme-free biosensors is their maintenance of functionality upon storage, as enzyme degradation is not a concern. This device achieved 90% of its original responsiveness following storage in PBS (pH 7) at 4 °C, out to two weeks time.

In another study, a GCE was functionalized with iron(III) tetra-pyridinoporphyrazine (FeTPyPz) in Nafion® for amperometric detection of paracetamol (acetaminophen) in solution.[122] FeTPyPz is a compound which mimics the action of the enzyme P450, which can metabolize paracetamol. This biomimetic sensor was coupled to a flow-cell for flow injection analysis (FIA) of aqueous samples containing paracetamol. FIA was compared with HPLC analysis for seven commercial formulations of paracetamol (*e.g.*, Tylenol®), indicating that detection by the sensor deviated by no more than 4.6% when compared to the HPLC method. The biomimetic sensor had a linear range of 1.0×10^{-5} to 5.0×10^{-2} M. When running samples at 30 min intervals for 4 h d^{-1} over a five-day period (320 injections total), there was little loss in response of the sensor (93% of initial signal) and no noticeable degradation of the Nafion® coating on the GCE. Furthermore, use of the sensor on samples of river water that were spiked with paracetamol resulted in percent recoveries of the drug that closely mirrored HPLC values (close to 100% recovery), indicating promise for this technique in environmental sampling.

Rahim *et al.* described fabrication of a mesoporous ceramic carbon electrode (CCE), composed of SiO_2/C-graphite, completed *via* sol–gel deposition of the silicon dioxide and carbon components, with integration of copper(II) phthalocyanine (CuPc) acting as a biomimetic enzyme (dopamine mono-oxygenase) for catalysis and amperometric detection of the redox reaction of dopamine.[123] Hydrogen peroxide in solution is utilized to oxidize the electrode CuPc to PcCu–OOH, which can in turn oxidize dopamine at the surface of the electrode. The oxidized dopamine can then be reduced by the electrode in an electrochemical reaction, leading to an amplification of signal that adds to the detection sensitivity of the design. The use of SiO_2 as an insulator promotes the exchange of electrons between the oxidized dopamine and the graphite in the electrode, as opposed to electron exchange between the CuPc and the electrode.[123] Additionally, the use of hydrogen peroxide in solution aids the CuPc construct in mimicking the dopamine monooxygenase enzyme, forming a Cu^{II}–O–O–H species that can oxidize the dopamine into a 1,2-quinone, before its reduction by the electrode surface potential. Figure 2.4 demonstrates the proposed mechanism of the reaction.

The sensor displayed high sensitivity ($0.63\ \text{nA dm}^3\ \mu\text{mol}^{-1}\ \text{cm}^{-2}$), with a limit of detection of $0.62\ \mu\text{mol dm}^{-3}$, a linear response range of 10–$140\ \mu\text{mol dm}^{-3}$, and a response time of ~ 1 s. The variation between different electrodes (relative standard deviation) was 38%, but repeatability between 10 measurements of a given electrode had an RSD of 1.37%. Furthermore, the electrodes retain sensitivity for up to nine months when stored at room temperature. Dopamine in 0.9% (m/v) physiological solution gave a response of $100\ (\pm 2)\%$, indicating an excellent ability to detect dopamine levels in solution.

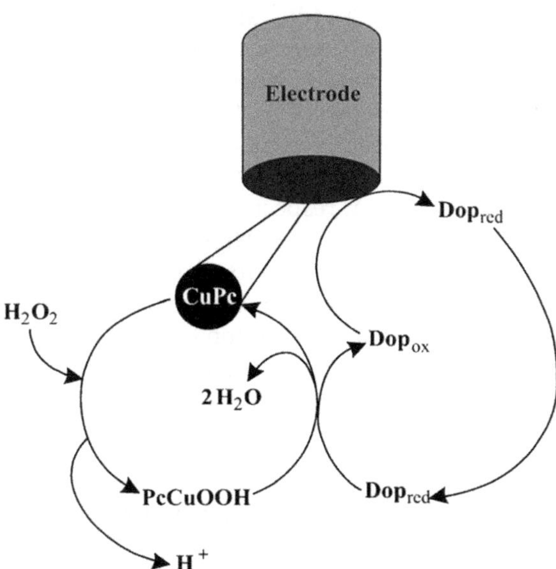

Figure 2.4 Proposed mechanism for the sensor SiO_2/C/CuPc response for dopamine. Copper phthalocyanine is represented by $CuPc_{red}$, $CuPc_{ox}$ is oxidized CuPc, and Dop_{red} and Dop_{ox} are the reduced and oxidized dopamine species, respectively.
(Reproduced from Rahim *et al.*,[123] with permission from Elsevier.)

Interference studies indicated that no interfering signals were incurred with catechol or resorcinol in solution, though ascorbic acid in concentrations greater than 4.0 μmol dm^{-3} interfered with the dopamine signal.

In the previous sections of this chapter, the use of superoxide dismutase for reactive oxygen species sensing was described. Zhou *et al.* created a biosensor to serve a similar function, monitoring the release of reactive oxygen species, and as such, monitoring electrode-adherent cellular release of the ROS.[124] An amperometric biosensor that mimicked the enzymatic activity of superoxide dismutase (SOD)–manganese(II) phosphate [$Mn_3(PO_4)_2$] complexes was created by utilizing a zeolite (ZSM-5) to facilitate cation exchange with Mn^{2+}. The Mn^{2+}-ZSM construct was protected by a coating of poly-(diallyldimethylammonium chloride) (PDDA), facilitating confinement of the electron transfer between the Mn^{2+} and ZSM-5. Mn^{2+}-exchanged zeolite structures with the PDDA coating (Mn^{2+}-ZSM/PDDA). The biosensor exhibited an estimated formal potential ($E^{\circ\prime}$) of 561 ± 6 mV (*vs.* Ag/AgCl, in saturated KCl), with the anodic and cathodic peak currents displaying a linear relationship with respect to the scan rate. The linear relationship indicates that the Mn^{2+} is incorporated into the crystalline zeolite structure and not limited by surface diffusion for electron exchange. During amperometric detection of the ROS O_2^-, the proposed mechanism suggests dismutation of O_2^- by Mn^{2+}, forming MnO_2^+ and H_2O_2, and subsequent reduction of the MnO_2^+ by another O_2^- species, yielding Mn^{2+} and O_2. Detection of O_2^- in solution (at either -0.1 V or 0.6 V, *vs.* Ag/AgCl) provided a linear detection range between concentrations of 5×10^{-7} to 1.2×10^{-3} M; a detection limit of 0.37 μM and response within 5 s were noted. There was limited interference observed from competing electro-active compounds when detection of O_2^- was tested, at a potential of -0.1 V, in the presence of O_2, H_2O_2, OH$\cdot$, $ONOO^-O_2$, ROO$\cdot$, Na^+, Ca^{2+}, K^+, Mg^{2+}, Zn^{2+}, Fe^{3+}, ascorbic acid (AA), uric acid (UA), dopamine (DA), and cysteine (Cys); cathodic currents of 6.5%, 9.83%, 7.28%, and 6.72% from OH$^\bullet$, $ONOO^-$, O_2, and 10 μM AA, respectively, while interfering currents from the other analytes were less than 1%. High stability was displayed by the electrodes, with three-times daily testing for four months with no reduction in response. Additionally, the inter-device response variation for five sensors was shown to be less than 4.2%. *In situ* detection of O_2^- released from HeLa cells that were adhered onto the sensor surface was *via* the cathodic current, detecting a release of ~ 1.1–10^{-5} M O_2^- from $\sim 2.5 \times 10^6$ cells, upon the addition of 10 mM zymosan (which induces O_2^- release), with a response within 5 s.

2.4   Molecularly Imprinted Polymers in Biosensing

2.4.1   Molecularly Imprinted Polymers Introduction

Molecularly imprinted polymers (MIPs) should be discussed, at least briefly in this chapter, as they demonstrate a biomimetic approach to analyte detection in sensing platforms. MIPs are cross-linked polymer networks, prepared by polymerizing material in the presence of a template molecule (target analyte), which can subsequently be washed out of the polymerized construct, leaving behind an imprinted site taking on the molecular shape and chemical functionality left by the template (Figure 2.5).[125] As such, MIPs possess artificial receptor sites, which can recognize and bind molecules of the target analyte upon their introduction to the MIP. In this regard, they mimic bioaffinity events, but can also be joined with catalytic mechanisms. When these polymer networks are coupled to a transduction mechanism, they can be utilized in biosensing applications. Optical (*e.g.*, fluorescence, surface plasmon resonance), electrochemical (*e.g.*, cyclic voltammetry), and mass-based (*e.g.*, QCM frequency shifts) transduction mechanisms are some of the means which may be utilized in MIP-based biosensing.

MIPs have typically been used for detection of small molecules, such as dopamine,[126] caffeine,[127] and nicotine.[128,129] However, there have been efforts to investigate larger proteins, such as hemoglobin[130,131] and Hev b1 latex allergen.[132] For example, MIPs of polyacrylamide hydrogels have been fabricated and templated for bovine hemoglobin (BHb) and examined for parameters governing template molecule removal, protein reloading, and molecular selectivity.[131] It was determined that BHb could be imprinted successfully, with an optimum wash of 10% (w/v):10% (v/v) of SDS:AcOH, which leads to greater than 90% re-loading efficiency of the BHb. Selectivity for BHb was clearly indicated when tested with analogues that are similar in structure and chemical functionality (cytochrome *c* and myoglobin).

(A) Non-covalent imprinting

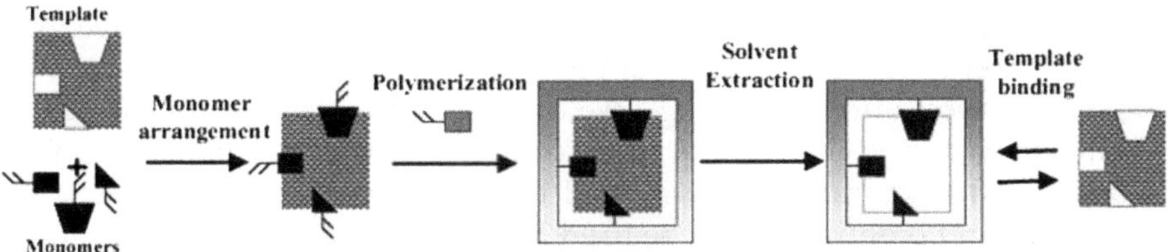

(B) Covalent imprinting

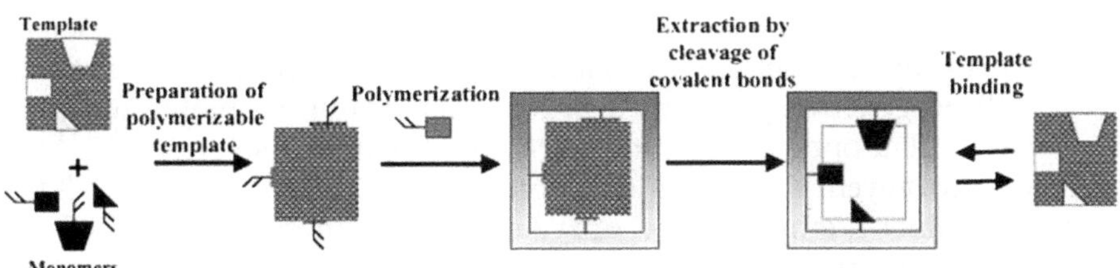

Figure 2.5　Schematic representation of (A) non-covalent and (B) covalent molecular imprinting procedures.
(Reproduced from Marazuela and Moreno-Bondi,[125] with permission from Springer.)

MIPs hold potential as biomimetic sensing units in a variety of areas. As an example, MIP-based sensors have been identified as a potential means of detection for environmental and food contaminants, with suggested growth in these sectors in the future.[133] MIPs show promise as a biomimetic sensing mechanism in the area of food analysis since they are artificially created and not as susceptible to loss of functionality compared to naturally derived biosensors, while possessing high specificity and stability under numerous conditions.[133] Their flexibility can lead to their use in various food analysis and process monitoring applications, such as detection of taste-producing compounds,[134] monitoring of ideal growth conditions for yeast,[135] or detection of herbicides.[136] In the case of toxin monitoring, detection of the toxic protein ricin has been developed using MIPs.[137] A number of different sensing applications may be addressed using MIPs as a transduction interface. Chapter 3 in this volume is dedicated to the development and applications of MIPs, so this will not be explored further in this chapter.

2.5 Whole Cell Biosensing

2.5.1 Whole Cell Biosensing Introduction

Whole cell-based biosensors fall under the umbrella of biomimetic biosensing in the respect that their transducer-coupled cellular responses (*e.g.*, cellular depolarization) to an external stimulus can mimic their natural stimuli responses, while facilitating conversion of those responses into a quantitative assessment of the stimulus. While the use of radiolabeling techniques, immunoassays, and other cell-labeling protocols have been utilized extensively in bench-top applications, there have not been as many biosensor applications utilizing whole cells as the sensing units. Part of challenge with this platform is maintenance of cellular viability and functionality, often requiring delicate cell culturing techniques. As such, biomolecules and biomimics may be used alternatively in sensing applications, as discussed in the previous sections of this chapter.

Nevertheless, there have been a few studies combining cells with transducers as a biosensing mechanism. A review by Monk and Walt highlighted some of the work that has been conducted using whole cells as the sensing mechanism in fiber optic biosensors.[138] Such work includes several studies that utilized genetically modified *E. coli* cells, exhibiting bioluminescence, coupled to fiber optic sensors for detection of toxins in solution[139–144] (*e.g.*, phenol, mytomycin C, cerulenin, or mercury) or volatile toxins[145,146] (*e.g.*, benzene, toluene, ethylbenzene, or xylene). Furthermore, a number of other bioluminescent microorganisms and mammalian cells have been joined to fiber optic sensors [*e.g.*, microalgae *Chlorella vulgaris*,[147] various yeast strains,[148] mouse fibroblasts,[149] and Gram-negative bacteria (*Alcaligenes eutrophus*)[150]]. These studies utilize whole cells as a sensing mechanism through optical signal transduction.

Additionally, there has been research conducted on mimicking larger biological systems, such as olfactory systems for odorant detection, and some work on mimicking of gustatory systems for tastant recognition. In these studies, whole cells expressing olfactory or gustatory receptors are utilized in order to detect specific odorant or tastant compounds. These whole cell sensing techniques are covered in more detail below, as they more pronouncedly mimic larger systems (olfactory and gustatory), rather than just functioning as isolated cellular units, more fully exemplifying the concept of biomimetic sensing.

2.5.2 Olfactory and Gustatory System Whole Cell Biosensing

Whole cells that express olfactory receptors may be coupled to transducers and utilized as olfactory system biomimics. Advances in these systems have been highlighted in several reviews.[151,152] These systems include the use of cells that can be excised from animals or cultured to express odorant receptors. For instance, human embryonic kidney (HEK-293) cells may be used to express the I7 odorant receptor and cultured on QCM transducers[153] or microelectrodes[154,155] for detection of aldehydes. Yeast (*Saccharomyces cerevisiae*) has also been used to express the human OR17-40 receptor, coupled to interdigitated gold microelectrodes, and utilized in helional detection.[156] As described in the following examples, there has also been work conducted with rats and insects for odorant detection.

A study by Wu *et al.* examined the use of light addressable potentiometric sensors (LAPS) coupled to cultured olfactory sensory neurons (OSNs) from Sprague–Dawley rats as a biomimetic odor detection system.[157] The cilia on the OSNs served as receptors for odorants, creating an action potential in the cells that could be detected on the laser illuminated areas of the LAPS chips. Illumination of the semiconductor chips by an He-Ne laser ($\lambda = 543.5$ nm, 5 mW) created electron–hole pairs and the chip was biased with a potential of 2.0 V to create a depletion region that would fluctuate with changes in surface potential (*e.g.*, action potentials of the OSNs) at the location of illumination, which allowed monitoring of odorant-stimulated OSNs. The laser could illuminate a specific cell with a 10 μm focal spot, allowing for cell-specific determination of odorant detection within a mixed population of cultured rat olfactory cells (Figure 2.6). The OSNs were tested using growth medium (DMEM) infused with odor mixtures (acetic acid, octanal, cineole, hexanal, and heptan-2-one) and agents that served as either an odor detection inhibitor (MDL12330A) or enhancer (LY294002). Inclusion of the odor detection inhibitor confirmed that the electrical signals being monitored on the LAPS chips before and after inhibitor administration were a result of odorant stimulation of the OSNs. Using this setup, the authors observed firing spikes over 5 s intervals that were more frequent during odorant mixture administration than those observed during control conditions or administration of the MDL12330A. These results indicated that stimulation by the tested odorant molecules could be accomplished and

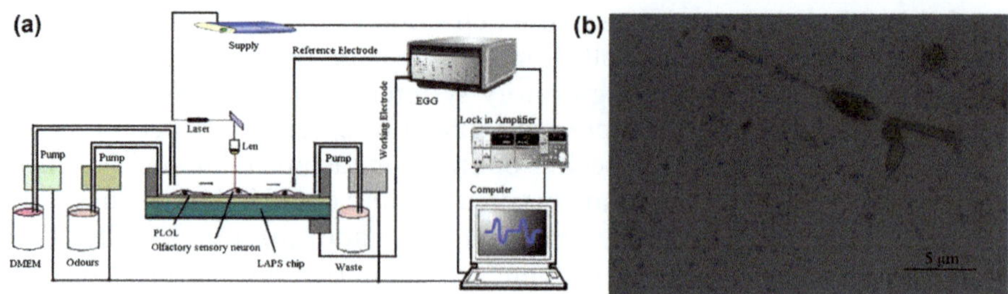

Figure 2.6   (a) Schematic diagram of the LAPS measurement set-up. (b) Results of rat OSNs cultured on LAPS chip for 3 days.
(Reproduced from Wu *et al.*,[157] with permission from Elsevier.)

measured. Furthermore, inclusion of the enhancer showed increase in the number of firing spikes per second when compared to control values or the odorant mixture alone. The sensor offers a means of monitoring electrophysiological responses of the OSNs to odorant mixtures, though additional work is needed to improve the signal-to-noise ratio ($S/N = 3$) and address observations that not all of the OSNs that were tested showed a response to the stimulants on the LAPS chips. The work shows promise for the study of olfactory transduction mechanisms or neural network research.

Liu *et al.* also utilized olfactory cells from Sprague–Dawley rats as the biomimetic odorant sensing mechanism.[158] In this case, the olfactory epithelium was dissected out of the donor animal and fixed onto a 36-channel microelectrode array (MEA) for monitoring of olfactory signals during odorant exposure. Acetic acid and butanedione were used as the odorants for the bioelectronic nose. The multiple channels of the MEA allowed for simultaneous monitoring of multiple cells across the affixed epithelial layer, providing data on the correlation and synchrony of transmembrane potentials during odorant exposure. This setup allows for the collection of data on the transmembrane potentials with respect to time, location, and frequency, providing insight into spatiotemporal differences between the spontaneous firing signals *versus* stimulant responses of the olfactory cells. This information may be useful in future bioelectronic nose designs, as well as providing information on the signal transduction patterns of the excised tissue. Differences in firing signals were observed following exposure to the acetic acid and butanedione molecules, as determined by power spectrum analysis, indicating recognition by the olfactory epithelial cells on the MEA.

In a study that was similar to the previous two that have been described, Liu *et al.* explored excised olfactory mucosa from Sprague–Dawley rats, affixed them to LAPS chips, and the olfactory action potentials were analyzed in response to presentation of acetic acid and butanedione (Figure 2.7).[159] The changes in photocurrent of the LAPS were used to measure the potential changes of the illuminated cells upon exposure to odorants. Utilizing this setup, differences in firing of action potentials were recorded when comparing the acetic acid and butanedione odorants, both of which changed the

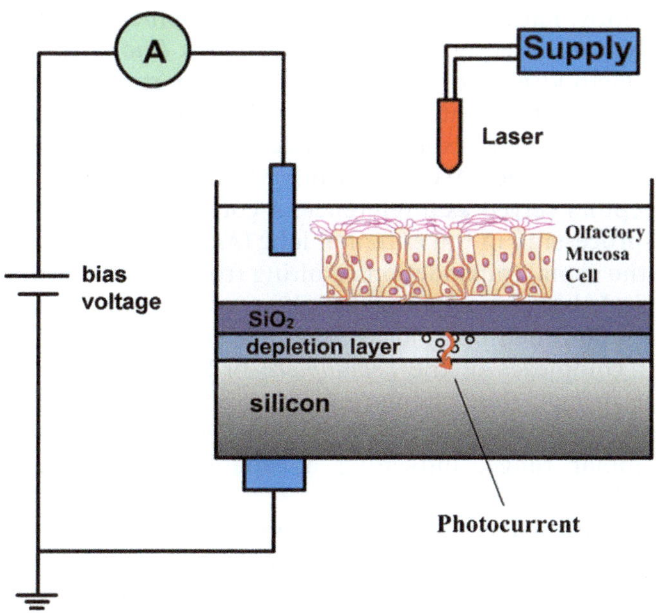

Figure 2.7 LAPS system of the olfactory mucosa tissue cells on the sensor surface. (Reproduced from Liu *et al.*,[159] with permission from Elsevier.)

firing frequency when compared to control conditions that lacked stimulation; approximately 40 spontaneous spikes in potential per minute were observed without stimulation. Generally speaking, it was noted that potential spikes were increased in frequency with exposure to butanedione, while exposure to acetic acid decreased the potential spiking frequency. However, following exposures to these odorants, firing frequencies returned to steady levels, in both cases, at rates that were notably higher than the rates prior to stimulant exposure. Questions still exist about the practicality of utilizing excised tissue in bioelectronic noses, particularly when compared with cultured cells, but this technology provides a setup that more closely mimics the naturally occurring olfactory setup for use in biosensing.

Olfactory receptors (Ors) from *Drosophila melanogaster* have also been studied with aims on creating a biosensor that mimics the olfactory system of insects.[160] Sf9 cells were transfected with genes to produce the *D. melanogaster* Ors, Or22a or Or67d, and confirmed *via* fluorescent antibody labeling. Fluorescence imaging of individual cells indicated an increase in Ca^{2+} in response to the addition of receptor-specific odorant ligands; ethyl butyrate and 11-*cis*-vaccenyl acetate were used to stimulate the Or22a and Or67d receptors, respectively. Furthermore, the Sf9 cells were successfully cultured on surface acoustic wave (SAW) chips for greater than 72 h, with strong adherence to the device occurring after 60 min. This finding holds promise for combined functionality of the transfected Sf9 cells with SAW resonators in future biosensing designs.

More recently, Racz *et al.* described the development of an infochemical detection system based on the moth *Spodoptera littoralis*.[161] Sf9 cells, expressing olfactory receptors, were made to be part of a "chemoreceiver" and were immobilized onto dual-SAW chips for detection of volatiles in various ratios. Sf9 cells expressing olfactory receptors were used on the sensing portion of the SAW chip, while Sf9 cells that did not express olfactory receptors served as a reference. A computational model, based on the signal processing in the antennal lobe (AL) of *Lepidoptera*, was created to decode the signal transduction resulting from odorant stimulation. This model included representations of olfactory receptor neurons, second-order projection neurons, local interneurons, glomeruli, and macroglomerular complexes of the *Lepidoptera* order of insect. Utilizing their biomimetic design, the "chemoreceiver" was able to detect octopamine hydroxide from a "chemoemitter" (expressing infochemical mixtures) in the micromolar range, indicating a capability to detect and classify pheromone blends.

Moreover, with respect to gustatory system mimics, a few studies have been conducted utilizing LAPS chips for potentiometric examination of tastant detection.[162,163] In one of these studies, circumvallate papillae were harvested from Sprague–Dawley rats, and the taste receptor cells were isolated and cultured on LAPS chips. Following isolation and culture, the taste receptor cells were examined for elicitation of responses from chemicals representing sour, bitter, sweet, and savory tastes, as well as for response to the neural transmitter ATP; these taste representative chemicals were HCl, $MgSO_4$, sucrose, and monosodium glutamate (MSG), respectively. The study with their biomimetic sensor indicated that it was difficult to use changes in extracellular potential firing rates to detect specific tastes, but taste density (analyte concentration) could be indicated by firing rates. There was also an observable set of taste cells that responded to ATP presentation, likely aiding in the transmission of taste signals between cells and to nerves. In the other study, a LAPS chip was utilized in a similar manner to examine acid-sensing (sour taste) in cells that were harvested from Sprague–Dawley rats. A Hodgkin–Huxley model was used to examine the changes in ionic currents, as a whole-cell patch-clamp type system. The experiments indicated that observable changes in extracellular potential firing rates and firing frequencies were pH-dependent in a number of the tested cells. However, some cells did not exhibit this type of response, but the authors suggested that these non-responsive cells may not have been acid-sensing cells; confirmation by immunocytochemistry protocols would be needed. In both of the LAPS chip studies, the use of whole cell biosensing was exhibited, demonstrating the potential of this platform to gain better understanding of the electrophysiology of taste receptor cells. In a commercial setting, this technology could theoretically be used in food flavor analysis.

Additionally, tastant sensing has been conducted with a sensor featuring S2 cells expressing the *Drosophila* G protein coupled receptor Gr5a.[164]

The Gr5a expressing S2 cells were cultured on an ion-sensitive field-effect transistor (ISFET) for label-free detection of sugar. When the cells grown on the biosensor were exposed to trehalose at varying concentrations, a concentration-dependent response was observable through voltage shifts of the ISFET; the voltage shifts are a function of changes in the ion content in the solution, as the trehalose interacts with the Gr5a receptors of the cells. This work provides insight into the transduction mechanisms (potentially ion channel signaling) in the gustatory system of *Drosophila*. Additionally, the biosensor shows promise as a platform that could introduce other cell types for whole cell biosensing diagnostic tools in other applications.

2.6 Summary

As has been discussed, biomimetic surfaces and materials have become a topic of much interest in biosensing. There are a number of ways to harness the biomimetic properties of materials, lending them to sensing applications. The use of lipid membranes have allowed for cellular membrane mimics, facilitating detection protocols in an environment similar to those found in cells. Furthermore, there have been a number of other approaches taken to immobilize biomolecules serving as sensing units in signaling or as a means of promotion of a more natural environment to enhance the binding activities of target cells. While these biomolecular approaches have been taken, there have also been efforts to mimic biomolecules, harnessing their recognition and catalytic properties, while doing so with a synthetic approach. Building upon this concept is the category of molecularly imprinted polymers. These constructs utilize templated polymer materials, as bioaffinity recognition materials for the imprinted material, allowing for specific recognition and allowing sensory transduction through a number of techniques. Lastly, the use of whole cells as the detection elements in sensing platforms has been utilized in a variety of applications, including mimics of olfactory and gustatory systems.

The advancements in nanomaterial fabrication techniques have also been harnessed for these biomimetic sensing technologies and should play a significant role in the future. The ability to harness the beneficial properties of materials at the nanoscale and interface with biological molecules at the scale at which they normally operate will likely drive future sensing developments. Furthermore, the ability to mimic natural materials while providing stability to sensing platforms should continue to make biomimicry an appealing method for the development of sensing technologies. The use lipid membranes, biomolecule immobilization techniques, and whole cell sensors are helpful approaches to ascertain information through the aforementioned detection platforms. Moreover, the utilization of more synthetic techniques such as biomolecular mimics or molecularly imprinted polymers seem like an appealing choice for biomimetic sensing platforms going forward.

References

1. W. Göpel and P. Heiduschka, *Biosens. Bioelectron.*, 1995, **10**, 853.
2. J. J. Ramsden, *Biosens. Bioelectron.*, 1998, **13**, 593.
3. N. Sugao, M. Sugawara, H. Minami, M. Uto and Y. Umezawa, *Anal. Chem.*, 1993, **65**, 363.
4. E. Reimhult, M. K. Baumann, S. Kaufmann, K. Kumar and P. R. Spycher, *Biotechnol. Genet. Eng.*, 2010, **27**, 185.
5. T. C. Chilcott, E. L. S. Wong, T. Böcking and H. G. L. Coster, *Physiol. Meas.*, 2008, **29**, S307.
6. H.-Z. Yu, S. Morin, D. D. M. Wayner, P. Allongue and C. H. de Villeneuve, *J. Phys. Chem. B*, 2000, **104**, 11157.
7. E. L. S. Wong, M. James, T. C. Chilcott and H. G. L. Coster, *Surf. Sci.*, 2007, **601**, 5740.
8. C. Karolis, H. G. L. Coster, T. C. Chilcott and K. H. Barrow, *Biochim. Biophys. Acta*, 1998, **1368**, 247.
9. T. Lehmann and J. Rühe, *Macromol. Symp.*, 1999, **142**, 1.
10. J. Li, Y. Du, P. Boullanger and L. Jiang, *Thin Solid Films*, 1999, **352**, 213.
11. U. Hilbig, O. Bleher, A. Le Blanc and G. Gauglitz, *Anal. Bioanal. Chem.*, 2012, **403**, 713.
12. R. Naumann, E. K. Schmidt, A. Jonczyk, K. Fendler, B. Kadenbach, T. Liebermann, A. Offenhäusser and W. Knoll, *Biosens. Bioelectron.*, 1999, **14**, 651.
13. T. Nakaminami, S. Ito, S. Kuwabata and H. Yoneyama, *Anal. Chem.*, 1999, **71**, 4278.
14. S. Godoy, B. Leca-Bouvier, P. Boullanger, L. J. Blum and A. P. Girard-Egrot, *Sens. Actuators B*, 2005, **107**, 82.
15. T. Jiao, B. D. Leca-Bouvier, P. Boullanger, L. J. Blum and A. P. Girard-Egrot, *Colloids Surf. A*, 2010, **354**, 284.
16. W. Bücking, G. A. Urban and T. Nann, *Sens. Actuators B*, 2005, **104**, 111.
17. R. Naumann, A. Jonczyk, C. Hampel, H. Ringsdorf, W. Knoll, N. Bunjes and P. Gräber, *Bioelectrochem. Bioenerg.*, 1997, **42**, 241.
18. W. Hoppe, W. Lohmann, H. Markl and H. Ziegler, *Biophysik, Zweite, Völlig Neubearbeitete Auflage*, Springer, Berlin, 1982, pp. 380–385.
19. I. Choi, S. Lee, S. Hong, Y. I. Yang, H. D. Song, J. Yi and T. Kang, *Sensors, IEEE*, 2009, **1–3**, 353.
20. J. S. Valentine, P. A. Doucette and S. Z. Potter, *Annu. Rev. Biochem.*, 2005, **74**, 563.
21. L. Banci, I. Bertini, A. Durazo, S. Girotto, E. B. Gralla, M. Martinelli, J. S. Valentine, M. Vieru and J. P. Whitelegge, *Proc. Natl. Acad. Sci. U. S. A.*, 2007, **104**, 11263.
22. H. A. Lashuel, *Sci. Aging Knowl. Environ.*, 2005, (38), pe28.
23. G. Woodhouse, L. King, L. Wieczorek, P. Osman and B. Cornell, *J. Mol. Recognit.*, 1999, **12**, 328.
24. P. C. Gufler, D. Pum, U. B. Sleytr and B. Schuster, *Biochim. Biophys. Acta*, 2004, **1661**, 154.

25. B. Schuster, P. C. Gufler, D. Pum and U. B. Sleytr, *IEEE Trans. Nanobiosci.*, 2004, **3**, 16.
26. B. Schuster and U. B. Sleytr, *Curr. Nanosci.*, 2006, **2**, 143.
27. G. Stark, *Membrane Transport in Biology*, ed. G. Giebisch, D. C. Tosteson and H. Ussing, Springer, Berlin, 1978, vol. 1, pp. 447–473.
28. P. Läuger, *Angew. Chem.*, 1985, **97**, 939.
29. B. Raguse, V. Braach-Maksvytis, B. A. Cornell, L. G. King, P. D. J. Osman, R. J. Pace and L. Wieczorek, *Langmuir*, 1998, **14**, 648.
30. H. Bao, Z. Peng, E. Wang and S. Dong, *Langmuir*, 2004, **20**, 10992.
31. S. R. Jadhav, Y. Zheng, R. M. Garavito and R. M. Worden, *Biosens. Bioelectron.*, 2008, **24**, 831.
32. R. F. Costello, I. R. Peterson, J. Heptinstall, N. G. Byrne and L. S. Miller, *Adv. Mater. Opt. Electron.*, 1998, **8**, 47.
33. S. R. Evans and I. Fritsch, *Electroanalysis*, 2004, **16**, 45.
34. A. Simon, A. Girard-Ergot, F. Sauter, C. Pudda, N. P. D'Hahan, L. Blum, F. Chatelain and A. Fuchs, *J. Colloid Interface Sci.*, 2007, **308**, 337.
35. J. S. Hansen, M. Perry, J. Vogel, J. S. Groth, T. Vissing, M. S. Larsen, O. Geschke, J. Emneús, H. Bohr and C. H. Nielsen, *Anal. Bioanal. Chem.*, 2009, **395**, 719.
36. J. B. Largueze, K. El Kirat and S. Morandat, *Colloids Surf. B*, 2010, **79**, 33.
37. B. Wicklein, M. Darder, P. Aranda and E. Ruiz-Hitzky, *ACS Appl. Mater. Interfaces*, 2011, **3**, 4339.
38. X. Song and B. I. Swanson, *Proc. SPIE*, 1999, **3537**, 280.
39. X. Song and B. I. Swanson, *Anal. Chem.*, 1999, **71**, 2097.
40. L. M. Freeman, S. Li, Y. Dayani, H. S. Choi, N. Malmstadt and A. M. Armani, *Appl. Phys. Lett.*, 2011, **98**, 143703.
41. L. L. Wang, A .K. Gaigalas and V. Reipa, *Biotechniques*, 2005, **38**, 127.
42. Q. He, Y. Tian, H. Möhwald and J. Li, *Soft Matter*, 2009, **5**, 300.
43. J. Y. Wong, J. Majewski, M. Seitz, C. K. Park, J. N. Israelachvili and G. S. Smith, *Biophys. J.*, 1999, **77**, 1445.
44. F. J. Carrion, A. De La Maza and J. L. Parra, *J. Colloid Interface Sci.*, 1994, **164**, 78.
45. S. McLaughlin, *Annu. Rev. Biophys. Biophys. Chem.*, 1989, **18**, 113.
46. M. Tanaka and E. Sackmann, *Nature*, 2005, **437**, 656.
47. R. P. Richter, R. Berat and A. R. Brisson, *Langmuir*, 2006, **22**, 3497.
48. X. Song, J. Shi and B. Swanson, *Anal. Biochem.*, 2000, **284**, 35.
49. S. Pavan and F. Berti, *Anal. Bioanal. Chem.*, 2012, **402**, 3055.
50. H. A. Godwin and J. M. Berg, *J. Am. Chem. Soc.*, 1996, **118**, 6514.
51. X. Bi, A. Agarwal, N. Balasubramanian and K. L. Yang, *Electrochem. Commun.*, 2008, **10**, 1868.
52. Y. M. Xu, H. Q. Pan, S. H. Wu and B. L. Zhang, *Chin. J. Anal. Chem.*, 2009, **37**, 783.
53. M. Lin, M. Cho, W. S. Choe, J. B. Yoo and Y. Lee, *Biosens. Biolectron.*, 2010, **26**, 940.

54. M. B. Villiers, S. Cortès, C. Brakha, J. P. Lavergne, C. A. Marquette, P. Deny, T. Livache and P. N. Marche, *Biosens. Bioelectron.*, 2010, **26**, 1554.

55. B. Viguier, K. Zòr, E. Kasotakis, A. Mitraki, C. H. Clausen, W. E. Svendsen and J. Castillo-Leòn, *ACS Appl. Mater. Interfaces*, 2011, **3**, 1594.

56. M. Mascini, A. Macagnano, G. Scortichini, M. Del Carlo, G. Diletti, A. D'Amico, C. Di Natale and D. Compagnone, *Sens. Actuators B*, 2005, **111–112**(suppl), 376.

57. C. Nakamura, Y. Inuyama, H. Goto, I. Obataya, N. Kaneko, N. Nakamura, N. Santo and J. Miyake, *Anal. Chem.*, 2005, 77, 7750.

58. I. Obataya, C. Nakamura, H. Enomoto, T. Hoshino, N. Nakamura and J. Miyake, *J. Mol. Catal. B*, 2004, **28**, 265.

59. S. Carnazza, C. Foti, G. Gioffrè, F. Felici and S. Guglielmino, *Biosens. Biolectron*, 2007, **23**, 1137.

60. H. Zhang, X. Li, I. Bay, R. Niu, Y. Jia, C. Zhang, L. Zhang and Y. Cao, *Biotechnol. Appl. Biochem.*, 2009, **53**, 185.

61. T. Z. Wu, *Biosens. Bioelectron.*, 1999, **14**, 9.

62. S. Jia, J. Fei, J. Zhou, X. Chen and J. Meng, *Biosens. Bioelectron.*, 2009, **24**, 3049.

63. O. Choi, B. C. Kim, J. H. An, K. Min, Y. H. Kim, Y. Um, M. K. Oh and B. I. San, *Enzyme Microb. Technol.*, 2011, **49**, 441.

64. X. Luo, M. Xu, C. Freeman, T. James and J. J. Davis, *Anal. Chem.*, 2013, **85**, 4129.

65. F. Yin, H. K. Shin and Y. S. Kwon, *Biosens. Bioelectron.*, 2005, **21**, 21.

66. A. M. Bond, *Modern Polarographic Methods in Analytical Chemistry*, Dekker, New York, 1980, pp. 8–83.

67. S. Lee, I. Choi, S. Hong, Y. I. Yang, J. Lee, H. D. Song, J. Yi and T. Kang, in *2009 IEEE Sensors*, (8th IEEE Conference on Sensors, Christchurch, New Zealand), IEEE, New York, 2009, pp. 422–424.

68. K. Gabrovska, T. Nedelcheva, T. Godjevargova, O. Stoilova, N. Manolova and I. Rashkov, *J. Mol. Catal. B: Enzym.*, 2008, **55**, 169.

69. F. Tian, W. Wu, M. Broderick, V. Vamvakaki, N. Chaniotakis and N. Dale, *Biosens. Bioelectron.*, 2010, **25**, 2408.

70. H. Vaisocherová, W. Yang, Z. Zhang, Z. Cao, G. Cheng, M. Piliarik, J. Homola and S. Jiang, *Anal. Chem.*, 2008, **80**, 7894.

71. N. D. Brault, C. Gao, H. Xue, M. Piliarik, J. Homola, S. Jiang and Q. Yu, *Biosens. Bioelectron.*, 2010, **25**, 2276.

72. K. A. Marx, T. Zhou, D. MacIntosh and S. J. Braunhut, *Mater. Res. Soc. Symp. Proc.*, 2004, **EXS-1**, 169.

73. R. Knerr, B. Weiser, S. Drotleff, C. Steinem and A. Göpferich, *Macromol. Biosci.*, 2006, **6**, 827.

74. N. Nath, J. Hyun, H. Ma and A. Chilkoti, *Surf. Sci.*, 2004, **570**, 98.

75. J. H. Lee, H. B. Lee and J. D. Andrade, *Prog. Polym. Sci.*, 1995, **20**, 1043.

76. S. Drotleff, U. Lungwitz, M. Breunig, A. Dennis, T. Blunk, J. Tessmar and A. Göpferich, *Eur. J. Pharm. Biopharm.*, 2004, **58**, 385.

77. U. Hersel, C. Dahmen and H. Kessler, *Biomaterials*, 2003, **24**, 4385.

78. Z. Zhang, S. Chen and S. Jiang, *Biomacromolecules*, 2006, **7**, 3311.

79. M. S. Mannoor, S. Zhang, A. J. Link and M. C. McAlpine, *Proc. Natl. Acad. Sci. U. S. A.*, 2010, **107**, 19207.

80. R. L. Buchanan and M. P. Doyle, *Food Technol. (Chicago, IL, U. S.)*, 1997, **51**, 69.

81. Y. Luo, G. Hu, A. Zhu, B. Kong, Z. Wang, C. Liu and Y. Tian, *Biosens. Bioelectron.*, 2011, **29**, 189.

82. J. M. McCord and I. Fridovich, *J. Biol. Chem.*, 1968, **243**, 5753.

83. J. M. McCord and I. Fridovich, *J. Biol. Chem.*, 1969, **244**, 6094.

84. H. A. Kontos and E. P. Wei, *J. Neurosurg.*, 1986, **64**, 803.

85. M. Y. Globus, O. Alonso, W. A. Dietrich, R. Busto and M. D. Ginsberg, *J. Neurochem.*, 1995, **65**, 1704.

86. C. E. Thomas, P. Bernardelli, S. M. Bowen, S. F. Chaney, D. Friedrich, D. A. Janowick, B. K. Jones, F. J. Keeley, J. H. Kehne, B. Ketteler, D. F. Ohlweiler, L. A. Paquette, D. J. Robke and T. L. Fevig, *J. Med. Chem.*, 1996, **39**, 4997.

87. M. Z. Wrona and G. Dryhurst, *Chem. Res. Toxicol.*, 1998, **11**, 639.

88. H. A. Sadek, P. A. Szweda and L. I. Szweda, *Biochemistry*, 2004, **43**, 8494.

89. J. Ellis, E. D. Castillo, M. Bayon, R. Grimm, J. F. Clark, G. Pyne-Geithman, S. Wilbur and J. A. Caruso, *J. Proteome Res.*, 2008, **7**, 3747.

90. H. Ji, H. Li, P. Martasek, L. J. Roman, T. L. Poulos and R. B. Silverman, *J. Med. Chem.*, 2009, **52**, 779.

91. S. Komatsu, R. Yamamoto, Y. Nanjo, Y. Mikami, H. Yunokawa and K. Sakata, *J. Proteome Res.*, 2009, **8**, 4766.

92. M. Müller, B. Kessler and K. Lunkwitz, *J. Phys. Chem. B*, 2003, **107**, 8189.

93. H. Zhang, J. J. Xu and H. Y. Chen, *J. Electroanal. Chem.*, 2008, **624**, 79.

94. E. Paleček, F. Jelen, C. Teijeiro, V. Fučik and T. M. Jovin, *Anal. Chim. Acta*, 1993, **273**, 175.

95. Y. Liu, P. Layrolle, J. Bruijn, C. Blitterswijk and K. Groot, *J. Biomed. Mater. Res.*, 2001, **57**, 327.

96. Y. Liu, E. Hunziker, P. Layrolle, J. Bruijn and K. Groot, *Tissue Eng.*, 2004, **10**, 101.

97. A. Oyane, M. Uchida and A. Ito, *Key Eng. Mater.*, 2004, **254–256**, 541.

98. M. Uchida, A. Oyane, H. Kim, T. Kokubo and A. Ito, *Adv. Mater.*, 2004, **16**, 1071.

99. A. Oyane, M. Uchida and A. Ito, *J. Biomed. Mater. Res.*, 2005, **72A**, 168.

100. A. Oyane, M. Uchida, Y. Ishihara and A. Ito, *Key Eng. Mater.*, 2005, **284–286**, 227.

101. A. Oyane, M. Uchida, K. Onuma and A. Ito, *Biomaterials*, 2006, **27**, 167.

102. M. Uchida, A. Ito, K. Furukawa, K. Nakamura, Y. Onimura, A. Oyane, T. Ushida, T. Yamane, T. Tamak and T. Tateishi, *Biomaterials*, 2005, **26**, 6924.

103. A. Oyane, K. Hyodo, M. Uchida, Y. Sogo and A. Ito, *Key Eng. Mater.*, 2006, **309–311**, 1181.

104. A. Oyane, Y. Yoshiro, M. Uchida and A. Ito, *Biomaterials*, 2006, **27**, 3295.

105. N. G. Tognalli, P. Scodeller, V. Flexer, R. Szamocki, A. Ricci, M. Tagliazucchi, E. J. Calvo and A. Fainstein, *Phys. Chem. Chem. Phys.*, 2009, **11**, 7412.
106. M. Hnilova, C. R. So, E. E. Oren, B. R. Wilson, T. Kacar, C. Tamerler and M. Sarikaya, *Soft Matter*, 2012, **8**, 4327.
107. A. J. Haes, S. L. Zou, G. C. Schatz and R. P. Van Duyne, *J. Phys. Chem. B*, 2004, **108**, 109.
108. X. Zhong, G. S. Qian, J. J. Xu and H. Y. Chen, *J. Phys. Chem. C*, 2010, **114**, 19503.
109. R. Lind, P. Connolly, C. D. W. Wilkinson, L. J. Breckenridge and J. A. T. Dow, *Biosens. Bioelectron.*, 1991, **6**, 359.
110. J. Chen, D. Du, F. Yan, H. X. Ju and H. Z. Lian, *Chem.–Eur. J.*, 2005, **11**, 1467.
111. C. Hao, F. Yan, L. Ding, Y. D. Xue and H. X. Ju, *Electrochem. Commun.*, 2007, **9**, 1359.
112. Y. Y. Wong, S. P. Ng, M. H. Ng, S. H. Si, S. Z. Yao and Y. S. Fung, *Biosens. Bioelectron.*, 2002, **17**, 676.
113. H. Matsui, *Proc. SPIE*, 2010, **7593**, 759310.
114. F. S. Damos, M. P. T. Sotomayor, L. T. Kubota, S. M. C. N. Tanaka and A. A. Tanaka, *Analyst*, 2003, **128**, 255.
115. S. S. Rosatto, L. T. Kubota and G. Oliveira Neto, *Anal. Chim. Acta*, 1999, **390**, 65.
116. A. Lindgren, J. Emnéus, T. Ruzgas, L. Gorton and G. Marko-Varga, *Anal. Chim. Acta*, 1997, **347**, 51.
117. F.-D. Munteanu, A. Lindgren, J. Emnéus, L. Gorton, T. Ruzgas, E. Csöregi, A. Ciucu, R. B. van Huystee, I. G. Gazaryan and L. M. Lagrimini, *Anal. Chem.*, 1998, **70**, 2596.
118. W. A. Alves, I. O. Matos, P. M. Takahashi, E. L. Bastos, H. Martinho, J. G. Ferreira, C. C. Silva, R. H. de Almeida Santos, A. Paduan-Filho and A. M. Da Costa Ferreira, *Eur. J. Inorg. Chem.*, 2009, 2219.
119. M. Zamocky and F. Koller, *Prog. Biophys. Mol. Biol.*, 1999, **72**, 19.
120. X. Chen, H. Xie, J. Kong and J. Deng, *Biosens. Bioelectron.*, 2001, **16**, 115.
121. I. Fridovich, *Annu. Rev. Biochem.*, 1995, **64**, 97.
122. M. C. Q. Oliveira, M. R. V. Lanza, A. A. Tanaka and M. D. P. T. Sotomayor, *Anal. Methods*, 2010, **2**, 507.
123. A. Rahim, S. B. A. Barros, L. T. Kubota and Y. Gushikem, *Electrochim. Acta*, 2011, **56**, 10116.
124. J. Zhou, Y. Luo, A. Zhu, Y. Liu, Z. Zhu and Y. Tian, *Analyst*, 2011, **36**, 1594.
125. M. D. Marazuela and M. C. Moreno-Bondi, *Anal. Bioanal. Chem.*, 2002, **372**, 664.
126. N. Gao, Z. Xu, F. Wang and S. Dong, *Electroanalysis (N. Y.)*, 2007, **19**, 1655.
127. B. S. Ebarvia, S. Cabanilla and F. Sevilla III, *Talanta*, 2005, **66**, 145.
128. D. Croux, A. Weustenraed, P. Pobedinskas, F. Horemans, H. Diliën, K. Haenen, T. Cleij, P. Wagner, R. Thoelen and W. De Ceuninck, *Phys. Status Solidi A*, 2012, **209**, 892.

129. C. T. Wu, P. Y. Chen, J. G. Chen, V. Suryanarayanan and K. C. Ho, *Anal. Chim. Acta*, 2009, **633**, 119.
130. S. M. Reddy, G. Sette and Q. Phan, *Electrochim. Acta*, 2011, **56**, 9203.
131. D. M. Hawkins, D. Stevenson and S. M. Reddy, *Anal. Chim. Acta.*, 2005, **542**, 61.
132. C. Sontimuang, R. Suedee and F. Dickert, *Anal. Biochem.*, 2011, **410**, 224.
133. A. J. Baeumner, *Anal. Bioanal. Chem.*, 2003, **377**, 434.
134. H. Z. H. M. Sun, J. T. S. Choy, D. R. Zhu and Y. S. Fung, *Sens. Actuators B*, 2008, **131**, 148.
135. K. Seidler, P. A. Lieberzeit and F. L. Dickert, *Analyst*, 2009, **134**, 361.
136. Z. Wu, C. Tao, C. Lin, D. Shen and G. Li, *Chem.–Eur. J.*, 2008, **14**, 11358.
137. M. Lulka, S. Iqbal, J. Chambers, E. Valdes, R. Thompson, M. Goode and J. Valdes, *Mater. Sci. Eng. C*, 2000, **C11**, 101.
138. D. J. Monk and D. R. Walt, *Anal. Bioanal. Chem.*, 2004, **379**, 931.
139. M. B. Gu, P. S. Dhurjati, Y. K. Van Dyk and R. A. LaRossa, *Biotechnol. Program*, 1996, **12**, 393.
140. M. B. Gu, C. C. Gil and J. H. Kim, *Biosens. Bioelectron.*, 1999, **14**, 355.
141. B. Polyak, E. Bassis, A. Novodvorets, S. Belkin and R. S. Marks, *Water. Sci. Technol.*, 2000, **42**, 305.
142. B. Polyak, E. Bassis, A. Novodvorets, S. Belkin and R. S. Marks, *Sens. Actuators B*, 2001, **B74**, 18.
143. M. B. Gu and G. C. Gil, *Biosens. Bioelectron.*, 2001, **16**, 661.
144. I. Biran, D. M. Rissin, E. Z. Ron and D. R. Walt, *Anal. Biochem.*, 2003, **315**, 106.
145. G. C. Gil, R. J. Mitchell, S. T. Chang and M. B. Gu, *Biosens. Bioelectron.*, 2000, **15**, 23.
146. G. C. Gil, Y. J. Kim and M. B. Gu, *Biosens. Bioelectron.*, 2002, **17**, 427.
147. C. Vedrine, J.-C. Leclerc, C. Durrieu and C. Tran-Minh, *Biosens. Bioelectron.*, 2003, **18**, 457.
148. I. Biran and D. R. Walt, *Anal. Chem.*, 2002, **74**, 3046.
149. L. C. Taylor and D. R. Walt, *Anal. Biochem.*, 2000, **278**, 132.
150. S. Leth, S. Maltoni, R. Simkus, B. Mattiasson, P. Corbisier, I. Klimant, O. S. Wolfbeis and E. Csoregi, *Electroanalysis*, 2002, **14**, 35.
151. K. C. Persaud, *IEEE Sens. J.*, 2012, **12**, 3108.
152. L. Du, C. Wu, Q. Liu, L. Huang and P. Wang, *Biosens. Bioelectron.*, 2013, **42**, 570.
153. H. J. Ko and T. H. Park, *Proc. 229th ACS Natl. Meeting*, 2005, **229**, U104.
154. S. H. Lee, S. H. Jeong, S. B. Jun, S. J. Kim and T. H. Park, *Electrophoresis*, 2009, **30**, 3283.
155. S. H. Lee and T. H. Park, *Biotechnol. Bioprocess Eng.*, 2010, **15**, 22.
156. M. Marrakchi, J. Vidic, N. Jaffrezic-Renault, C. Martelet and E. Pajot-Augy, *Eur. Biophys. J. Biophys. Lett.*, 2007, **36**, 1015.
157. C. Wu, P. Chen, H. Yu, Q. Liu, X. Zong, H. Cai and P. Wang, *Biosens. Bioelectron.*, 2009, **24**, 1498.

158. Q. Liu, W. Ye, L. Xiao, L. Du, N. Hu and P. Wang, *Biosens. Bioelectron.*, 2010, **25**, 2212.
159. Q. Liu, W. Ye, H. Yu, N. Hu, L. Du, P. Wang and M. Yang, *Sens. Actuators B*, 2010, **146**, 527.
160. M. D. Jordan and R. A. J. Challiss, *Procedia Comput. Sci.*, 2011, 7, 281.
161. Z. Rácz, M. Cole, J. W. Gardner, M. F. Chowdhury, W. P. Bula, J. G. E. Gardeniers, S. Karout, A. Capurro and T. C. Pearce, *Int. J. Circ. Theor. Appl.*, 2013, **41**, 653.
162. P. Chen, B. Wang, G. Cheng and P. Wang, *Biosens. Bioelectron.*, 2009, **25**, 228.
163. P. Chen, X. D. Liu, B. Wang, G. Cheng and P. Wang, *Sens. Actuators B*, 2009, **139**, 576.
164. H. C. Lau, T. E. Bae, H. J. Jang, J. Y. Kwon, W. J. Cho and J. O. Lim, *Jpn. J. Appl. Phys.*, 2013, 52, 04CL02.

Hydrogel-Based Molecularly Imprinted Polymers for Biological Detection

HAZIM F. EL-SHARIF,[a] DEREK STEVENSON,[a] KEITH WARRINER[b] AND SUBRAYAL M. REDDY*[a]

[a] Department of Chemistry, Faculty of Engineering and Physical Sciences, University of Surrey, Guildford, Surrey, GU2 7XH, UK; [b] Department of Food Science, University of Guelph, Guelph, ON, Canada, N1G 2W1
*Email: s.reddy@surrey.ac.uk

3.1 Introduction

Molecularly imprinted polymers (MIPs) are smart materials designed to selectively bind certain molecules with a degree of chemical affinity. Molecular imprinting has rapidly become an effective method of imprinting highly specific and selective recognition sites in synthetic polymers. As such, MIPs have generally been referred to as plastic antibodies or synthetic receptors. Over the years, MIPs have increased in popularity to fabricate sensors,[1,2] in separation methods,[3,4] in protein crystallization,[5,6] and in catalysis.[7] MIPs are highly cross-linked polymer networks that are intrinsically stable, robust, and are able to facilitate their application in extreme environments. The functional mechanism is similar to antibodies or enzymes, where detection is based on shape, orientation, and interaction between the molecule and the MIP, much like a lock and key.[3,8,9] The imprinting process occurs when polymer networks are generated by either a self-assembly *in situ* imprinting

RSC Detection Science Series No. 3
Advanced Synthetic Materials in Detection Science
Edited by Subrayal Reddy

Published by the Royal Society of Chemistry, www.rsc.org

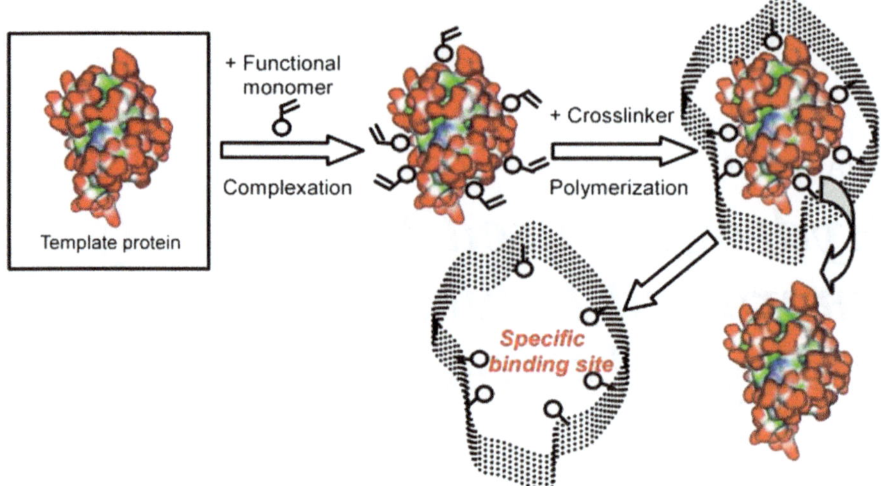

Figure 3.1   The molecular imprinting process. Reversible interactions between template and functional monomers are formed by pre-association in polymerization.[64]

approach or a post-polymerization imprinting approach. This is achieved by allowing functional and matrix monomers to pre-associate and polymerize with a template in a solution mixture (see Figure 3.1). The association between the template and monomer is fixed in place by polymerization in the presence of a controlled amount of a cross-linker monomer, which can impart the robustness required for the polymer.[10] Removing the template from the bulk polymer leaves behind an idealized cavity capable of recognizing and rebinding the same template. The reversible interactions between template and functional monomers usually involves one or more of the following mechanisms: (a) reversible covalent bonds; (b) covalently attached polymerizable binding groups that are activated for non-covalent interaction by template cleavage; (c) electrostatic interactions; and (d) hydrophobic or van der Waals interactions.[11] In the non-covalent imprinting approach the degree of association between the monomer and the target is predominantly based on simple molecular interactions such as hydrogen bonding, hydrophobic, and electrostatic interactions. Alternatively, in the covalent approach the degree of association between the functional monomer and the target is mostly based on covalent interactions before polymerization. Target molecules can be extracted from the bulk polymer by cleaving the covalent linkage, and then that same target can be rebound by covalently reattaching the molecule or by virtue of intermolecular semi-covalent hydrogen bonding interactions.

3.2   Traditional Imprinting

Historically, molecular imprinting has been used for the detection of small molecules, generally of low molecular weight and non-biological origin.

These have included drug molecules such as atenolol, caffeine, and nicotine, and herbicide or pesticide molecules such as atrazine.[12–17] Such molecules have been extensively exploited as imprinting templates, leading to significant achievements in solid-phase extraction (SPE), sensors, membranes, and enzyme-like catalysis.[3,7,18–21] Traditionally, SPE and chromatography were the main developing areas for the application of MIP technology.[12] The main perceived advantage of MIPs over biological antibodies for SPE was the ease with which they can be obtained and the consequent lower cost and speed.[12] However, MIPs are not without flaws: one of the main drawbacks in MIP approaches to SPE is the difficulty in removing the entire template analyte molecule. Complete template removal is imperative for accuracy of results, since leaching of residual analyte into actual samples would cause major contamination issues. Moreover, MIPs are usually prepared with comparatively large amounts of template, which exacerbates the latter leaching problem. Thus, structural analogues (of the target analyte) can be imprinted as the template to tackle such problems.

As an example of small molecule imprinting, many herbicides such as atrazine have been imprinted using non-covalent strategies.[13] Atrazine is a common broad-spectrum herbicide that has been used extensively over the last 30 years. Gas and liquid chromatography have both been used extensively for the detection and quantification of pesticides at trace levels, but these techniques can be time consuming and costly.[13] Lavignac *et al.* used a number of approaches under both equilibrium and non-equilibrium conditions to investigate MIP performance for atrazine using high-performance liquid chromatography (HPLC). The affinity and selectivity of the polymers were initially evaluated under non-equilibrium conditions through calculated imprinting factors (6.68 and 0 for MIP and NIP, respectively, where NIP = non-imprinted polymer). Relative imprinting factors using structural analogues of atrazine (triazine, ametryn, simazine, chlorotoluron, propazine, prometryn, alachlor, and metribuzine) also revealed high selectivities for the imprinted polymers. In order to determine affinity constants and binding site concentrations, isotherm data were re-plotted in the form of a Scatchard plot (see Eqn 3.1).[13] This is a linearized form of the Langmuir equation, of which the transformation has shown to distort experimental error, and only assumes single affinity constant binding site populations. B_{max} is the apparent maximum number of binding sites, K_d the equilibrium dissociation constant, F the concentration of free analyte, and B the concentration of bound analyte:

$$\frac{B}{F} = \frac{B_{max} - B}{K_d} \tag{3.1}$$

However, Lavignac *et al.* expressed that due to the heterogeneous distribution of binding sites in MIP matrices, MIP–ligand binding studies for simple organic molecules, such as pesticides, herbicides, and drugs, have generally reported non-linear concave curves.[13] Therefore, alternative approaches indicative of binding site cooperativity that are commonly observed in biological ligand–receptor systems were used for MIP–ligand binding analysis. The Hill equation (see Eqn 3.2), where Y is the binding site occupancy and n_h is the Hill coefficient, relates to a linear Scatchard plot when n_h is equal to 1.0, and is indicative of ligand binding with no cooperativity to one site:

$$\log\frac{Y}{1-Y} = n_h \times \log[F] - n_h \times \log K_d \qquad (3.2)$$

Variations in n_h, *i.e.* if greater than 1.0, present a sigmoidal graph indicating receptor/ligand having multiple binding sites with positive cooperativity. Such would be expected of MIP–ligand binding due to the heterogeneous distribution of binding sites. However, if n_h is less than 1.0 it can also be indicative of multiple binding sites, nonetheless with different affinities for template or negative cooperativity. It was observed that both selectivity and affinity were dependent on the concentration of the ligand and that, unusually, selectivity and affinity were better at higher atrazine concentrations (70 µmol). It was concluded that this phenomenon resulted from the formation of atrazine complexes during the pre-polymerization stage and during rebinding during higher concentrations. This allowed MIP polymer particles to demonstrate an improved atrazine affinity when the conditions favored complex formation.

Numerous traditional MIPs are now commercially available for binding toxins (*e.g.* bisphenol A), pesticides (*e.g.* catechol), and antibiotics (*e.g.* penicillin G).[22] Companies such as Biotage, PolyIntell, Raptor Detection, and Semorex offer to develop custom-made MIPs.[22] One of the leading authorities and commercial pioneers of MIPs for process scale separations, analytical chromatography, and sample preparation is Biotage (formally known as MIP Technologies). For the last decade, Biotage have offered tailor-made SupelMIP SPE cartridges for individual analytes and analyte classes.[23,24] SupelMIPs consist of highly cross-linked polymers that can be used within the whole pH range (1–14). They exhibit minimal swelling and shrinking, and offer selectivity for the extraction of individual trace analytes in complex matrixes. Over the years, SupelMIP design and selectivity have provided faster and simpler sample prep methods, better mass spectrometry (MS) compatibility by reducing ion suppression, and have allowed lower detection limits and improved sensitivity.[23,24]

Although the imprinting of small molecules has its importance in the analytical world and has seen many advances over the years, there have been efforts to investigate larger and more complex biomacromolecule structures such as proteins.[9,25–27] As such, MIPs have been regarded as 'antibody mimics' and have shown clear advantages (see Table 3.1) over real antibodies for sensor technology.[9,26,28] Although antibodies exhibit a high degree of selectivity, any biological recognition element is inherently unstable, with limited shelf-life even when stored under optimum conditions.[28] However, MIPs are highly cross-linked, intrinsically stable, robust, and are able to facilitate their application in extreme environments.[7] Accordingly, chemical

Table 3.1 Characteristic comparison between MIPs and antibodies.

Characteristic	Antibody	MIPs
Selection	*In vivo*	*In vitro*
Size (kDa)	~150–160	Not applicable
Production	Animal or recombinant (ethical considerations)	Synthetic (large scale, low costs)
Stability	Several weeks at 4 °C	Years at room temperature
Binding site	Monoclonal: homogeneous; polyclonal: heterogeneous	Strategy dependent; heterogeneous or homogeneous
Target molecules	Immunogenic macromolecules	Strategy dependent
Application conditions	Physiological	Strategy and template dependent

sensors and biosensors are ever increasing in use within the field of modern analytical chemistry. Their opportunities include clinical diagnostics, environmental analysis, food analysis, and production monitoring, as well as the detection of illicit drugs, genotoxicity, and chemical warfare agents.[7,9,10,20,25,29,30] The incorporation of MIP-based bio-mimetic receptor systems capable of binding target molecules with affinities and specificities on a par with natural receptors has been accomplished.[9,26]

The remainder of this chapter is now dedicated to the advances made in biomacromolecular imprinting within hydrogel-based MIPs. Their progress towards commercial alternatives and readily available diagnostic tools in biosensing applications will be discussed. The design characteristics in which MIPs are utilized in biosensing applications are outlined in terms of MIP coupling to transduction mechanisms.

3.3 Biomacromolecular Imprinting within Hydrogels

The constant demand for new and innovative biological recognition methods that rely on alternatives to antibodies, enzymes, and various biological reagents has led to the development of receptor-like synthetic smart materials such as MIPs. The imprinting of large biomacromolecules, such as pathogens or proteins, does however present a variety of challenges. Proteins, for example, are relatively labile, and have changeable conformations which are sensitive to various factors, *e.g.* solvent environments, pH, and temperature.[31–33] At present, biological reagents of animal origin such as antibodies and enzymes are mainly employed for protein recognition purposes.[31] To date, antibodies are the most successful affinity tools used in modern analysis such as diagnostics, purification, and therapeutics.[31] Alternative affinity tools have been based on nucleic acids (aptamers) and polypeptides (engineered binding proteins).[28] However, such biological reagents can sometimes be difficult to reproduce or are expensive, have low shelf lives, and have ethical issues surrounding their origin.[28,31] Therefore, receptor-like synthetic smart material MIPs have been intensively studied as

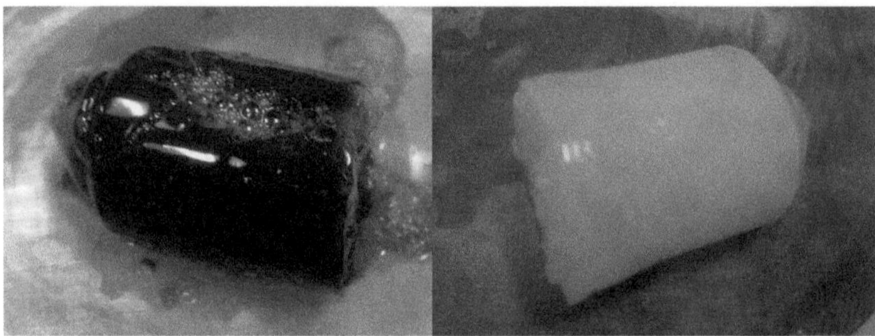

Figure 3.2   An example of MIP (*left*) and NIP (*right*) hydrogels made of polyacrylamide. The MIP retains its reddish color due to the imprinting of bovine
hemoglobin.

substitutes for natural receptors and have been shown to offer many
attractive features.[1,2,5,6,34]

Recently, the design of aqueous phase and environmentally sensitive
'smart hydrogels' have been employed.[35,36] Imprinting methods for preparing MIPs in aqueous media are being extensively researched in order to
provide a viable alternative solution that would sustain protein stability.[35,36]
However, traditional methods for preparing MIPs in aqueous media are also
associated with problems such as a low binding efficiency and the possibility
of protein denaturation.[32] Hydrogels have the ability to sense changes of pH,
temperature, or the concentration of metabolite and release their load as a
result of such a change. Hydrogels are insoluble cross-linked macromolecular networks of polymer chains swollen in water and other biological
fluids (see Figure 3.2).[36] They are hydrophilic and have a highly absorbent
nature, enabling them to contain over 99% water. Hydrogels are usually
made by co-polymerization of two or more monomers, with the disadvantage
of having relatively low mechanical strength. This can be overcome by either
cross-linking or by crystallization. The degree and nature of cross-linking
and the crystallinity of the polymer are responsible for its properties in the
swollen state. The degree to which it imbibes water without the loss of shape
is important in many natural hydrogels. Hydrogels have become excellent
carriers for the release of drugs and active macromolecules in their equilibrium swollen state as dynamic swollen systems.[36] The bio-compatibility of
these gels is attributed to their ability to mimic natural tissue due to their
high water content and their special surface properties.

3.3.1   Protein Imprinting

Proteins or polypeptides are organic compounds built from a linear chain
series of up to 20 different L-α-amino acids; these are joined together by the
peptide bonds between the carboxyl and amino groups of adjacent amino
acid residues. Proteins are essential parts of organisms and participate in
virtually every process within cells.[31] The extent to which proteins are

commonly researched in MIP applications, and proteomics in general, is in response to their important roles within the medical diagnostic area. Proteins involved in cell growth and cell signalling are widely researched as potential targets for cardiac disease and muscle injury markers for novel therapies.[37] For example, when screening for food allergens, it is often specific proteins that trigger allergic responses. There are numerous different allergen proteins present in individual foods, all of which need to be specifically detected during analysis, *e.g.* egg and sesame allergies are attributed to lysozyme and globulins, respectively.[38] However, current measurements are not in place to enforce such legislation relating to the content of food allergens due to a lack of measurement methods for specific food allergens and lack of certified reference material (CRM) in order to enforce legislation.[38]

Over the years, protein imprinting has generally involved one of the three following approaches.[39] The first is bulk imprinting, where protein templates are wholly imprinted in the bulk of polymer matrices and recognized as a whole by functional monomers. The second approach is surface imprinting, where protein templates are partially imprinted or stamped on bulk polymer surfaces with recognition and binding sites at specific orientations. The last approach is epitope imprinting, which is where a part or fragment of the protein template is imprinted, allowing for the recognition of whole proteins. As mentioned previously, the imprinting of large biomacromolecules is not without its challenges. Owing to the large size of proteins (6000 Da to several million Da) it is essential to control the size and number of pores that are generated (in bulk and on surface) during MIP synthesis, together with the density of the MIP network.[39] Hydrogels of polyacrylamides and/or other acrylamide derivatives, and advanced polymeric materials such as sol–gel composites, have been extensively studied and are particularly suitable for controlling the latter factors for effective MIP design and synthesis.

3.3.1.1 Polyacrylamide-Based Hydrogels

Polyacrylamide (polyAA) gels are known to possess the necessary parameters to successfully produce an analyte-specific MIP, *i.e.* being very inert, offering hydrogen bonding capabilities, and being biocompatible.[31,36] As such, polyAA hydrogels have presented themselves as excellent candidates for hydrogel-based MIPs (HydroMIPs). PolyAA is a nitrogen-containing member of the acrylate family of polymers, and is a suitable imprinting matrix for biological molecules as it is water soluble, cheap, easily produced, and can be tailored to possess attractive structural parameters. These hydrogels are formed in aqueous solution by copolymerization of acrylamide and small amounts of a bifunctional cross-linker, which can impart the robustness required for the polymer. The cross-linker is usually N,N'-methylene-bisacrylamide (bisAA), a dimer of acrylamide.[34] This is a vinyl addition polymerization initiated by a free radical polymerization (FRP) system. FRP has three distinct stages: initiation, propagation, and termination. The polymerization of acrylamide is initiated by the addition of ammonium persulfate (APS) (which on dissolving in water forms free radicals) in the presence of tetramethylenediamine (TEMED). The TEMED, through its ability to accelerate the rate of formation of free radicals, acts as a catalyst for the polymerization.

The copolymerization of acrylamide with bisAA produces a mesh-like network in three dimensions (see Figure 3.3). This consists of elongated acrylamide chains with random interconnections formed from the bis-acrylamide cross-linker, resulting in a gel with a characteristic porosity; this will depend on the polymerization conditions and monomer concentrations. The association between protein and acrylamide monomers is fixed in place when the polymerization of acrylamide occurs. Once this polymerization has occurred and the protein has been imprinted, it is then able to be extracted or eluted from the bulk polymer, leaving behind a cavity capable of recognizing and re-absorbing or rebinding the same protein. The specifics behind bonding interactions in the molecular imprinting process involving proteins and polyAA matrices are dependent on the degree of pre-association between the template protein molecules and polyAA. These associations are most likely based on simple molecular interactions such as hydrogen bonding and other non-covalent bonding.[32]

Several studies have been conducted and pioneered in the field of aqueous phase molecular imprinting, using polyacrylamide HydroMIPs for the selective imprinting of proteins.[1,2,32–34] HydroMIP polyacrylamide gels templated for a range of proteins have been assessed for template removal, reloading, and selectivity.[2,5,33,34] As such, they are capable of nano-selectively recognizing biomolecules in medicine, food, and the environment.[31] Hawkins *et al.* explored in detail a variety of template bovine hemoglobin (Hb) removal strategies, including varying ratios of a sodium dodecyl sulfate surfactant:acetic acid (SDS:AcOH) (at pH 2.8).[34] Hb is one of the most commonly imprinted proteins.[2,5,33,34] It is an iron-containing oxygen transport metalloprotein[34] and is a globular protein with a molecular weight around 64 500 Daltons (Da), with a composite tetramer structure made up of four subunit protein chains. Structural changes in any of these four molecular subunits can result in the manifestation of hereditary diseases such as sickle cell anemia, thalassemia, and hemoglobinopathies.[31] Hawkins *et al.* concluded that an optimum 10% (w/v) ratio was reported to be the most effective for protein removal, resulting in a 90% imprinting efficiency of re-loaded protein selectively bound within the MIP. This was attributed to the combination of low pH and surfactant content. Proteins generally have an assortment of negative and positive charges due to charged R groups, while also retaining hydrophobic domains due to nonpolar R groups.[32] The surfactant–acid combination disrupts both covalent and hydrogen bonds, and exposes hydrophobic areas as well as protonation of the template protein, resulting in a positive net charge. At concentrations above 8 mM, SDS starts to form micelles; this occurs when the concentration

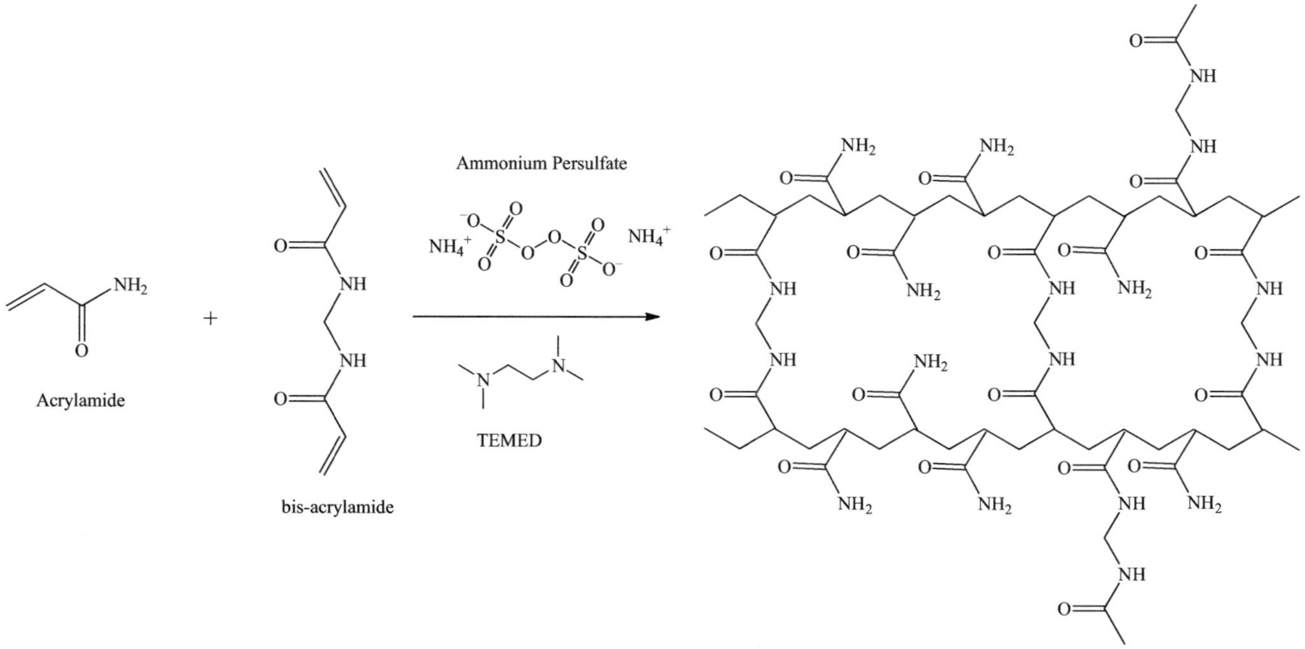

Figure 3.3　The polymerization of acrylamide and bis-acrylamide (cross-linker) to form polyAA matrices by FRP; APS as initiator and TEMED as catalyst.

of surfactant is greater than the critical micelle concentration (CMC). These negatively charged SDS micelles attract the positive protein and bind to the main peptide chains at a ratio of $1:2$ (SDS:amino acid residue). As a result, an overall negative charge is imparted upon the protein; this is directly proportional to the mass of the protein (1.4 g SDS/g protein). This results in the uncoiling or unfolding of the peptide chains and induces conformational changes (*i.e.* reduces secondary, tertiary, or quaternary structures to linearized primary amino acids), where protein is ultimately denatured and solubilized, thus exiting the cavities within the HydroMIP complex. Surfactant strategies are increasing in popularity for template protein removal protocols for various applications.[34]

The molecular imprinting of Hb using semi-interpenetrating polymer network (semi-IPN) hydrogels has also been investigated.[40,41] The molecularly imprinting methods utilized a mild aqueous media [1% (v/v) acetic acid] of chitosan [2–4% (w/v)] and acrylamide in the presence of bisAA as the cross-linking agent. Xia *et al.*[40] investigated the morphology of the semi-IPN and their selective capabilities using thermal analysis, X-ray diffraction, differential scanning calorimetry (DSC), and environmental scanning electron microscope (ESEM). Langmuir analysis showed that an equal class of adsorption was formed in the hydrogel. The adsorption equilibrium constant (4.27 g mL^{-1}) and maximum adsorption capacity (36.53 mg g^{-1} wet hydrogel) revealed high Hb template recognition over the non-imprinted polymer. Zeng *et al.*[41] further studied the direct electrochemistry and electrocatalysis of the Hb semi-IPN hydrogels based on polyacrylamide and chitosan. UV–VIS spectroscopy showed that the semi-IPN provided a favorable microenvironment around the Hb to retain the enzymatic bioactivity and native structure of the Hb. The Hb–PAM–chitosan film exhibited a three-dimensional network porous structure. This would qualify the Hb–PAM–chitosan film for good conductivity of electrons. Direct electron transfer of Hb was achieved by casting Hb–PAM–chitosan films onto glassy carbon electrode surfaces. The immobilized Hb showed good bio-electrocatalytic activity; current values were linear with increasing H_2O_2 concentration ranging between 5–420 µM. The unique semi-IPN hydrogel would have wide potential applications in direct electrochemistry, biosensors, and biocatalysis.

Another less commonly imprinted metalloprotein is catalase (Cat); this is possibly due to its larger size and difficulty of imprinting.[32,42,43] Cat is a common enzyme with a molecular weight of 250 000 Da and is found in nearly all living organisms. The cellular role of catalase is the removal of hydrogen peroxide, as it is a harmful byproduct of many normal metabolic processes. In order to prevent damage, it must be quickly converted into other, less dangerous, substances. Thus, catalase is frequently used by cells to rapidly catalyze hydrogen peroxide decomposition. A catalase deficiency may increase the likelihood of developing type II diabetes and peroxisomal disorder (acatalasia). Genetic polymorphisms in catalase and its altered expression and activity are associated with oxidative DNA damage and subsequently the individual's risk of cancer susceptibility. To this end, studies into catalase crystallization facilitated by MIPs have been recently researched.[6] Saridakis *et al.* described a new initiative of MIP design for producing protein crystals essential for determining high-resolution 3D structures of several key proteins, including catalase.[5,6] It was hypothesized that MIPs could represent ideal nucleants for the formation of large single crystals of otherwise hard to crystallize proteins because they are designed to specifically attract a template protein. The researchers evaluated the ability of polyAA HydroMIPs imprinted with seven proteins to induce nucleation of their own cognate proteins, as well as other proteins. With the exception of the catalase MIP, the resulting polymers were able to induce the formation of crystals of nine tested proteins, including, in some cases, non-cognate proteins with similar molecular weights. These MIP cavity applications using surfactant protocols in protein crystallization have demonstrated attractive features in terms of crystal formation time, yield, and metastability without the use of known nucleants. One example of the easily MIP-induced crystallized proteins was trypsin. This is a non-metalloprotein serine protease enzyme found in the digestive system of many vertebrates, where it hydrolyses proteins. It is often referred to as a proteolytic enzyme, or proteinase,[34] and can be purified very easily, and hence it has been used widely in various biotechnological processes.[6] Trypsin is commonly used in biological research during proteomics experiments to digest proteins into peptides for mass spectrometry analysis, *e.g.* in-gel digestion. Trypsin is particularly suited for this, since it has a very well-defined specificity.[34]

Later on, Reddy *et al.* compared the functionality and selectivity of hydrogel-based MIPs using acrylamide (AA), *N*-hydroxymethylacrylamide (NHMA), and *N*-isopropylacrylamide (NiPAm) for investigation of trypsin and BHb crystallization in comparison to mass-based quartz crystal microbalance (QCM) detection.[5] QCM is primarily an ultra-sensitive mass sensing device, with the ability to measure mass changes as low as 1 µg cm^{-2} on a piezoelectric quartz crystal resonator in real-time (approximately 100 times more sensitive than a 0.1 mg electronic fine balance).[44,45] To detect binding of protein to the MIP, changes in the resonant frequency of the QCM from the added mass were calculated using the Sauerbrey equation (Eqn 3.3):

$$\Delta f = \frac{-2\Delta m f_0^2}{A\sqrt{\rho_q \mu_q}} \tag{3.3}$$

In this equation, Δf is the shift in frequency (Hz), f_0 is the base frequency (Hz) of the QCM prior to functionalization of the electrode, Δm is the change in mass (g) of the QCM, A is the surface area of the electrode (cm^2), ρ_q is the density of the QCM (2.65 g cm^{-3}), and μ_q is the shear modulus of the quartz (2.947×10^{11} g cm^{-1} s^{-2}). The principle is that if a rigid layer is evenly deposited on one or both of the electrodes, the resonant frequency will decrease proportionally to the mass of the adsorbed layer. This was developed for oscillation in air and only applies to rigid masses attached to the crystal.[44,45] However, in a fluid environment, the fluid viscosity η_L and the fluid density ρ_L also impart an impact on the change in frequency, as dictated by the Kanazawa and Gordon equation (Eqn 3.4):

$$\Delta f = -f_0^{3/2} \left(\frac{\eta_L \rho_L}{\pi \rho_q \mu_q} \right)^{1/2} \tag{3.4}$$

The polyAA, poly-NHMA, and polyNiPAm MIPs showed varying abilities to crystallize proteins when imprinted with trypsin or BHb and tested for their ability to crystallize various other non-template proteins [macrophage migratory inhibition factor (MIF), trypsin, lysozyme, thaumatin, and BHb]. While BHb imprinted MIPs showed no cross-selectivity in crystallization, some degree of MIP cross-selectivity was illustrated for trypsin MIPs and other similar sized proteins ($\sim$15 kDa), such as lysozyme. QCM measurements were also comparable to the resulting MIP selectivity trends. It was therefore suggested that protein crystallization could be used to examine the cross-selectivity of a MIP dependent upon the size of the protein being imprinted and the functional groups of the imprinted polymer.

Other imprinting methods that have proven effective in imprinting small proteins such as trypsin and lysozyme have incorporated photo-grafting surface-modified polystyrene beads as matrices.[46] Lysozyme is an important index in the diagnosis of various diseases. As such, lysozyme has become one of the most commonly imprinted proteins due to its relatively small size and imprinting ease.[32] Lysozyme is a glycoside hydrolase enzyme that has 129 amino acid residues and a molecular weight of approximately 14 700 Da.[47] Lysozyme catalyzes the hydrolysis of 1,4-β-linkages between *N*-acetylmuramic acid and *N*-acetyl-D-glucosamine residues in peptidoglycan and between *N*-acetyl-D-glucosamine residues in chitodextrins. This catalytic activity is non-specifically targeted to the bacterial cell membranes and related to general non-specific organism defense. Large amounts of lysozyme can be found and easily purified from egg-white. It is also very easy to crystallize, which is not the case for most of the other proteins.[6] This feature of lysozyme is widely used for its purification. Lysozyme is part of the innate immune system, existing as a natural form of protection from pathogens like *Salmonella*, *E. coli*, and *Pseudomonas* species.[47] Qin *et al.* illustrated a simplistic fabrication method for lysozyme MIP beads in aqueous media using mesoporous chloromethylated polystyrene beads (MCP beads).[46] Grafting of lysozyme imprinted copolymers with acrylamide and bis-acrylamide, using dithiocarbamate iniferter (initiator transfer agent terminator) as support, was achieved through surface initiated living-radical polymerization. This technique allowed for the efficient control of the grafting process and suppressed solution propagation. Fourier transform infrared spectroscopy, elemental analysis, nitrogen sorption analysis, and scanning electron microscopy concluded that lysozyme–MIP beads had more well-distributed surface pores, without any visible gel formation in solution, compared to traditional MIPs prepared by initiated radical polymerization.

MIP sensors have also been widely fabricated for detection of bovine serum albumin (BSA).[32,43,48] BSA is a soluble monomeric globular protein with a molecular weight of 66 776 Da, similar to that of Hb. BSA is a carrier for steroids, fatty acids, and thyroid hormones and stabilizes extracellular fluid volumes. BSA is generally used because of its stability, its lack of effect in many biochemical reactions, and its low cost since large quantities of it can be readily purified. As a result, BSA has many uses as a carrier protein, as a stabilizing agent in enzymatic reactions, and in gel shift assays.[48] It is commonly used to determine the quantity of other proteins, by comparing an unknown quantity of protein to known amounts of BSA. This attribute serves as an excellent cross-selective template study for Hb. In light of this, several studies have investigated the optimum conditions of separation selectivity based on BHb and BSA MIPs formed on surfaces of amino-silica and polyacrylamide hydrogel MIPs.[37,48,49] Gai *et al.* investigated the interaction of functional acrylamide monomers and template BHb protein in different molar ratios, solution pH, and ionic strength of polyacrylamide hydrogel MIPs.[48] MIP adsorption selectivity was also investigated using BHb and BSA independently and competitively. The molar ratio range of acrylamide to BHb (100 : 1 to 1200 : 1) was found to be optimum at 600 : 1 for monomer–protein interaction, indicating that a stable monomer–protein complex was formed at this molar ratio. The adsorption experiments indicated that BHb–MIP had better selective adsorption and recognition properties compared to BHb, especially in the presence of BSA as a competing protein. This was attributed to the MIPs recognition of the synergistic effect of shape complementarity and multiple hydrogen bonding interactions in the MIP cavities rather than just the similarity of either protein's molecular weight or size separation. Thus they concluded that the MIPs could serve as selective separating materials for target proteins from protein mixtures of similar molecular weights.[47]

Moreover, the use of fluorescently labeled and autofluorescing BSA MIPs have also demonstrated potential in optical sensing.[50] Hawkins *et al.* prepared polyacrylamide hydrogel MIPs templated with FITC-conjugated BSA and BHb and imaged using confocal and two-photon confocal microscopy. Using these methods, visual confirmation of the binding properties of the MIPs in comparison to a control NIP was possible. Furthermore, the integrity of the MIP polymer network after elution of the FITC-labeled BSA was also confirmed by enhancing the residual fluorescence signal. The use of two-photon microscopy also confirmed the effectiveness of the 10% SDS/10% AcOH (w/v) in eluting the template protein from the MIP. In controlled experiments, protein denaturation and unfolding caused unquenching of the autofluorescence in the molecule's protoporphyrin IX complex, giving rise to the ability of imaging the protein molecules during MIP elution.

Owing to the wide use, success, and suitability of polyAA as an imprinting matrix for a range of biological and non-biological molecules, other acrylamide-based derivatives are being explored and considered for further optimization (see Table 3.2).[5,6] One often used derivative,

Table 3.2 List of possible functional monomers for polyacrylamide-based hydrogels.

Monomer	Structure	Molar Mass (g mol^{-1})	Chain Length
Acrylamide (AA)	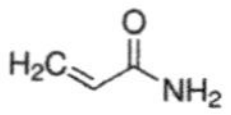	71.081	4
N,*N*-Dimethylacrylamide (DEA)	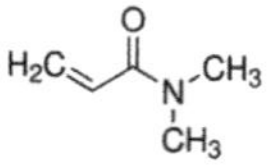	99.13	5
N-(Hydroxymethyl)acrylamide (NHMA)	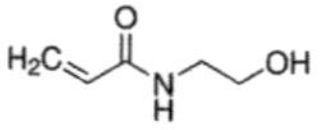	101.10	6
N-Iso-propylacrylamide (NiPAm)	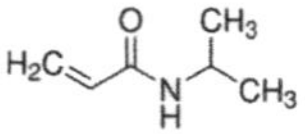	113.16	6
N-[Tris(hydroxymethyl)methyl]acrylamide (TrisHA)	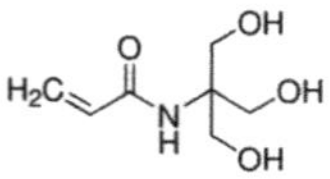	175.18	7
2-Acrylamido-2-methyl-1-propanesulfonic acid (AMPS)	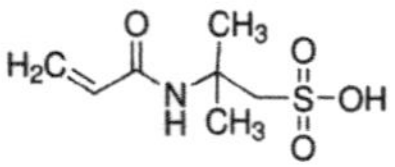	207.25	8
(3-Acrylamidopropyl)trimethylammonium chloride (APTAC)	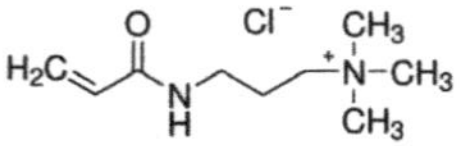	206.71	9

poly(*N*-hydroxymethylacrylamide) (polyNHMA), has demonstrated good potential for high-throughput DNA analysis by microchannel electrophoresis.[51] When polyNHMA is used both as a separation matrix and as a dynamic coating in bare silica capillaries, the matrix can resolve over 620 bases of a contiguous DNA sequence within 3 h.[51] Another popular and potential hydrogel is poly(*N*-isopropylacrylamide) (polyNiPAm). This is a thermosensitive water swellable hydrogel.[51] At 33 °C these hydrogels undergo a reversible lower critical solution temperature (LCST) phase transition from a swollen hydrated state to a shrunken dehydrated state, losing around 90% of their mass. Since polyNiPAm gels expel their liquid contents at near body temperature, it has widely been investigated for possible applications in controlled drug delivery.[31] As with acrylamide, the mechanism of polymerization of each of these functionalized acrylamides is also vinyl addition polymerization initiated by FRP. Thus, similarly to polyAA, NHMA and NiPAM form hydrogels in aqueous solution by their copolymerization with small amounts of a bifunctional cross-linker, usually bis-AA, which can impart the robustness required for the polymer.

One particular study compared the functionality and selectivity of hydrogel-based MIPs using acrylamide, NHMA, and NiPAm for investigation in mass-based QCM detection.[5] Thin films of the three different types of MIPs imprinted for bovine hemoglobin (BHb) were deposited onto QCM crystals and analyzed for their selectivity against an analogue protein bovine serum albumin (BSA). Best selectivity was attributed to NHMA MIP hydrogels, where only target BHb induced a change frequency. PolyAA MIPs, although selective for target BHb, also exhibited some degree of cross-selectivity for similarly sized BSA protein. Interestingly, the NiPAm MIP exhibited a near-zero frequency response to template BHb and cognate BSA, indicating that NiPAm is equally unselective for both proteins as is the control non-imprinted polymer. The lack of response from NiPAm to either BHb or BSA suggests that there is a resistance for either protein to bind to the polymer. The hydrophilic shell of the protein appears to be important in contributing to this lack of binding with the hydrophobic polymer. The differences in selectivity are attributed to the hydrophilicity of the polymers. NHMA is by far the most hydrophilic of the three, rendering NiPAm the most hydrophobic due to its isopropyl group. It is suggestive that the hydrophobic interaction with the protein limits the MIPs efficacy in protein binding, and could explain the lack of specificity in binding between target BHb and analogue BSA displayed by acrylamide, which lacks the hydroxyl group present in NHMA.

Alternative and readily used cross-linkers, other than bis-AA, can vary the physical or chemical gel properties. These include 1,4-diallylpiperazine (DAP), *N,N'*-bisacrylylcystamine (BAC), and *N,N'*-diallyltartardiamide (DATD) (see Table 3.3). BAC and DATD are both disruptable cross-linkers which enable gels to be solubilized. Alternative cross-linkers may be more or less reactive in polymerization than bisAA. For example, DAP can be substituted for bisAA on a weight basis without changing polymerization protocols.

Table 3.3 List of possible cross-linkers for polyacrylamide-based hydrogels.

Monomer	Structure	Molar Mass (g mol^{-1})	Chain Length
N,N'-Methylene-bis-Acrylamide (bis-AA)	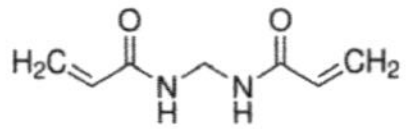	154.17	9
N,N'-(1,2-Dihydroxyethylene)bisacrylamide (DHEBA)	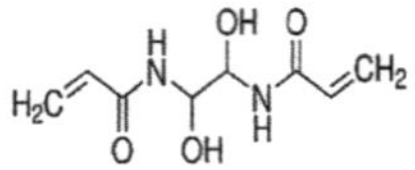	200.19	10
1,4-Bis(acryloyl)piperazine (BAP)	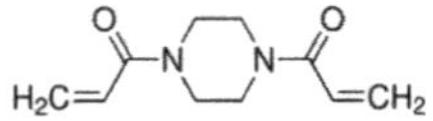	194.3	10
1,4-Diallylpiperazine (DAP)	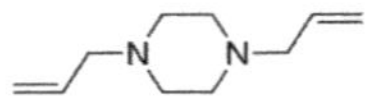	166.3	10
N,N'-Diallyltartardiamide (DATD)	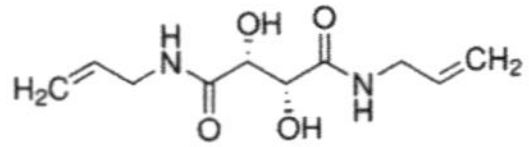	228.25	12
N,N'-Bisacrylylcystamine (BAC)	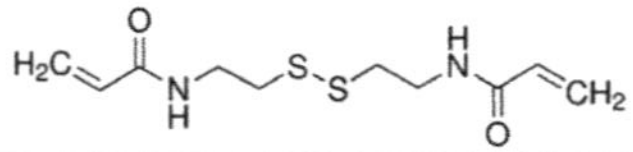	260.38	14

Changing the cross-linker type or composition has offered several advantages in the realm of gel electrophoresis. These have included reduced background for silver staining, increased gel strength, and higher resolution gels. Therefore, alterations in cross-linkers will in turn alter the polymer matrix and therefore could have an impact on protein imprinting efficiency for MIP hydrogels.

Elastomeric electrophoresis gels have been produced using a combination of dextran and polyacrylamide cross-linked with N,N'-(1,2-dihydroxyethylene)bisacrylamide (DHEBA).[52] It was observed that the effective pore sizes were smaller, compared with conventional polyacrylamide gels cross-linked with DHEBA at the same total monomer and cross-linker concentrations. Enhancement of the resolving power of the gel was also observed by using the elastomeric gels. Sontimuang *et al.*[53] demonstrated the use of DHEBA as a cross-linker in MIPs using a copolymer of vinylpyrrolidone and methacrylic acid (MAA-NVP-DHEBA) for the imprinting of allergen protein hevein (Hev b1) from latex gloves. A biosensor was developed using photopolymerization of the MIPs ($\lambda = 254$ nm) onto capacitive microelectrode surfaces. Hev b1 detection occurred within minutes and with a detection limit of 10 ng mL^{-1}. Different hevein allergenic proteins isolated from natural rubber latex from the rubber tree (Hev b1, Hev b2, and Hev b3) were distinguished by the MIPs, depending on the dimension and conformation of these proteins and with a selectivity factor of 4. Non-Hev b proteins, such as lysozyme, ovalbumin, and BSA, were also tested but only exhibited a selectivity factor of 2. Moreover, the MIP-based sensor exhibited good operational stability of up to 180 days when used continuously at room temperature.[53]

3.3.1.2 Polymethacrylic Acid Hydrogels

Methacrylic acid (MAA) is an organic compound that can also undergo vinyl addition polymerization initiated by a free radical generating system much like polyacrylamide. To date, polyMAA imprinted polymers in the form of particles are reportedly made by various polymerization methodologies.[54] Previously, Pérez-Moral *et al.* conducted a comparative study of imprinted polymer particles prepared by different polymerization methods.[55] In this comparative study, MAA polymers with a similar composition were synthesized by bulk, suspension, emulsion, two-step swelling, and precipitation polymerizations. Rebinding capacities of their template propranolol under identical analytical conditions was assessed from organic and aqueous solutions by radio ligand binding. Specific rebinding in organic solution (toluene + 0.5% acetic acid) was in the order of: precipitation (50%) > suspension (40%) > bulk (35%) > core-shell (15%) > two-step swelling (10%). However, specific rebinding from an aqueous solution (25 mM sodium citrate + 0.5% acetic acid + 2% ethanol, pH 4.6) revealed a different trend: two-step swelling (20%) ≈ suspension (19%) ≈ bulk (19%) > core-shell (15%) > precipitation (0%).

In attempts to improve rebinding capacities, and adjust to aqueous solution compatibility, MAA hydrogels have been synthesized. MAA hydrogels have commonly been synthesized by copolymerizing with acrylamide or hydroxyethyl methacrylate (HEMA) and small amounts of a bifunctional bisacrylamide or ethylene glycol dimethacrylate (EGDMA) cross-linker.[56] The copolymerization retains the hydrogels elasticity and decreases the sequence length and tacticity of the methacrylic chain segment. By enhancing the macromolecular chain structure the efficiency of the carboxyl groups that associate with templates are optimized. Several specific polyMAA MIPs for the drug reserpine have been developed and characterized.[54] Evaluation of the various polymers by binding assays indicated that the optimum ratio of functional monomer to template was 4:1. Furthermore, the imprinting effect of the MIPs was assessed by the chromatographic method, which demonstrated that the MIPs had better chromatographic behavior and selectivity than those of the corresponding NIPs. The preparation of reserpine MAA MIPs and elucidation of imprinting and recognition mechanisms may serve as useful references for other drug-based MIPs.

Imprinting of proteins in a copolymer system of polyMAA incorporated with polyacrylamide and 2-(dimethylamino)ethyl methacrylate (DMA) has been studied. Ou *et al.* investigated the electrostatic functional groups in a system of polyacrylamide incorporated with MAA and DMA for the possibility of imprinting of lysozyme.[56] As mentioned previously, lysozyme is an important index in the diagnosis of various diseases, and an increase of lysozyme concentration in cerebrospinal fluid has been described in patients with tuberculosis meningitis, neurosyphilis, and fungal meningitis.[56] It was postulated that the electrostatic functional groups would assist the recognition of protein and can be realized through the optimum distribution of positive charges, negative charges, and hydrogen bonds. MIPs were prepared with 0.573 M acrylamide, 0.573 M MAA, and 0.573 M DMA at a total solution concentration of 20% (w/w). These were able to adsorb 83% more lysozyme than the non-imprinted polymer. At high MIP preparation concentrations of 20% and 40% total monomers (w/w), significant imprinting efficiencies of 1.83 and 3.38, respectively, were achieved. However, the results indicated that approximately 27% (w/w) of the lysozyme template was not able to be extracted. It was suggested that since amino acids (tryptophan, tyrosine, cysteine, and histidine) are susceptible to free radical modification by radical OH groups, radical intermediates could decay or react with another free radical to form covalent bonds.[56] The addition of MAA reduced the extent of copolymerization and in turn increased the amount of extracted template. This was attributed to the complexation of acid groups with lysozyme, preventing the attack of free radicals of the template protein.

Hirayama *et al.* also used a combination of either MAA and acrylamide or *N,N*-(dimethylamino)propylacrylamide (DMAPAA) to form hydrogel lysozyme MIPs.[57,58] Surface-modified silica beads imprinted with lysozyme were applied to a QCM sensor. The modified silica polymer thin-film layer was able to grasp lysozyme selectively compared to the non-imprinted polymer. The amount of specifically bound lysozyme was highest for lysozyme–MIPs prepared with 0.46 mol% MAA monomer and 5.98 mol% DMAPAA. It was hypothesized that the specified binding of lysozyme to MIP was based on electrostatic interactions between basic groups in lysozyme and the COO^- group of acrylic acid, and the acidic groups in the lysozyme molecule interacting with the amino group of DMAPAAm. For lysozyme rebinding, the distance between the COO^- group or that between the amino group of DMAAm was an important factor for improving the binding capacities. The specificity of the MIPs was tested using hemoglobin, revealing that only lysozyme specifically bound to the polymer MIP particles. The rebinding ability did not alter when tested for reliability by binding and releasing lysozyme repeatedly. The development of biosensor strategies for the detection of proteins is imperative for applications in proteomics, medical diagnostics, and pathogen detection.[59] However, protein-detecting arrays remain underdeveloped due to the lack of highly selective and specific binding agents that interact with protein surfaces through complementary interactions.[59] One possible solution, however, is within differential receptor array systems that occur in nature.[59–62] These differential systems use non-specific and/or weakly interacting agents to routinely conduct pattern-based recognition fingerprints for various bioanalytes.[59]

An electronic sensor that works in a similar way is a chemometric tool that decodes statistical information and classifies standards for recognition, such as principal component analysis (PCA).[61,63] PCA can be used to explain the variance–covariance structure of a set of variables through linear combinations, and is often used as a dimensionality reduction technique.[63] Takeuchi *et al.* demonstrated the use of a chemometric strategy (PCA; multivariate analysis) for molecular recognition and classification of five proteins using plural imprinted acrylic acid and DMA polymers.[64,65] Six different protein-imprinted polymers were synthesized using three template proteins, cytochrome *c* (Cyt), ribonuclease A (Rib), and α-lactalbumin (Lac), and acidic or basic functional monomers of acrylic acid and DMA, respectively. The resulting MIPs produced unique fingerprints when rebound with both corresponding and non-template (albumin and myoglobin) proteins. Three-dimensional PCA scores of the binding assay MIP data (see Figure 3.4) revealed that a clear protein distinction was possible, and that protein-imprinted polymer arrays can be applied to protein profiling by pattern analysis of binding activity for each polymer.[64–66]

A final example of MAA-based hydrogel MIPs is provided by the work of Adrus and Ulbricht, who reported the first photopolymerized MIP hydrogel for protein recognition.[67] Temperature-responsive poly(*N*-isopropylacrylamide) (NiPAm) hydrogels were imprinted with lysozyme *via in situ* photo-initiated cross-linking polymerization. These were tailored by tuning the ionic content through methacrylic acid concentrations as template binding co-monomer, while keeping the ratio between the bisAA cross-linker and NiPAm fixed. Moderate salt concentrations (0.3 M NaCl) were found to be suited for template removal without phase separation or irreversible collapse

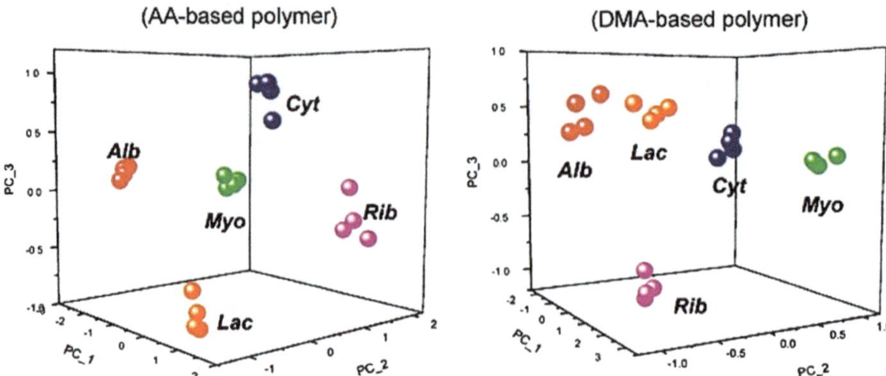

Figure 3.4   PCA score plots showing the discrimination of four trials of five different proteins based upon bound amounts of acrylic acid (AA)-based and 2-(dimethylamino)ethyl methacrylate (DMA)-based polymers. Alb: albumin; Cyt: cytochrome *c*; Lac: lactalbumin; Myo: myoglobin; Rib: ribonuclease A; Alb and Myo are non-templated proteins.[64,65]

of the hydrogel at sufficient degrees of monomer conversion. Polymer swelling and protein (lysozyme and cytochrome *c*, template and non-template, respectively) binding were investigated for imprinted and non-imprinted gels at temperatures above and below the lower critical solution temperature (LCST) of 32 °C. The degree of swelling below LCST increased with increasing MAA content for both NIP and MIP gels. By increasing template-monomer content, increasing ionic group content is the reason for high water uptake.[67] In contrast, mainly below LCST and also at lower MAA content (2 wt%) the equilibrium swelling ratio of the MIP hydrogels was higher than that of their NIP counterparts. Lysozyme imprinted gels showed a much higher affinity, selectivity, and binding capacity for their template lysozyme compared to non-template cytochrome *c* and the non-imprinted NIP controls. Protein binding capacity was strongly reduced above the LCST of 32 °C from ~350 to ~50 mg g^{-1}, and to zero for non-imprinted gels. Specific lysozyme binding to the imprinted gels caused a large concentration-dependent hydrogel deswelling. Increasing lysozyme concentrations up to 1.2 g L^{-1} continuously induced shrinking of the MIP hydrogels. Specific interactions between template-protein and imprinting sites and subsequent conformational change within the hydrogel network may have caused the specific shrinking. It was proposed that protein binding to imprinted sites caused local conformation and solvation changes, and ultimately led to different solvation in the imprinted sites. The NIP hydrogels, however, only experienced a slight volume decrease and only at high lysozyme concentration. This was attributed to a non-specific effect of ion pairing, and was later modulated by the content of the co-monomer MAA. It can be concluded that protein recognition and signal transduction can be combined within one material since selective protein binding to MIP hydrogels leads to a pronounced deswelling. Adrus and Ulbricht proposed that this could enable additional functions of such hydrogels, for instance the development of protein micro-sensors based on measuring the swelling pressure.[67]

3.3.1.3   Poly(2-hydroxyethyl methacrylate) Hydrogels

Since Wicheterle and Lim's first synthesis of poly(2-hydroxyethyl methacrylate) (pHEMA) hydrogels in the 1960s,[68] they have rapidly become one of the most widely used hydrogels to date.[36] Because pHEMA gels have soft, flexible, and water-absorbing plastic properties they have been extensively researched and used as materials for ureters, cardiovascular implants, breast implants, nasal cartilage replacements, contact lenses, corneal implants, drug release devices, tissue repair surgery, and in numerous dental applications such as root canal fillers and soft denture liners.[35,36,69–71] The superior biocompatibility and modification ease that pHEMA exhibits for tailor-made hydrogels to suit specific roles has allowed for such a diverse range of applications. Owing to a controlled release rate and increased surface area, created by the size and shape of the pHEMA fibres' swelling capacities, pHEMA has been used to adhere to cells and act as absorbents of proteinaceous materials, creating consistent healthy equilibriums when applied to natural tissue.[70–73] A particular study used pHEMA hydrogels absorbed with active ingredients that are applied to human tissue, such as humectants and proteins. Approximately 28–80% of the hydrophilic and hydrophobic fluids were absorbed by pHEMA hydrogels. This may promote rapid in-growth of cells and capillaries when pHEMA gels are applied to tissue mass for long periods of time.[70–73]

Hydrogels of pHEMA function by rotating around the central carbon; in water the polar hydroxyethyl side turns outward and the material becomes flexible. Polymerization occurs under the influence of light and free-radical initiators, whereby most pHEMA chains are cross-linked into a complex three-dimensional network by another cross-linking compound. The properties of pHEMA can be modified by this cross-linking to a certain degree, but its hydrophilicity can be increased by the introduction of a second hydroxyl group using various monomers. A variant of pHEMA is poly(2-hydroxypropyl methacrylate) (pHPMA), which also exhibits similar properties and advantages. Copolymers of HEMA and HPMA have been utilized to create hydrogel matrices for drug delivery systems (see Figure 3.5). The drug diffusion is controlled by the type and ratios of co-monomers used together with certain cross-linking agents.[70–73]

The free-radical polymerization of pHEMA with an ethylene glycol dimethacrylate (EGDMA) cross-linker is a strategy that has been widely used.[47] One particular study investigated EGDMA and several other cross-linking monomers for lysozyme, ribonuclease A, and myoglobin using combinatorial approaches to MIP synthesis. In this instance these approaches have proved necessary in order to obtain an optimized polymer for given target proteins, such as myoglobin (Mb).[32,47] Mb is also an iron-containing and

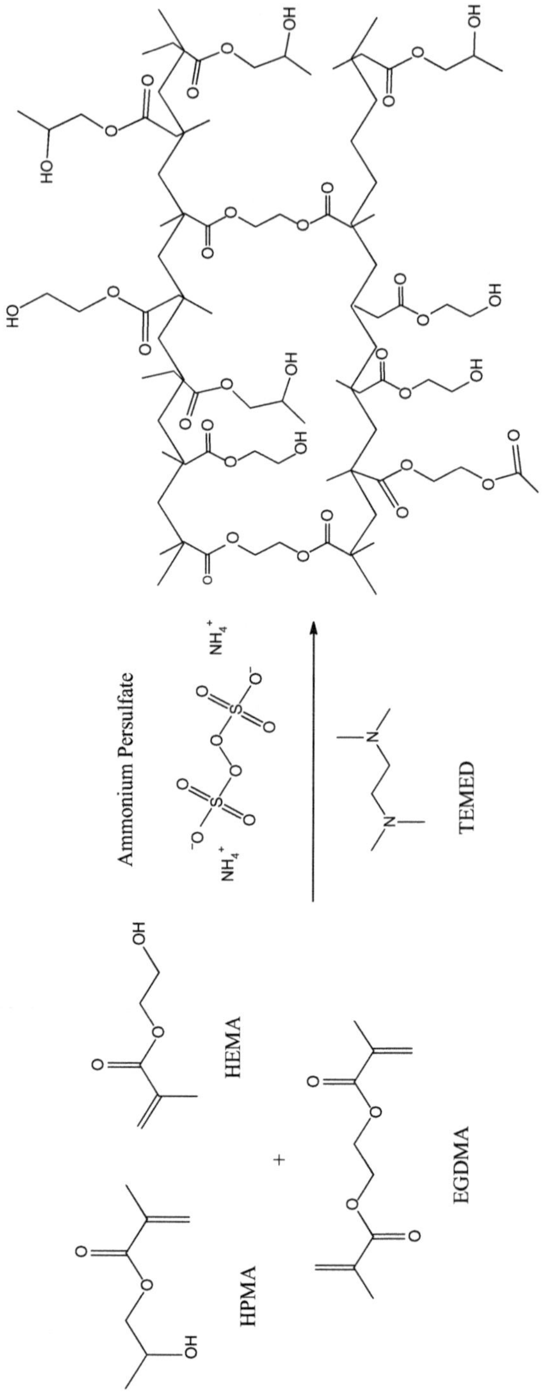

Figure 3.5 The copolymerization of PHEMA and PHPMA.

oxygen binding metalloprotein that is found in the muscle tissue of vertebrates. Mb is structurally similar to a single subunit of Hb with a molecular weight of 16 700 Da, which is about a quarter that of Hb. Mb is the putative protein that in high concentrations in urine is indicative of rapid breakdown of muscle (*e.g.* rhabdomyolysis). It can, however, cause acute renal failure due to its toxicity to the renal tubular epithelium.[47] Because of this, Mb is a sensitive marker for muscle injury, making it a potential marker for myocardial infarction in patients with chest pain.[31] MIPs specific for Mb could serve as antibodies capable of replacing their natural counterparts and potentially have an application in diagnostic analysis. Hung-Yin *et al.* reported a fast method combining molecular imprinting and microcontact printing (µCP) techniques (a form of soft lithography) to prepare a MIP thin film. Microcalorimetry information of the interaction between monomers and protein stamps was used to design optimal MIP compositions. Both techniques were combined to create high-performance MIPs. Isothermal titration of monomers to proteins stamps was also investigated to screen the functional monomer for MIPs.[47] MIPs were photopolymerized, allowing for rapid parallel MIP synthesis of different compositions, also requiring small volumes of monomers (*ca.* 4 µL). By initially selecting the cross-linker on the basis of having a minimal recognition for Mb, and using this as a starting point for functional monomer selection, successful production of Mb MIPs with exceptionally high selectivity was achieved. The affinity of the polymers for Mb, when prepared with a variety of different cross-linkers and no functional monomer, was evaluated. Tetraethylene glycol dimethacrylate (TEGDMA) gave the most selective lysozyme binding, while poly(ethylene glycol) 400 dimethacrylate (PEG400DMA) was most selective for ribonuclease A and myoglobin. The Mb MIPs showed a promising absorbance compared to polymers of similar composition, but formed in the absence of the Mb template (non-imprinted polymers). The combinatorial technique also avoided potential solubility problems, with the protein targets offering a convenient, microscale platform to optimize MIP compositions.

Calcification tendencies correlating to topography and porosity of pHEMA materials have been investigated using varying concentrations of water and the co-monomer ethoxyethyl methacrylate at different strengths of cross-linking agent ethylene glycol dimethacrylate.[69] Lou *et al.* prepared four groups of pHEMA hydrogels using either bulk polymerization or solution polymerization at subcritical diluent contents, *i.e.* '*homogeneous hydrogels*', and at initial water concentration above critical levels, *i.e.* '*heterogeneous hydrogels*' or phase separation sponges. A series of 100 wt% HEMA formulations and 80/20 wt% [HEMA/ethoxyethyl methacrylate (EEMA), a co-monomer] formulations generated pHEMA hydrogels with various porous structures, facilitating the investigation of the relationships between structural features of pHEMA hydrogels and their calcifying tendencies. Scanning (SEM) and transmission electron microscopy (TEM) revealed the topography of pHEMA hydrogels with surface irregularities and some porosity, respectively. Calcification and pHEMA calcium ions diffusion ability (30–40 µm and 200 µm for hydrogels that exhibited substantial surface irregularities and micro channels) were detected by light microscopy, SEM, and energy dispersive analysis of X-rays (EDAX) after incubation of the materials in a metastable calcifying solution for 48 days. It was concluded that surface defects were the major contributors to calcium deposition, with a tendency to form at higher solvent volume fractions (40–45 wt%) which increased the porosity. These optimizations have facilitated MIP HEMA gel composition strategies based on rational porosity levels in order to maximize MIP efficiencies in terms of analyte rebinding.[70–73]

In an attempt to improve the imprinting efficiency of pHEMA hydrogels, Sreenivasan pre-associated a metal ion prior to polymerization for the imprinting of polyaromatic hydrocarbons (PAHs).[72] Naphthalene, a model PAH, was imprinted; it was proposed that due to the lack of functional groups (*e.g.* OH, COOH, NH_2, *etc.*), inadequate and inefficient affinity sites would hinder MIP functionality for such PAH molecules. In turn, the addition of metal ions, such as silver (Ag^+), could impart enhanced interaction between the functional monomers and the PAHs. Metal ion 'electron acceptors' can complex with many organic compound 'electron donors' and various types of association products can be formed by weak charge transfer interaction.[72] As a result of incorporating Ag^+ ions (in the form of silver nitrate, 100 mg) during the synthesis, the equilibrium uptake of the template naphthalene molecule rose from 3.3 ± 0.4 µg/100 mg (without using Ag ions) to 66.0 ± 2 µg/100 mg, resulting in a MIP to NIP ratio of 27.50 : 1. The adsorption of a similar PAH (anthracene) was also tested upon the Ag-induced naphthalene imprinted pHEMA MIP to assess the MIPs' affinity towards the original template. It appeared that the adsorption capacity of molecules lacking in functional groups could be increased by the inducement of the metal ion methodology without hindering its selectivity or affinity.

As previously described, traditional MIPs have been prepared as solid particles (*i.e.* powders for solid phase extraction and chromatography), or in the last decade as thin films (either hydrogel layers or micropatterned structures). More recently, however, nanoscale MIP fabrication has led to higher density MIP structures (with higher surface-to-volume ratios) in the quest of higher sensitivities.[11,74] One example of MIP nanospheres (MIPNs) has been achieved using block copolymer self-assembly techniques synthesized by controlled/"living" free radical polymerization (CRPs).[75] In the last decade, CRPs have attracted considerable interest in the MIP field for the controlled synthesis of advanced MIPs with tailor-made structures and improved binding properties.[75–78] In one case, iniferter-induced CRP was attempted to attach pHEMA chains onto MIP particles of bupivacaine, but only very short pHEMA oligomer chains (i.e. trimers of HEMA) were grafted. The resulting MIP only showed molecular recognition in organic solvent/ buffer mixtures instead of pure aqueous solutions.[78]

Zhao *et al.* prescribed a new approach to prepare MIPNs using a diblock copolymer, poly[(*tert*-butyl methacrylate)-*block*-HEMA] (PtBMA-*b*-pHEMA).[75]

Cyclohexane, a block selective solvent, 2-acrylamido-6-(carboxylbutyl-amido)pyridine (ACAP), and cross-linkable methacryloyl side groups were also introduced into the polymer to produce spherical micelles. Combining molecular imprinting and block copolymer self-assembly techniques, the resulting MIPNs had uniform nanoscopic sizes, good dispersibility in the solvent, and higher rebinding capacities and selectivities than traditional bulk MIPs. Hydrolysis of the PtBMA shell allowed for the water dispersibility of the nanospheres and the ability to penetrate through cell membranes. It was revealed that the molar density of ACAP functional groups in MIPNs was about 1.6-fold of that in the traditionally prepared bulk MIPs (7.1×10^{-4} mol g^{-1} *vs.* 4.5×10^{-4} mol g^{-1}), while the rebinding capacity of the MIPNs was about 2.4-fold higher at 8 mM template concentration (450 to 190 µmol g^{-1}). The higher binding capacities were attributed to the nanoscopic recognition sites in the core-shell structure of the MIPNs.

More recently, in an attempt to suppress nonspecific hydrophobic interactions between MIP and template molecules, approaches using post-modification procedures of preformed MIPs have been investigated, either by surface grafting of hydrophilic polymer layers or by chemical modification of the surface functional groups of the MIP particles.[76,77] It was hypothesized that the presence of the significant hydrophobic interactions between MIPs and template in pure aqueous media has led to their water incompatibility, ultimately leading to high hydrophobically driven non-specific bindings for both the MIPs and their control NIPs obscuring their inherent specific recognition ability.[76] Moreover, when analyzing macromolecular biological samples such as proteins, it is often seen that surface accumulation can occur.[76,77] This in turn may block imprinted binding sites within MIP particles, rendering them inaccessible to the template molecules.

In light of this, Pan *et al.* demonstrated the first controlled synthesis of water-compatible MIP microspheres with ultrathin hydrophilic pHEMA shells.[76] Initially, MIP microspheres were prepared with surface-immobilized reactive dithioester groups *via* surface-initiated reversible addition–fragmentation chain transfer (RAFT) precipitation polymerization (RAFTPP). The pre-polymerization mixtures [template 2,4-dichlorophenoxyacetic acid (2,4-D), functional monomer 4-vinylpyridine (4-VP), cross-linker EGDMA, and a porogenic solvent mixture of methanol and water (4/1 v/v)] were stirred for 2 h at 25 °C in order to allow the formation of a more stable complex between the functional monomer and template molecules. The 2 h pre-association modification in the polymerization procedure led to higher yields and larger particle sizes. Polymerization was then performed with cumyl dithiobenzoate (CDB) as the chain transfer agent and azobisisobu-tyronitrile (AIBN) as the initiator at 60 °C for 24 h with a reactant composition of 2,4-D/4-VP/EGDMA/AIBN/CDB being 1/4/20/0.88/1.76 (molar ratio) and the volume percentage of the porogenic solvent being about 99%. After purification by Soxhlet extraction, light pink MIP particles (due to the presence of dithioester groups) were then grafted with pHEMA brushes and

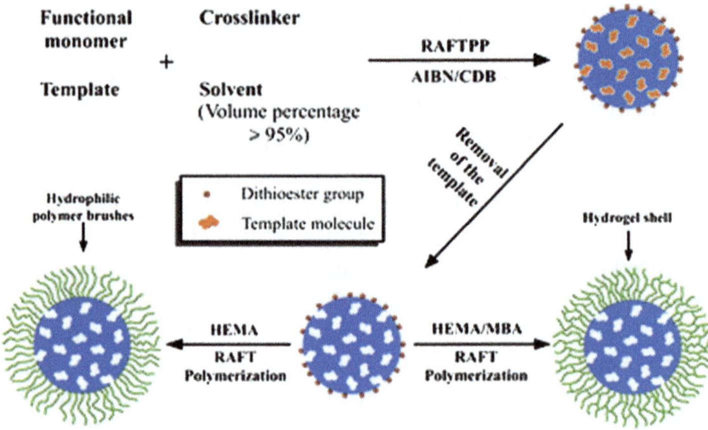

Figure 3.6 Schematic illustration for a controlled synthesis of MIP microspheres with ultrathin hydrophilic polymer shells of pHEMA brushes and a pHEMA hydrogel layer via RAFT polymerization.[76]

lightly cross-linked pHEMA hydrogel layers *via* RAFT polymerization. A detailed schematic illustration of this procedure can be seen in Figure 3.6.

The RAFT polymerization technique is one of the most powerful CRPs due to its good control over polymer structures, its applicability to a wide range of monomers, and mild reaction conditions.[76,79] The introduction of hydrophilic grafted brushes onto MIP microspheres was hypothesized to improve their surface hydrophilicity and led to pure water-compatible binding properties. Characterization and verification of microspheres using SEM revealed increasing MIP and NIP number-average diameters (D_n) depending on their composition: ungrafted (2.971 and 3.350 µm), grafted with PHEMA brushes (2.989 and 3.368 µm), and those grafted with a PHEMA hydrogel shell (2.999 and 3.375 µm) (see Figure 3.7). The presence of ultrathin hydrophilic polymer shells on the MIP microspheres was also confirmed by SEM, Fourier transform infrared spectroscopy (FTIR), fluorescent labeling treatment, contact angle studies, and water dispersion stability tests. Polydispersity indices (U) were also calculated and decreased in the same trend as D_n. The binding properties of the MIP/NIP microspheres in a mixture of methanol and water were found to be higher for MIPs than NIPs, suggesting the presence of selective binding sites in the obtained MIPs. Despite this, both grafted MIP and NIP microspheres bound fewer templates than corresponding ungrafted microspheres, revealing the obvious occurrence of surface modification for the grafted MIP/NIP microspheres. However, once switching to equilibrium binding experiments in pure aqueous solution, template binding capacities of the facile surface-grafted MIP/NIP microspheres decreased with increasing grafting time (i.e. the chain length of the polymer brushes) and had an overall higher % rebinding. Furthermore, grafted hydrogel MIP microspheres showed much higher template binding capacities in pure water than grafted brush MIPs, which correlated

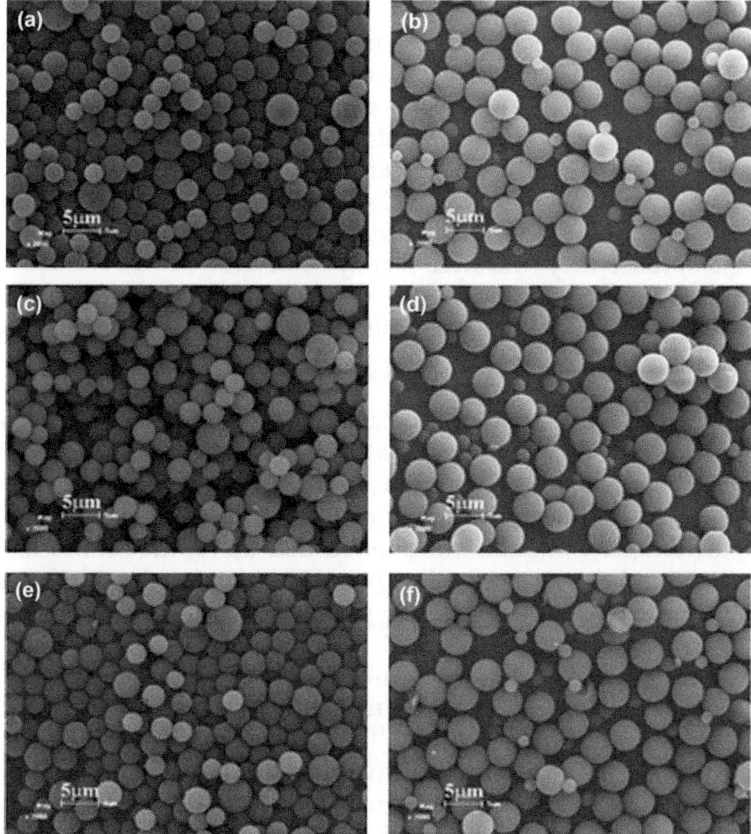

Figure 3.7   SEM images of ungrafted (a) MIP and (b) NIP microspheres, (c) MIP and (d) NIP microspheres grafted (12 h) with pHEMA brushes, and (e) MIP and (f) NIP microspheres grafted (12 h) with a pHEMA hydrogel shell.

well with their surface hydrophilicity. These results strongly verified that hydrophilic polymer shell-grafting approaches via surface-initiated RAFT polymerization is an efficient preparation method for pure water-compatible MIPs. Moreover, the significant effect of chain lengths in the grafted polymer brushes and the presence of cross-linking in the grafted polymer shells on the surface hydrophilicity and water-compatibility of the MIP microspheres were of great importance for the rational design of water-compatible MIPs when using controlled surface-grafting approaches.

3.3.1.4   Polysiloxane Sol–Gels

By and large, MIPs have been predominantly derived from organic polymers, often requiring organic solvents for their synthesis.[18,32,35,36,80] Recently, however, inorganic polymers have been increasingly used for MIP synthesis.[81–83] Polysiloxanes are mixed inorganic–organic silicone polymers consisting of an inorganic silicon–oxygen backbone with attached organic (carbon, hydrogen, and oxygen) side groups. Silicones are inert, synthetic compounds with a variety of forms and uses, including silicone oil, silicone grease, silicone rubber, and silicone resin.[81–84] Generally, polysiloxanes are the result of the sol–gel process; this concerns the transition of a liquid 'sol' system into solid 'gel' network, and some other byproducts like water or alcohol. The sol–gel process has been described as a two-step process, the first being hydrolysis, followed by a condensation polymerization of an alkoxide (see Eqn 3.5). Alkoxides are ideal chemical precursors for sol–gel synthesis because they hydrolyze readily in water.[84]

$$Si(OR)_4 + H_2O \rightarrow HO\text{–}Si(OR)_3 + Si(OR)_4 \rightarrow [(RO)_3Si\text{–}O\text{–}Si(OR)_3] + ROH$$

$$(3.5)$$

A well-studied alkoxide in molecular imprinting is tetraethyl orthosilicate (TEOS);[85,86] TEOS matrices fabricated by the two-step sol–gel method have resulted in protein-specific imprinted polysiloxane porous scaffolds (see Figure 3.8). Evaluation of the ability of these scaffolds to preferentially recognize the template biomolecule showed that up to three times more template was bound than a competitor protein, which is a level of preferential binding similar to values reported in the molecular imprinting literature for both organic and inorganic materials.[85]

Sol–gel imprinting is rapidly earning worldwide attention due to its versatility and directness in synthesizing inorganic ceramic materials at low temperatures and mild conditions.[47,81–84] High purity, homogeneity, controlled porosity, nanoscale structuring, and the flexibility of incorporating organic moieties are some of the most remarkable features offered by this method for generating highly sensitive and selective matrices to incorporate analyte molecules for sensing applications.[83,84] These materials under sol–gel technologies have found numerous applications in different fields, such as the glass industry, ceramics, thin films, and different biological and chemical sensors.[81,82]

Sol–gel methods have been used to prepare porous silica scaffolds molecularly imprinted for proteins and peptides.[86] Lee *et al.* have previously

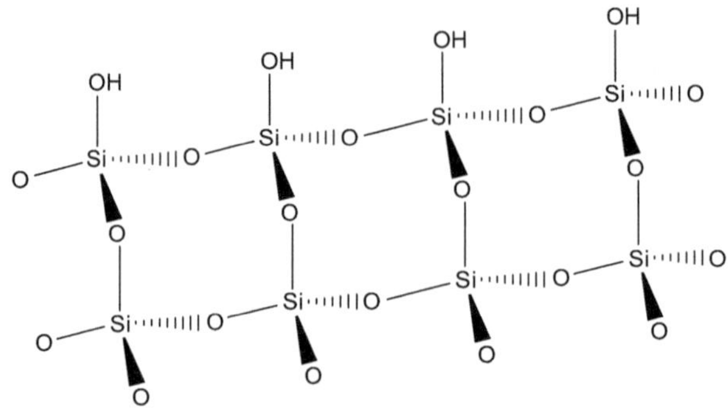

Figure 3.8   Schematic detailing a TEOS sol–gel matrix structure.

imprinted both lysozyme and RNase A using sol–gel processing in macro-porous polysiloxane (silica) scaffolds based on TEOS. The quantity of surface-accessible protein (62% of loaded protein) was varied by changing the amount of protein loaded into the sol. The amount of protein per unit surface area ranged from 0.3 μg m^{-2} for low loading of RNase to 152 μg m^{-2} for high loading of lysozyme. The amount of template was also tested against competitors individually and in mixtures, where up to 3.6 and three times more template bound compared with the competitor, respectively. This concurred with typical MIP values for both organic and inorganic materials.[85] This approach was later optimized using the 'epitope approach', which involved imprinting a peptide fragment of a target protein to promote specific adsorption of the entire protein; this is similar in which an antigen binds to an antibody *via* the epitope.[34,86] Brown and Puleo imprinted both whole lysozyme and a 16-residue lysozyme C peptide into porous silica scaffolds using the same sol–gel process.[86] After removing the template, the scaffolds were exposed to lysozyme and/or a competitor of comparable size, RNase A. The study concluded that although a preferential binding trend was present for the peptide-imprinted scaffolds, much less preference was seen in the presence of competitor molecules. Thus this implies that imprinting with the entire protein molecule was most effective for complex recognition. Optimum conditions for the template protein separation selectivity were also studied based on bovine hemoglobin (BHb) MIPs formed on the surfaces of amino-silica.[81,84,85] These studies have investigated the relationship between fundamental parameters for preparation of MIPs and their effect on the separation selectivity. Sol–gel MIPs were prepared *via* polymerization on the surface of aminopropyl silica followed by immobilization of the proteins, using a mixture of trimethoxy(octyl)silane, 3-aminopropylsilane, and tetra-ethoxysilane at a 42.5 : 42.5 : 5 : 15 ratio. These were polymerized at pH 7.0, before template protein removal by a solution of 0.1 M NaOH and 10% (v/v) HAc–10% (w/v) SDS. The results indicated that the MIPs obtained by this method showed the best separation selectivity for template lysozyme.

3.3.2 Biomarker Imprinting

The continued discovery of cancer biomarkers necessitates improved methods for their detection. Cancer/tumor markers, or simply 'biomarkers', are generally proteins or transcription factors that have previously been identified as being switched on in cancer, leading to unchecked growth of cells.[37,87–89] Biomarkers signifying the presence of any tumor/cancer may be defined on the basis of gene products uniquely expressed or overexpressed in tissue, serum, or urine.[89] In cancer therapy, antibodies can be raised against specific proteins present on the surface of tumors. Several markers and antibody counterparts have been recognized and established, such as the Her2/neu breast cancer gene and Herceptin antibody. However, biomarkers, and/or their counterpart antibodies which are currently a standard detection test, have a number of key limitations. Although antibodies exhibit a high degree of selectivity, any biological recognition element is inherently unstable with limited shelf-life even when stored under optimum conditions. This has justified the evaluation of new biomarkers/antibodies.[87]

One example is prostate specific antigen (PSA), and has commonly been used as a biomarker for initial diagnosis, monitoring of response to treatment, and prediction of prostate cancer (PC) risk and of treatment outcome.[88,89] PC is the second most common cause of cancer-related death in men. However, PSA is a prostate-specific and not a prostate cancer-specific marker; therefore it lacks both sensitivity and specificity to accurately detect the presence of PC, requires adjustment for age and prostate volume, is frequently raised in non-cancer conditions such as benign hypertrophy and prostatitis and, so far, has been controversial as a screening tool.[88] Alternatively, engrailed-2 (EN2) protein, a homeodomain-containing transcription factor, has been found to be expressed in ovarian cancer, melanoma, and PC cell lines, and secreted into the urine by PC and not by normal prostate tissue or stroma.[37,88,89] If the use of EN2 is proven effective, then a less invasive, less technical, and cheap urine test (much like a pregnancy test-kit) that could incorporate a MIP-based antibody could be deemed more suitable for early diagnosis.

Highly selective MIPs for the cancer marker sarcosine have been prepared and used as solid-phase extraction (SPE) sorbent materials.[90] Hashemi-Moghaddam *et al.* have prepared a very simple MIP procedure using methacrylic acid as functional monomer and a mixture of acetonitrile/water (4/1, v/v) as porogen. This method overcame the problems usually related to the imprinting of biological polar compounds. Moreover, potentiometric sensors based on surface molecular imprints have also been investigated using self-assembled monolayers to design sensing elements for the detection of cancer biomarkers.[91] Wang *et al.* have developed gold-coated silicon chips using thiol molecules (hydroxyl-terminated alkanethiol) chemically bound to the metal substrate and self-assembled into highly ordered monolayers. Using this method, biomolecules of purified carcinoembryonic antigen (CEA) and human amylase were co-adsorbed and then removed, creating footprint cavities in the monolayer matrix. Potentiometrically measured changes upon re-adsorption of the biomolecules in the chip had a sensitive detection range of 2.5–250 ng mL^{-1}. The developed CEA assay was validated against a standard immunoassay, and was tested for its specificity by cross-reacting with hemoglobin, ultimately having no response to either hemoglobin or a non-imprinted control sensor. This approach demonstrated the potential of generating a new general inexpensive, highly sensitive, specific, and simple assay methodology for the detection of protein cancer biomarkers and to protein-based macromolecular structures.[91]

3.3.3 Virus Imprinting

Although best management practices can minimize the introduction of disease causing agents, it is not possible to exclude all due to the open nature of farming. Therefore, emergency preparedness is critical to prevent the spread of the disease causing agent and implement corrective actions. In a typical scenario, an outbreak is initially identified by symptoms exhibited by animals/plants through chance observation or *via* inspection. Samples are collected and sent to a central laboratory for confirmatory testing. In the time elapsed from sample collection to results being available, which can be a week in some cases, the infection or toxic material could spread over a wide area. A classic example of the impact of a delayed response was illustrated in the foot-and-mouth outbreak in the UK that occurred in the 1990s. Here, there was a 48 h delay for performing confirmatory testing, in which time the infection could not be contained and spread throughout the animal herds within the UK, leading to billion pound losses. Consequently, there are incentives for diagnostic tests that can be applied on-site that will enable rapid detection of toxic agents or infectious materials. Once identified, emergency preparedness plans could be rapidly implemented, thereby containing the agent at source. It is generally acknowledged that outbreaks of infectious diseases in agriculture and release of toxic chemicals into water courses can have high, long-lasting, economic impacts. To specifically estimate the impact of rapid diagnostic devices is difficult, given that screening forms part of an overall prevention strategy. Yet, there is a clear demand for on-site diagnostic devices, given that the sector is worth over $5 bn a year and is growing. Currently, diagnostic devices for on-site testing are unavailable or lack the required sensitivity and are expensive. The development of MIP sensors could provide tools for farmers, inspectors, and researchers to rapidly confirm a toxic or infectious agent. The low cost of the sensors ($5 per test) along with ease of use, high sensitivity (ng virus), and rapid analysis time (1 min) should make the sensors commercially viable for routine use.[92–94]

The early detection of biological or chemical toxic agents, *i.e.* viruses, within the environment, animal production facility, or water sources is critical in emergency response management. Whether the toxic agent is naturally disseminated or deliberately released, early detection is key to enable containment and recovery strategies to be implemented. This has led to the demand for on-site testing devices that can be used by non-technical personnel in the field when investigation of viral outbreaks or toxic release is required.[94,95] On-site diagnostic (OSD) and point-of-care (POC) tools for in-field use are an essential part of emergency management, given that early detection of an environmental hazard leads to rapid containment and corrective action. However, development of sample preparation methods that function to concentrate targets and remove interferences has received less attention. Moreover, current on-site diagnostics are heavily dominated by immuno-sensors such as the lateral flow immunoassay and require multiple step protocols and technical expertise. Although antibodies exhibit a high degree of selectivity, any biological recognition element is inherently unstable with limited shelf-life even when stored under optimum conditions.[62] In addition, antibodies are difficult and expensive to produce, hence typically linked to commonly occurring analytes (for example, *Salmonella*) for which there is a large demand. In contrast, for hazards that are less frequently encountered but can cause significant disruption, such as tomato spotted wilt virus (TSWV), the time and resources to produce antibodies is often lacking. For example, the swine flu outbreak of 2009 resulted in an increased demand for immuno-sensors to detect the virus.

However, antibody production lacks the flexibility to produce large volumes required to develop sensors in the short term. Indeed, immunoassays for swine flu have yet to be marketed, but are currently under development. The instability, cost, and inflexibility of antibodies have led to interest in artificial receptors, commonly referred to as plastic antibodies. Although a range of strategies have been evaluated, those based on MIPs have received the most attention with commercial solid-phase extraction (SPE) based on the technology becoming commercially available. Molecular imprints have been demonstrated for many classes of molecules, including drugs and pesticides, amino acids, peptides, sugars, and hormones, with the imprinting of small organic molecules now well established and considered routine.[12,96]

Recently, the imprinting of microbes such as human rhinovirus and various plant viruses such as tobacco mosaic virus (TMV), pepper mild mottle virus (PMMoV), and tomato spotted wilt virus (TSWV) have been reported.[97–99] These plant viruses are widely spread and infectious diseases that affect a wide range of plantations such as vegetables, flowers, and herbs.[98] One of the main reasons for using TMV as a model template, for example, is that it is very robust and can be subjected to harsh environmental conditions (90 °C and pH values between 3.5 and 9) without losing its conformation and activity.[98] TMV or *tobamovirus* is a group IV positive-sense single-strand RNA [(+)ssRNA] plant virus of the *Virgaviridae* family.[100] It has a tubular shape with an external diameter of 18 nm, internal diameter of 4 nm, and a length of 300 nm. Its structure consists of a central, helical RNA strand 6400 bases long, coated with a 2130 protein shell (capsomer) per capsid unit. PMMoV is also known as a tobamovirus of the same family as TMV, also retaining a similar structure to TMV in being helical and rod-shaped (312 nm long and 18 nm in diameter), with an isoelectric point (p*I*) of 3.38–3.71.[100]

PMMoV is one of the key viral pathogens in bell peppers and responsible for significant crop losses in North America and Europe.[101] Symptoms caused by this pathogen may vary, based on the specific host cultivar; however, general symptoms usually include various degrees of plant mottling, chlorosis, curling, dwarfing, and distortion of the vegetable, leaves, and even whole plants. Specific vegetation symptoms include a reduction in size, mottling, color changes, and an obvious distorted appearance. Controlling PMMoV is important for pepper production worldwide; however, recent research shows that PMMoV may be transmitted to humans and cause fever and abdominal pains.[102] Metagenomic studies have shown that

PMMoV is stable throughout the human digestive system, and consequently identified viable PMMoV amounts in the stool of human subjects (up to 10^9 virions per gram of dry fecal weight).[102] In contrast, TSWV is known as a *tospovirus*, and is a group V negative-sense single-strand RNA [(−)ssRNA] plant virus of the *Bunyaviridae* family. TSWV has been found to be ubiquitous in the environment and can infect well over 1000 plant species, causing significant economic damage to many agronomic and horticultural corps.[101,103,104] Viral infection results in spotting and wilting of the plant, although can vary depending on species, reduced vegetative output, and eventually death. No antiviral cures have been developed for plants infected with a TSWV and infected plants should be removed from a field and destroyed in order to prevent the spread of the disease.[101,103,104]

Over the years, virus imprinted MIPs have shown increased binding affinities and have demonstrated that virus imprinting can induce selective binding of target viruses based on the virions shape.[93,98,100,105] This has been a significant attribute in virology, given that virus particles cannot be observed with light microscopy and are difficult if not impossible to cultivate in the laboratory.[62] Detection by polymerase chain reaction (PCR) or cell culture experiments can also be difficult or time consuming. It can be envisaged that virus imprinted films hold strong potential for future sensing devices.

Although imprinted polymers hold promise as alternatives to antibodies, the ability to transduce the binding of analyte into a measurable signal is problematic. It is possible to use detection platforms such as QCM or surface plasmon resonance (SPR) to measure mass changes during the binding of a target analyte to imprinted films.[97] While the QCM system lends itself to portability, the SPR approach is currently unsuitable for on-site diagnostic devices due to instability of the optics. The SPR technique can also be unsuitable for thin film polymer characterization due to an incompatibility in the refractive index operation range of the device. Successful monitoring of plant viruses directly in the plant sap within minutes has been achieved, based on biomimetic polymer MIPs in combination with highly sensitive QCM sensor techniques.[106] Dual polarization interferometry (DPI) has also been identified as a suitable and highly sensitive optical technique that is compatible with thin film polymers for in-lab optimization of MIP binding properties prior to QCM application.[2] Further viral imprinting innovations are the integration of MIPs with microfluidic devices using electroosmotic forces to minimize non-specific binding effects.[62,93,107] Electroosmotic flow is induced by the application of an electric field that generates a zone of mobile ions referred to as the double layer (double layer capacitance) that induces flow of liquid. This technique has been used to minimize non-specific binding of DNA to the walls of microfluid channels modified with DNA probes. Electoosmotic flow also results in convection within the microfluidic channel that can rapidly present the target to the bio-recognition element, leading to detection times in the order of seconds.[62,93,107]

3.4   Applications of Molecular Imprinting

An exciting new potential field of MIP technology is in bio-recognition elements in chemical and physical signal generation of biosensors.[10,108] Biosensors are analytical tools consisting of biologically active materials used in close conjunction with a device that will convert a biochemical signal into a quantifiable electrical signal. Biosensors have many advantages, such as simple and low-cost instrumentation, fast response times, minimum sample pre-treatment, and high sample throughput. Consequently, their utilization is ever increasing within the field of modern analytical chemistry, with opportunities growing in clinical diagnostics environmental and food analysis, genotoxicity, the detection of illicit drugs, and chemical warfare agents.[29] The focus is on the complementary intersection between molecular recognition, nanotechnology, and molecular imprinting to improve the analytical performance and robustness of devices.

Biosensors for large bio-macromolecules, such as proteins, are currently expensive to develop because they require the use of expensive antibodies. However, MIPs are becoming ever more promising as viable synthetic receptors, and offer the potential replacement of antibodies. Clear advantages over real antibodies have been recognized with MIP applications for sensor technology since they are highly cross-linked, intrinsically stable, robust, and are able to facilitate their application in extreme environments.[3,109,110] MIPs have proven to produce selective extractions that rival immunoaffinity-based separations, but without the tediously lengthy time-consuming process. They also have the advantage over immunoassays in that imprints can be made of compounds which are difficult, if not impossible, to raise with antibodies and can resist elevated temperatures and pressures.[62] Investigating the development of highly selective capture reagents such as synthetic antibodies ultimately could provide an inexpensive, fast, and efficient diagnostic method for highly sensitive analytical procedures within clinical diagnostics, environmental analysis, food analysis, production monitoring, and the pharmaceutical area.[3,10,80] Therefore potential MIP applications within the field of modern analytical chemistry are most likely in chemical and biosensors, as their opportunities are increasing in clinical diagnostics, environmental analysis, food analysis, and production monitoring.[3,10,80]

3.4.1   Immuno-Based Sensors

Lateral flow immunoassay tests have been a popular POC technique since the mid-1980s, and have since been employed in numerous areas, including clinical, veterinary, and environmental applications.[62] One familiar example is the home pregnancy test kit. This test works by binding the hormone marker for pregnancy [human chorionic gonadotropin (hCG) $\sim$25.7 kDa], from either blood or urine, to an antibody and an indicator. These are 99% accurate in lab tests and can bind up to 10–25 mIU mL^{-1}. Some cancerous tumors also produce this hormone; therefore, elevated levels (0–5 mIU mL^{-1}) measured in the absence of pregnancy can lead to a cancer diagnosis.

Test kit development has traditionally relied on antibodies. However, as previously mentioned, antibodies have drawbacks in terms of production expenses, lack of long-term stability, and ethical issues surrounding the use of animal-based antibodies. Hence, they are typically linked to commonly occurring analytes (for example, *Salmonella*) for which there is a large demand. In contrast, for hazards that are less frequently encountered but can cause significant disruption, such as tomato spotted wilt virus (TSWV), the time and resources to produce antibodies is often lacking. For example, the swine flu outbreak of 2009 resulted in an increased demand for immuno-sensors to detect the virus. However, antibody production lacks the flexibility to produce large volumes required to develop sensors in the short term. Indeed, immunoassays for swine flu have yet to be marketed but are currently under development. Although a range of strategies have been evaluated, those based on MIPs have received the most attention, with commercial solid-phase extraction (SPE) based on the technology becoming commercially available. Molecular imprints have been demonstrated for many classes of molecules, including drugs and pesticides, amino acids, peptides, sugars, and hormones, with the imprinting of small organic molecules now well established and considered routine.[12,96]

3.4.2 Transducers Used in MIP-Based Sensors

One important part of bio-sensing is the transducer, which monitors the reaction between bio-selector and analyte. Among various physical transducers (electrochemical, piezoelectric, *etc.*), mass sensitive devices such as surface acoustic wave (SAW), SPR, and QCM have become popular for sensing applications.[44,45,111,112] Although the sensor format can be used to detect almost any target of interest, acoustic, electrical, and optical sensors based on HydroMIPs for protein and virus detection could be incorporated into a microfluidic (handheld) device with non-specific binding being minimized through electroosmotic flow. The sensors could provide tools for researchers and provincial inspectors to sample for hazardous materials in the field (for example, plant virus), thereby strengthening the effectiveness of emergency response strategies. In addition, the sensors should be affordable to use by producers and processors to routinely screen for potential chemical or biological hazards within food production systems. In general, MIPs are well suited for biosensor technologies that rely on chemical and physical signal generation. MIPs have been useful in several studies that investigated drug loading, controlled slow release of therapeutic agents, spherical nano-bead production, and emulsion polymerization.[3,11,42,80]

MIPs may well provide an alternative solution to current automated high-throughput screening assay techniques. There is even the possibility of developing MIP decaffeinating filters which are reusable and thus more economical. With specific reference to HydroMIPs, QCM sensing technologies have implications for the development of rapid protein diagnostics indicative of, for example, cancer and cardiac disease states. The development of QCM systems for use in fluids has rapidly opened a new world of applications, including bio-sensing and micro-rheology.[44] Moreover, polysiloxane (sol–gel) MIPs provide an easy approach for polymerizing sensitive layers of recognition element directly over bio-sensing devices.[81] Thus, sol–gel MIPs have found uses in several analytical applications, and could even be effectively used as detection systems for endocrine disrupters in the near future.[81,82,113,114]

3.5 Molecular Imprinting Challenges

Although MIPs have been successfully applied and proven their worth to a large range of molecules and bio-macromolecules, there are still current drawbacks. The question of how selective they really are and how this can be improved still remains. One of the most dominant drawbacks in MIP technology is the unprecedented degree of influence that a variation in pH, ionic strength, and sample matrix concentration all have on the gel properties.[33,36] In turn, this can affect the three-dimensional shape and chemical characteristics of the template molecule during polymerization. This is particularly true when imprinting large bio-macromolecules such as proteins. As previously mentioned, proteins are relatively labile and have changeable conformations which are sensitive to solvent environments, pH, and temperature, all of which present a variety of challenges.[32,33,65,110] It has been postulated that problems with macromolecules could be caused by the use of charged functional monomers causing non-specific electrostatic interactions.[32]

The second challenge, critical step, and decisive manoeuver for optimal MIP performance is the template removal process.[115] If there are remaining template molecules in the MIPs, less cavities will be available for rebinding, which decreases efficiency. Furthermore, if template bleeding occurs during analytical applications, errors will arise. Despite the relevance to the MIP performance, template removal has received scarce attention and is currently the least cost-effective step of the MIP development.[115,116] Attempts to reach complete template removal may involve the use of highly drastic conditions in conventional extraction techniques, resulting in the damage or the collapse of the imprinted cavities.

Moreover, as with antibodies, MIPs have also shown a degree of cross-selectivity, in that they can bind molecules similar to the native template and cause non-specific binding.[32,33,65,110] It is thought that this is due in part to an excess of functional monomer molecules being randomly distributed and frozen within the imprinted cavity during polymerization that have an effect on the selectivity of rebound molecules. Initial template interactions with functional monomers largely determine the recognition properties of the matrix. Thus more sophisticated monomers capable of forming better, stronger, and more stable interactions that offer better positioning and complementary functionality are widely being sought.[32,56,80]

References

1. S. M. Reddy, G. Sette and Q. Phan, *Electrochim. Acta*, 2011, **56**, 9203–9208.
2. S. M. Reddy, D. M. Hawkins, Q. T. Phan, D. Stevenson and K. Warriner, *Sens. Actuators B*, 2013, **176**, 190–197.
3. K. Mosbach, *Anal. Chim. Acta*, 2001, **435**, 3–8.
4. N. Masqué, R. M. Marcé and F. Borrull, *Trends Anal. Chem.*, 2001, **20**, 477–486.
5. S. M. Reddy, Q. T. Phan, H. El-Sharif, L. Govada, D. Stevenson and N. E. Chayen, *Biomacromolecules*, 2012, **13**, 3959–3965.
6. E. Saridakis, S. Khurshid, L. Govada, Q. Phan, D. Hawkins, G. V. Crichlow, E. Lolis, S. M. Reddy and N. E. Chayen, *Proc. Natl. Acad. Sci. U. S. A.*, 2011, **108**, 11081–11086.
7. C. Alexander, H. S. Andersson, L. I. Andersson, R. J. Ansell, N. Kirsch, I. A. Nicholls, J. O'Mahony and M. J. Whitcombe, *J. Mol. Recognit.*, 2006, **19**, 106–180.
8. K. Haupt, A. V. Linares, M. Bompart and B. T. S. Bui, *Mol. Imprinting*, 2012, **325**, 1–28.
9. B. T. S. Bui and K. Haupt, *Anal. Bioanal. Chem.*, 2010, **398**, 2481–2492.
10. P. A. Lieberzeit, R. Samardzic, K. Kotova and M. Hussain, *Proc. Eng.*, 2012, **47**, 534–537.
11. A. Poma, A. P. F. Turner and S. A. Piletsky, *Trends Biotechnol.*, 2010, **28**, 629–637.
12. D. Stevenson, *Trends Anal. Chem.*, 1999, **18**, 154–158.
13. N. Lavignac, K. R. Brain and C. J. Allender, *Biosens. Bioelectron.*, 2006, **22**, 138–144.
14. R. C. Advincula, *Korean J. Chem. Eng.*, 2011, **28**, 1313–1321.
15. C. Alexander, H. S. Andersson, L. I. Andersson, R. J. Ansell, N. Kirsch, I. A. Nicholls, J. O'Mahony and M. J. Whitcombe, *J. Mol. Recognit.*, 2006, **19**, 106–180.
16. Y. Fuchs, O. Soppera, A. G. Mayes and K. Haupt, *Adv. Mater.*, 2013, **25**, 566–570.
17. S. Kunath, N. Marchyk, K. Haupt and K. Feller, *Talanta*, 2013, **105**, 211–218.
18. A. G. Mayes and M. J. Whitcombe, *Adv. Drug Delivery Rev.*, 2005, **57**, 1742–1778.
19. S. Asman, N. A. Yusof, A. H. Abdullah and M. J. Haron, *Molecules*, 2012, **17**, 1916–1928.
20. C. Dai, J. Zhang, Y. Zhang, X. Zhou and S. Liu, *PLoS One*, 2013, **8**, e78167.
21. N. A. Yusof, N. D. Zakaria, N. A. M. Maamor, A. H. Abdullah and M. J. Haron, *Int. J. Mol. Sci.*, 2013, **14**, 3993–4004.
22. V. J. B. Ruigrok, M. Levisson, M. H. M. Eppink, H. Smidt and J. van der Oost, *Biochem. J.*, 2011, **436**, 1–13.
23. C. Widstrand, B. Boyd, J. Billing and A. Rees, *Am. Lab.*, 2007, **39**, 23–24.

24. C. Widstrand, E. Yilmaz, B. Boyd, J. Billing and A. Rees, *Am. Lab.*, 2006, **38**, 12.
25. Q. Deng, J. Wu, X. Zhai, G. Fang and S. Wang, *Sensors*, 2013, **13**, 2994–3004.
26. S. Gupta, M. Saxena, N. Saini, R. Mahmooduzzafar, R. Kumar and A. Kumar, *PLoS One*, 2012, 7, e43527.
27. T. Renkecz, G. Mistlberger, M. Pawlak, V. Horvath and E. Bakker, *ACS Appl. Mater. Interfaces*, 2013, **5**, 8537–8545.
28. V. J. B. Ruigrok, M. Levisson, M. H. M. Eppink, H. Smidt and J. van der Oost, *Biochem. J.*, 2011, **436**, 1–13.
29. S. Pradhan, M. Boopathi, O. Kumar, A. Baghel, P. Pandey, T. H. Mahato, B. Singh and R. Vijayaraghavan, *Biosens. Bioelectron.*, 2009, **25**, 592–598.
30. J. Haginaka, C. Miura, N. Funaya and H. Matsunaga, *Anal. Sci.*, 2012, **28**, 315–317.
31. M. J. Whitcombe, I. Chianella, L. Larcombe, S. A. Piletsky, J. Noble, R. Porter and A. Horgan, *Chem. Soc. Rev.*, 2011, **40**, 1547–1557.
32. D. E. Hansen, *Biomaterials*, 2007, **28**, 4178–4191.
33. H. F. El-Sharif, Q. T. Phan and S. M. Reddy, *Anal. Chim. Acta*, 2014, **809**, 155–161.
34. D. M. Hawkins, D. Stevenson and S. M. Reddy, *Anal. Chim. Acta*, 2005, **542**, 61–65.
35. M. E. Byrne, K. Park and N. A. Peppas, *Adv. Drug Delivery Rev.*, 2002, **54**, 149–161.
36. M. E. Byrne and V. Salian, *Int. J. Pharm.*, 2008, **364**, 188–212.
37. J. Z. Hilt and M. E. Byrne, *Adv. Drug Delivery Rev.*, 2004, **56**, 1599–1620.
38. J. Heick, M. Fischer and B. Poepping, *J. Chromatog. A*, 2011, **1218**, 938–943.
39. Y. Ge and A. P. F. Turner, *Trends Biotechnol.*, 2008, **26**, 218–224.
40. Y. Q. Xia, T. Y. Guo, M. D. Song, B. H. Zhang and B. L. Zhang, *Biomacromolecules*, 2005, **6**, 2601–2606.
41. X. Zeng, W. Wei, X. Li, J. Zeng and L. Wu, *Bioelectrochemistry*, 2007, **71**, 135–141.
42. Y. Lv, T. Tan and F. Svec, *Biotechnol. Adv.*, 2013, **31**, 1172–1186.
43. E. Verheyen, J. P. Schillemans, M. van Wijk, M. Demeniex, W. E. Hennink and C. F. van Nostrum, *Biomaterials*, 2011, **32**, 3008–3020.
44. U. Latif, S. Can, O. Hayden, P. Grillberger and F. L. Dickert, *Sensors Actuators B*, 2013, **176**, 825–830.
45. T. M. A. Gronewold, *Anal. Chim. Acta*, 2007, **603**, 119–128.
46. L. Qin, X. He, W. Zhang, W. Li and Y. Zhang, *J. Chromatogr. A*, 2009, **1216**, 807–814.
47. H. Lin, C. Hsu, J. L. Thomas, S. Wang, H. Chen and T. Chou, *Biosens. Bioelectron.*, 2006, **22**, 534–543.
48. Q. Gai, F. Qu and Y. Zhang, *Sep. Sci. Technol.*, 2010, **45**, 2394–2399.
49. M. E. Brown and D. A. Puleo, *Chem. Eng. J.*, 2008, **137**, 97–101.
50. D. M. Hawkins, A. Trache, E. A. Ellis, D. Stevenson, A. Holzenburg, G. A. Meininger and S. M. Reddy, *Biomacromolecules*, 2006, 7, 2560–2564.

51. M. N. Albarghouthi, B. A. Buchholz, P. J. Huiberts, T. M. Stein and A. E. Barron, *Electrophoresis*, 2002, **23**, 1429–1440.
52. K. Aizawa, *Polym. Adv. Technol.*, 2000, **11**, 481–487.
53. C. Sontimuang, R. Suedee and F. Dickert, *Anal. Biochem.*, 2011, **410**, 224–233.
54. M. P. Davies, V. De Biasi and D. Perrett, *Anal. Chim. Acta*, 2004, **504**, 7–14.
55. N. Pérez-Moral and A. G. Mayes, *Anal. Chim. Acta*, 2004, **504**, 15–21.
56. S. H. Ou, M. C. Wu, T. C. Chou and C. C. Liu, *Anal. Chim. Acta*, 2004, **504**, 163–166.
57. K. Hirayama and K. Kameoka, *Japan Analyst*, 2000, **49**, 29–33.
58. K. Hirayama, Y. Sakai and K. Kameoka, *J. Appl. Polym. Sci.*, 2001, **81**, 3378–3387.
59. H. C. Zhou, L. Baldini, J. Hong, A. J. Wilson and A. D. Hamilton, *J. Am. Chem. Soc.*, 2006, **128**, 2421–2425.
60. K. Toko, *Mater. Sci. Eng. C*, 1996, **4**, 69–82.
61. E. A. Baldwin, J. Bai, A. Plotto and S. Dea, *Sensors*, 2011, **11**, 4744–4766.
62. A. González Oliva, H. J. Cruz and C. C. Rosa, *Sens. Update*, 2001, **9**, 283–312.
63. J. Zeravik, A. Hlavacek, K. Lacina and P. Skládal, *Electroanalysis*, 2009, **21**, 2509–2520.
64. T. Takeuchi, D. Goto and H. Shinmori, *Analyst*, 2007, **132**, 101–103.
65. T. Takeuchi and T. Hishiya, *Org. Biomol. Chem.*, 2008, **6**, 2459.
66. K. D. Shimizu and C. J. Stephenson, *Curr. Opin. Chem. Biol.*, 2010, **14**, 743–750.
67. N. Adrus and M. Ulbricht, *Polymer*, 2012, **53**, 4359–4366.
68. O. Wichterle and D. Lim, *Nature*, 1960, **185**, 117–118.
69. X. Lou, S. Vijayasekaran, R. Sugiharti and T. Robertson, *Biomaterials*, 2005, **26**, 5808–5817.
70. J. Pilar, J. Kriz, B. Meissner, P. Kadlec and M. Pradny, *Polymer*, 2009, **50**, 4543–4551.
71. B. Hacioglu, K. A. Berchtold, L. G. Lovell, J. Nie and C. N. Bowman, *Biomaterials*, 2002, **23**, 4057–4064.
72. K. Sreenivasan, *J. Appl. Polym. Sci.*, 2008, **109**, 3275–3278.
73. L. Li and L. J. Lee, *Polymer*, 2005, **46**, 11540–11547.
74. M. Bompart, K. Haupt and C. Ayela, *Mol. Imprinting*, 2012, **325**, 83–110.
75. Z. Li, J. F. Ding, M. Day and Y. Tao, *Macromolecules*, 2006, **39**, 2629–2636.
76. G. Pan, Y. Ma, Y. Zhang, X. Guo, C. Li and H. Zhang, *Soft Matter*, 2011, **7**, 8428–8439.
77. B. Dirion, Z. Cobb, E. Schillinger, L. I. Andersson and B. Sellergren, *J. Am. Chem. Soc.*, 2003, **125**, 15101–15109.
78. J. Oxelbark, C. Legido-Quigley, C. S. A. Aureliano, M. Titirici, E. Schillinger, B. Sellergren, J. Courtois, K. Irgum, L. Dambies, P. A. G. Cormack, D. C. Sherrington and E. De Lorenzi, *J. Chromatogr. A*, 2007, **1160**, 215–226.

79. G. Pan, Y. Zhang, X. Guo, C. Li and H. Zhang, *Biosens. Bioelectron.*, 2010, **26**, 976–982.
80. M. P. Davies, V. De Biasi and D. Perrett, *Anal. Chim. Acta*, 2004, **504**, 7–14.
81. A. Lukowiak and W. Strek, *J. Sol-Gel Sci. Technol.*, 2009, **50**, 201–215.
82. B. D. MacCraith, C. McDonagh, A. K. McEvoy, T. Butler, G. O'Keeffe and V. Murphy, *J. Sol-Gel Sci. Technol.*, 1997, **8**, 1053–1061.
83. S. N. Tan, W. Wang and L. Ge, in *Comprehensive Biomaterials*, ed. P. Ducheyne, Elsevier, Oxford, 2011, pp. 471–489.
84. A. Mujahid, P. A. Lieberzeit and F. L. Dickert, *Materials*, 2010, **3**, 2196–2217.
85. K. Lee, R. R. Itharaju and D. A. Puleo, *Acta Biomater.*, 2007, **3**, 515–522.
86. M. E. Brown and D. A. Puleo, *Chem. Eng. J.*, 2008, **137**, 97–101.
87. W. J. Catalona, J. P. Richie, F. R. Ahmann, M. A. Hudson, P. T. Scardino, R. C. Flanigan, J. B. Dekernion, T. L. Ratliff, L. R. Kavoussi, B. L. Dalkin, W. B. Waters, M. T. Macfarlane and P. C. Southwick, *J. Urol.*, 1994, **151**, 1283–1290.
88. R. Morgan, A. Boxall, A. Bhatt, M. Bailey, R. Hindley, S. Langley, H. C. Whitaker, D. E. Neal, M. Ismail, H. Whitaker, N. Annels, A. Michael and H. Pandha, *Clin. Cancer Res.*, 2011, **17**, 1090–1098.
89. S. E. McGrath, A. Michael, H. Pandha and R. Morgan, *FEBS Lett.*, 2013, **587**, 549–554.
90. H. Hashemi-Moghaddam, M. Rahimian and B. Niromand, *Bull. Korean Chem. Soc*, 2013., **34**, 2330–2334.
91. Y. Wang, Z. Zhang, V. Jain, J. Yi, S. Mueller, J. Sokolov, Z. Liu, K. Levon, B. Rigas and M. H. Rafailovich, *Sens. Actuators B*, 2010, **146**, 381–387.
92. O. Hayden, P. A. Lieberzeit, D. Blaas and F. L. Dickert, *Adv. Funct. Mater.*, 2006, **16**, 1269–1278.
93. B. Rodoni, *Virus Res.*, 2009, **141**, 150–157.
94. A. Fereres, *Virus Res.*, 2009, **141**, 111–112.
95. D. Kapczynski and D. Swayne, in *Vaccines for Pandemic Influenza: Current Topics in Microbiology and Immunology*, ed. R. W. Compans and W. A. Orenstein, Springer, Berlin, 2009, pp. 133–152.
96. O. Ramstrom, L. Ye, M. Krook and K. Mosbach, *Chromatographia*, 1998, **47**, 465–469.
97. L. D. Bolisay, J. N. Culver and P. Kofinas, *Biomaterials*, 2006, **27**, 4165–4168.
98. O. Hayden, P. A. Lieberzeit, D. Blaas and F. L. Dickert, *Adv. Funct. Mater.*, 2006, **16**, 1269–1278.
99. F. L. Dickert and O. Hayden, *Anal. Chem.*, 2002, 74, 1302–1306.
100. M. He, C. He and N. Ding, *Virus Res.*, 2012, **163**, 374–379.
101. M. Snippe, R. Goldbach and R. Kormelink, *Adv. Virus Res.*, 2005, **65**, 63–120.
102. P. Colson, H. Richet, C. Desnues, F. Balique, V. Moal, J. Grob, P. Berbis, H. Lecoq, J. Harle, Y. Berland and D. Raoult, *PLoS One*, 2010, 5, e10041.
103. F. Yang, W. Liu, G. Zhang and F. Wan, *Journal of Environmental Entomology*, 2011, **33**, 241–249.

104. N. Sankarakumar and Y. W. Tong, *J. Mater. Chem. B*, 2013, **1**, 2031–2037.
105. A. Namvar and K. Warriner, *Biosens. Bioelectron.*, 2007, **22**, 2018–2024.
106. F. Dickert, O. Hayden, R. Bindeus, K. Mann, D. Blaas and E. Waigmann, *Anal. Bioanal. Chem.*, 2004, **378**, 1929–1934.
107. F. L. Dickert, O. Hayden, P. Lieberzeit, C. Haderspoeck, R. Bindeus, C. Palfinger and B. Wirl, *Synth. Met.*, 2003, **138**, 65–69.
108. B. B. Prasad and I. Pandey, *Sens. Actuators B*, 2013, **181**, 596–604.
109. D. R. Kryscio and N. A. Peppas, *Acta Biomater.*, 2012, **8**, 461–473.
110. J. O. Mahony, K. Nolan, M. R. Smyth and B. Mizaikoff, *Anal. Chim. Acta*, 2005, **534**, 31–39.
111. G. N. M. Ferreira, A. da-Silva and B. Tomé, *Trends Biotechnol.*, 2009, **27**, 689–697.
112. C. Steinem and A. Janshoff, in *Encyclopedia of Analytical Science*, ed. P. Worsfold, A. Townshend and C. Poole, Elsevier, Oxford, 2nd edn, 2005, pp. 269–276.
113. S. V. Aurobind, K. P. Amirthalingam and H. Gomathi, *Adv. Colloid Interface Sci.*, 2006, **121**, 1–7.
114. P. C. A. Jerónimo, A. N. Araújo, M. Conceição and B. S. M. Montenegro, *Talanta*, 2007, **72**, 13–27.
115. R. A. Lorenzo, A. M. Carro, C. Alvarez-Lorenzo and A. Concheiro, *Int. J. Mol. Sci.*, 2011, **12**, 4327–4347.
116. A. Ellwanger, C. Berggren, S. Bayoudh, C. Crecenzi, L. Karlsson, P. K. Owens, K. Ensing, P. Cormack, D. Sherrington and B. Sellergren, *Analyst*, 2001, **126**, 784–792.

Nanoparticle Technologies in Detection Science

NIAMH GILMARTIN[a] AND CAROL CREAN*[b]

[a] School of Biotechnology and Biomedical Diagnostics Institute, National Centre for Sensor Research, Dublin City University, Dublin 9, Ireland;
[b] Department of Chemistry, Faculty of Engineering and Physical Sciences, University of Surrey, Guildford, Surrey, GU2 7XH, UK
*Email: c.crean@surrey.ac.uk

4.1 Introduction

Improvements in analytical devices such as enhanced performance, smaller sample volumes and decreased power consumption have accelerated the growth in nanomaterials research. Nanostructured materials display unique properties not traditionally observed in their bulk counterparts. Enhanced electrical/optical properties and increased surface area render them suitable for applications such as nanoelectronics, photovoltaics and chemical/biological sensing.[1] It is possible to achieve enhanced sensitivity, an improved response time and smaller size by using nanomaterials and here we report some of the success that has been achieved in this area with respect to nanoparticle technologies in biosensor detection science.

As defined by a recent European Commission (2013) review,[2] nanomaterials are "a natural, incidental or manufactured material containing particles, in an unbound state or as an aggregate or as an agglomerate and where, for 50% or more of the particles in the number size distribution, one or more external dimensions is in the size range 1–100 nm". In comparison, single atoms are about a third of a nanometer in diameter, whereas common

RSC Detection Science Series No. 3
Advanced Synthetic Materials in Detection Science
Edited by Subrayal Reddy
© The Royal Society of Chemistry 2014
Published by the Royal Society of Chemistry, www.rsc.org

macroscale objects such as a sheet of paper is about 100 000 nanometres thick. Nanomaterials are larger than individual atoms/molecules but smaller than bulk materials, and therefore have characteristic properties that neither completely obey quantum nor classical physics. Their small size results in large surface-to-volume ratios and is also responsible for superior electronic and optical properties when compared with the bulk. They are of interest for sensing applications where the nanomaterial's large surface area facilitates interaction with an increased number of target molecules when compared to their bulk samples,[3–7] leading to the development of biosensors with enhanced sensitivities and improved response times.[8,9] This enhanced sensitivity is important for clinical diagnostics as the concentration of targets can be very low in biological samples. A good example of this is DNA sensors, which generally rely on the polymerase chain reaction (PCR) for signal amplification. By developing biosensors with improved sensitivity it will eliminate the need for PCR and thus simplify DNA biosensors.[8] Nanomaterials can be incorporated into many types of biosensor configurations to develop magnetic, optical, electrical or electrochemical biodevices for the detection of many biological molecules, including nucleic acids, antibodies, proteins, toxins and bacteria.[10–15]

4.1.1 Gold Nanoparticles

One of the most extensively investigated nanomaterials for sensing, detecting and imaging are gold nanoparticles (AuNPs) due to their unique tunable optical properties. Colloidal gold has been used to colour since the 4th or 5th century BC. More recently, it was the use of colloidal gold in electron microscopy immunostaining that increased its popularity in the 1970s. The biggest rise in popularity, however, can be attributed to advances in synthetic methods and characterization tools from the 1990s. Improvements in the control over their size and shape are due to the continuously evolving synthesis techniques of gold nanoparticles.[16] AuNPs are generally synthesized by reducing Au(III) precursors in the presence of stabilizing ligands, the most popular of which uses citrate as both reductant and stabilizer for $HAuCl_4$. The AuNP surface and core properties can be manipulated for sensing and imaging (see Figure 4.1); however, the reproducibility, reliability and long-term stability in manufacturing AuNP-based biosensing assays are challenges that need to be overcome.

The size confinement effect is responsible for the unique optical characteristics of AuNPs providing new electronic and optical properties. The strong vibrant colour of their colloidal solution is a distinct feature of AuNPs that is caused by the surface plasmon resonance (SPR) absorption.[17] SPR is the coherent oscillation of conduction band electrons induced by electromagnetic field interactions and this effect is absent in individual atoms and the bulk form.[18] Many research groups over the last two decades have studied the optical properties of AuNPs with various sizes and shapes. As predicted by theory, the SPR absorption peak red shifts to longer

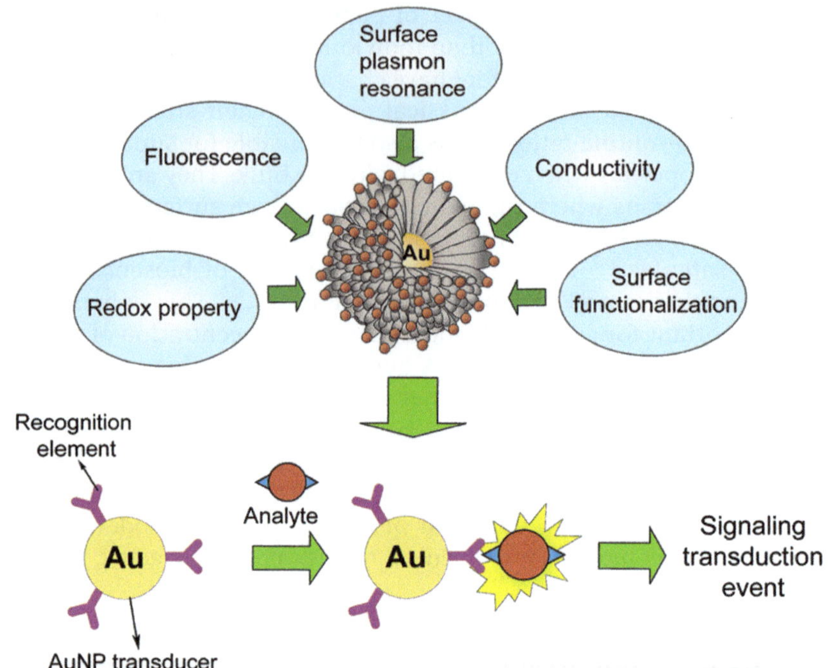

Figure 4.1 Examples of the physical properties of AuNPs and an illustration of a AuNP-based detection system.
(Reproduced from Saha *et al.*[16] with permission from the American Chemical Society.)

wavelengths with increasing particle size, with SPR absorptions also dependent on particle shape, surface modification, aggregate morphology, dielectric properties and refractive index of the surrounding medium.[18] The SPR peak can be tuned by changing the size and shape of AuNPs. AuNPs can also be used as probes for single-molecule surface-enhanced Raman scattering (SERS) detection, giving SERS enhancement factors on the order of 10^{15}.[19] Such optical enhancement may enhance the detection sensitivity of current AuNP-based biosensors. In addition to the optical properties of AuNPs, advantage can be taken of the electrical and redox activity of these NPs in biosensing, examples of which will be shown later in the chapter.

4.1.2 Carbon Nanotubes

In bionanotechnology, carbon nanotubes (CNTs) are one of the most studied carbon nanomaterials. The range of applications researched includes imaging, therapeutics, bioelectronics and sensing. CNTs, the allotrope of carbon discovered by Iijima in 1991,[20] consist of graphene sheets rolled into cylinders of sp^2 hybridized atoms (Figure 4.2). CNTs exist as single- (SWNTs), double- (DWNTs) and multi-walled (MWNTs) structures. The diameter for

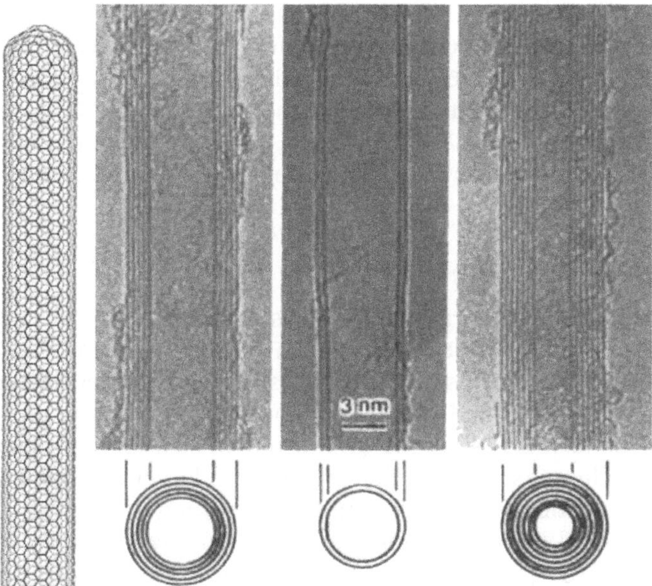

Figure 4.2   A SWNT can be visualized as a rolled-up sheet of graphite capped by half a C_{60} molecule (*left*). CNTs can also exist as DWNTs and MWNTs. TEM images reveal the number of walls present (*right* shows 5, 2 and 7 layers). (Reproduced from Iijima[20] with permission from Macmillan.)

SWNTs is usually less than 2 nm, whereas diameters for MWNTs range between 2–100 nm, depending on the number of shells present.[1] CNTs may be microns long and this combined with their narrow diameters leads to excellent material properties, such as a large surface area and high aspect ratio. SWNTs have a number of unique properties which include electrical, photoluminescence and Raman scattering. The tube helicity and diameter determine the electrical properties of a CNT, with approximately one-third of all tubes being metallic and two-thirds semiconducting.[21]

CNTs are generally synthesized using one of three techniques: chemical vapour deposition, electric arc discharge or laser ablation, all of which result in SWNTs with slightly varying properties.[21] At the moment, production of exactly one type of CNT is limited to the number of walls on the CNT, with some SWNT batches even containing DWNTs and MWNTs, among other types of nanostructured carbon. Exact production of a single type of chiral or semiconducting SWNT, without contaminants, is unfortunately not yet possible and considerable batch to batch variation is also common. Post-synthetic methods have been attempted to separate CNT mixtures. SWNTs have been separated by diameter and chirality using density gradient ultracentrifugation,[22] ion exchange chromatography,[23] gel chromatography[24] and conjugated polymer extraction.[25] Such methods can lead to SWNT fractions that are 99% enriched with either metallic or semi-conducting SWNTs within a narrow diameter range. While such separation techniques appear promising, the majority of current CNT research focuses on working with CNT mixtures since no definitive large-scale purification of a single CNT type exists. Chemical modification of CNTs can lead to disruption of the native structure with a concurrent change in properties. The debate around toxicity of CNTs continues, despite their potential in a variety of biomedical applications and a large degree of research to that end.

4.1.3   Quantum Dots

Quantum dots (QDs), inorganic nanocrystals with a size of approximately 1–10 nm, are composed of III–V (GaN, GaP, GaAs, InP or InAs) or II–VI (ZnO, ZnS, CdS, CdSe or CdTe) semiconductors, with the latter category dominating. Similar to AuNPs and CNTs, their small dimensions result in quantum mechanical behaviour, which for QDs results in size-tunable absorption and emission wavelengths (the smaller the QD, the larger the band gap energy, and thus the smaller the emission and absorption wavelengths compared to the same material in bulk) (Figure 4.3). High-quality monodisperse QDs are synthesized by the pyrolysis of organometallic precursors in organic solutions containing hydrophobic coordinating ligands. Aqueous solutions will yield hydrophilic QDs, but these NPs do not have the quality of those made from hot organic solvents.

QD colour tunability and narrow symmetric emission bands (*e.g.*, ~20–40 nm for CdSe) makes them ideal for multiplexed optical sensing. Multiplexing allows several different parameters to be assessed with just one measurement, which is important in bioapplications and will be discussed later in the chapter. QDs have properties of very high photostability and brightness in addition to large extinction coefficients over a wide wavelength range, which allows excitation of different QDs by a single-wavelength

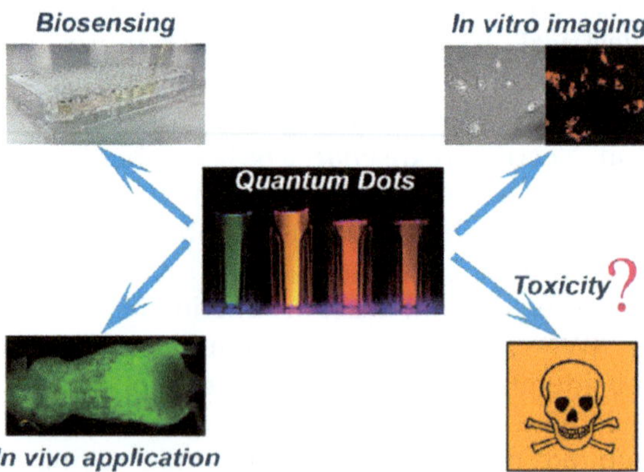

Figure 4.3   Applications and concerns of quantum dots; *middle image* shows the colour tunability of these NPs.
(Reproduced from Wang *et al.*[26] with permission from the American Chemical Society.)

excitation source.[27] QDs have many advantages over common fluorophores (such as narrow emission band and lack of photobleaching); therefore their use in biological applications has increased greatly over the last 10 years.[28]

4.2 Bioconjugation

Preparing NP bioconjugates is not trivial and there are many challenges associated with it, including the wide variability in their structure, meaning there is no one strategy to fit all NP modifications. Some NPs are uniform and consist of only one type of material, for example AuNPs, while others have core/shell structures (*e.g.* QDs) or are binary alloys. Most metal, semi-conductor and carbon NPs are hydrophobic and therefore require surface modification to become hydrophilic and biocompatible. This may often comprise attaching hydrophilic ligands, which should also provide chemical functional groups that can act as sites for subsequent bioconjugation. This chemistry has important implications for NP colloidal stability and ligand–NP coverage is often highly heterogeneous.[34]

Biomolecules can be conjugated to the surface of NPs directly or by using cross-linking molecules. Bioconjugation usually adds biorecognition sites (*e.g.*, antibodies, aptamers) or bioderived activities such as catalysis (*e.g.*, enzymes, DNAzymes). Depending on the NP or biomolecule in question, there can be large differences in size with respect to each other. NPs may display size heterogeneity or polydispersity and can therefore have a wide range of surface areas and volumes, leading to different surface loadings of the biocomponent. The surface of inorganic NPs does not always support the formation of new covalent bonds and therefore limits the choice of available conjugate chemistries to those that build from coordinated ligands. Effects from nanoscale curvature or differences in reactivity between carbon nano-tube tips and sidewalls further complicate conjugation. Proteins have numerous functional groups available for coupling (*e.g.*, carboxyls, amines), which can result in a distribution of protein orientations on the NP surface and mixed protein activity. The NP, its interfacial chemistry and the properties of the biomolecule all contribute to the interactions during conjugation and must be considered when dealing with the choice of conjugation chemistry.[34] Many factors determine the conjugation strategies available, including NP size, shape, material, surface chemistry and available functional groups, and also the biomolecule size and chemical composition. Conjugation strategies are generally either covalent (coupling directly to the NP surface or to a surface ligand) or non-covalent in nature (including electrostatic attachment, other forms of adsorption and encapsulation) (see Figure 4.4).

4.2.1 Covalent and Dative Chemistry

Directly attaching biomolecules to a metallic NP surface is often achieved by dative bonds, while attachment to non-metallic NPs or ligands on the NP surface is realized by covalent bonds. Dative bonds form when two electrons

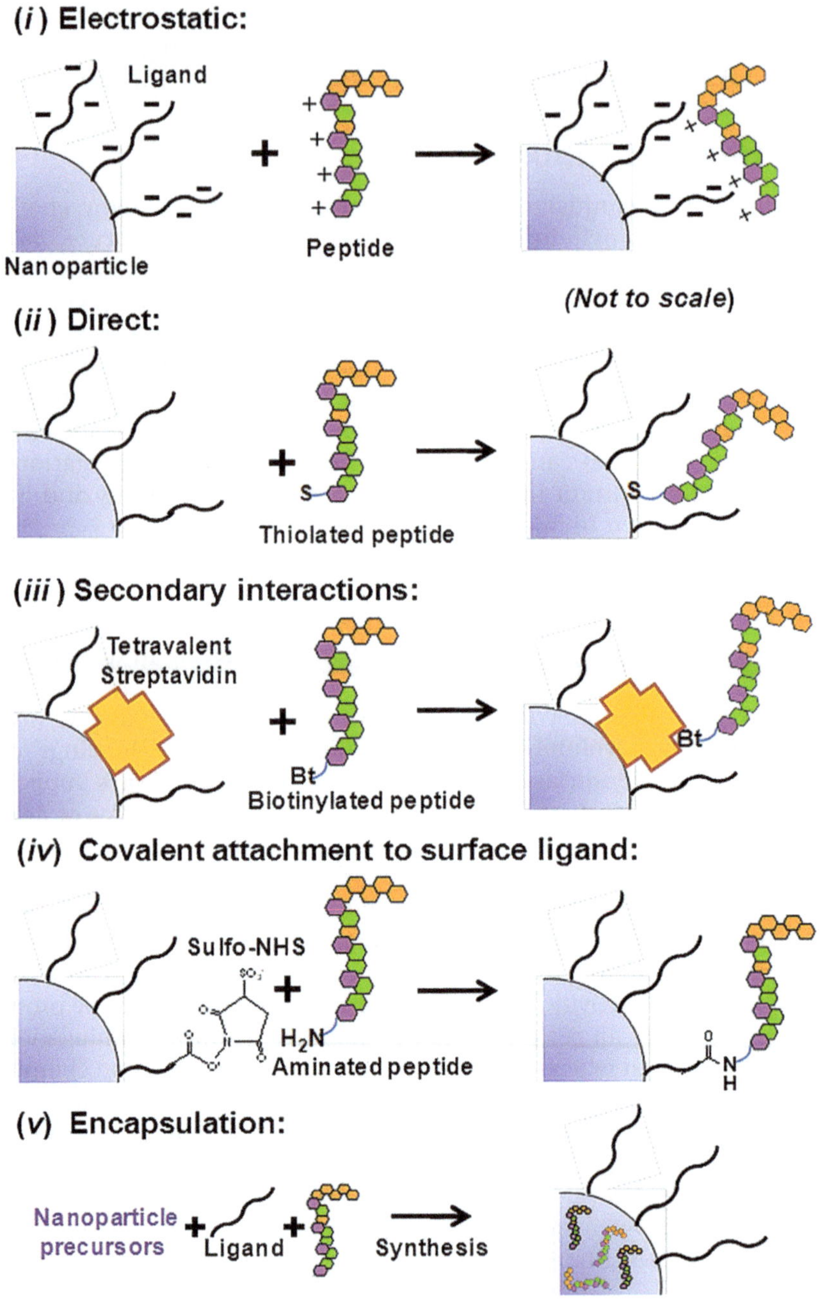

Figure 4.4 Five typical methods of bioconjugation to NPs using a peptide. (Reproduced from Sapsford *et al.*[34] with permission from the American Chemical Society.)

in the bond originate from a single atom (the NP metal surface) and differ from covalent bonds by longer bond lengths and greater polarity. Examples of dative bonding include metal ion chelation and Au–thiol chemisorption, where the thiol sulfur atom contributes a lone pair of electrons to the empty orbitals of gold atoms at an interface.[29] Dative bonds are not as strong as true covalent linkages and are affected by changes in pH, oxidation and displacement by other similar molecules.[29] Dative bonds can be strengthened, however, by increasing the number of interactions, *e.g.* the use of multidentate thiols increases the overall attachment strength of a ligand to both Au and semiconductor NPs or surfaces.[29] While not considered strictly as covalent bonds, certain electron-donating amino acids, particularly histidine, can form stable coordinate complexes with transition metals such as Ni(II), Cu(II), Zn(II) and Co(II). Covalent linkages also involve cross-linkers acting as a bridge between the NP and the biomolecule.

4.2.2 Non-covalent Attachment

Most non-covalent attachment involves weak coordination bonding, with bond stability dictated by equilibrium dissociation constants. Multidentate interactions can improve the stability of these NP bioconjugates and are often referred to as self-assembly since the chemistry frequently requires only stoichiometric mixing of both constituents. This approach is favoured because it offers facile attachment without the need for additional reagents.[30] Electrostatic attachment between oppositely charged NPs and biomolecules is the simplest and most widely used non-covalent approach. Complications with this method can include ionic strength of the surrounding medium, the concentration of reagents and the type and magnitude of charge of components.[31] Functionalization, for example of the strong negative charge of most nucleic acids (due to the phosphate backbone), takes place with positively charged NPs. Engineering NPs and proteins to contain charges can also permit attachment. For example, a maltose binding protein dimer was engineered to display a positively charged leucine region, allowing self-assembly to QDs coated with negatively charged ligands, which oriented the protein binding pocket away from the NP surface.[32] Hydrophobic–hydrophobic, hydrophobic–hydrophilic or host–guest interactions are other types of non-covalent attachment strategies.[33] Secondary interactions between functional groups on the NP surface and other moieties also facilitate non-covalent interaction. The well-known biotin–avidin interaction can be used for attachment if both the NP and biomolecule are functionalized with one of the pair in addition to other enzyme–substrate or receptor–ligand interactions.

4.2.3 Commonly Utilized Chemistries: Functional Groups and Conjugation Reactions

The chemical handles used for conjugation on proteins and peptides include the ε-amino group (on lysine side chains), N-terminal primary amines, the

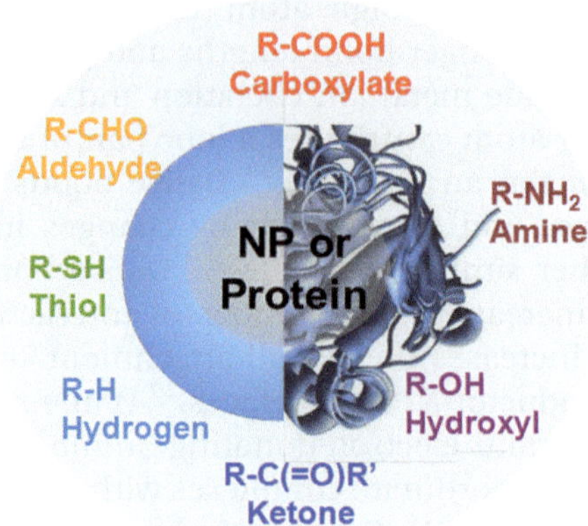

Figure 4.5   Common NP and protein functional groups that can be used for conjugation.
(Reproduced from Sapsford *et al.*[34] with permission from the American Chemical Society.)

guanidinium group (arginine side chains), the cysteine thiol group, carboxyl groups on aspartic acid or glutamic acid, or the C-terminus and the phenol on tyrosine (Figure 4.5).[29] Alternative reaction sites such as hydroxyl groups or aldehydes are found on carbohydrates and glycosylated proteins. Ribose sugars and phosphates are available for functionalization on nucleic acids. Modifying amines with *N*-hydroxysuccinimidyl (NHS) esters, carbodiimide mediation (EDC)[29] of carboxyls with amines, maleimide attachment to thiols, and diazonium modification of tyrosine phenols are the most common chemistries used on these groups. Recombinant modification of proteins allows cysteine residues (*i.e.*, thiols) to be engineered for site-specific functionalization. In addition, protein N-termini can be modified using N-terminal transamination and inserting polyhistidine residues into proteins allows interactions with nitrilotriacetic acid (NTA)-modified surfaces. One of the most comprehensive guides to bioconjugation chemistries is by Hermanson.[29]

4.2.4   Bioconjugation Strategies for AuNPs, QDs and CNTs

4.2.4.1   AuNPs

The bioconjugation chemistries of AuNPs are limited, despite their being one of the most studied and widely applied NP material to date. This chemistry is mainly driven by the NP surface, the stabilizing ligand or some functional intermediary.[34] Similar to most NP materials, one strategy for the bioconjugation of AuNPs is to add specific chemical handles by coating with ligands or polymer. This is invariably achieved by forming an intermediate monolayer of bifunctional thiol ligands on the surface of the AuNP.[34] Monolayers of organosulfur compounds such as alkanethiols, dialkyl disulfides, dialkyl sulfides, alkyl xanthates and dialkyl thiocarbamates spontaneously form on gold surfaces, with alkanethiols and dialkyl disulfides the most commonly used. Bifunctional linkers allow further chemical modification, *e.g.* thioalkyl acids or amines, with standard carbodiimide or maleimide coupling. Thiol-terminated biomolecules can also be directly bound to the AuNPs. One of the most well-known examples of this comes from the Mirkin group, where polyvalent AuNP–oligonucleotide conjugates were developed and which have been recently reviewed.[35] A third and common strategy is the adsorption of biomolecules on hydrophilic NPs. Depending on the biomolecule, different degrees of binding affinity can be found due to varying numbers of inherent amine, carboxyl, hydroxyl, imidazole, phosphate and thiol groups. This approach is commonly used to prepare probes for SERS-based assays. Protein A or streptavidin can be adsorbed onto AuNPs, allowing simple modification for many antibodies and biotinylated biomolecule conjugates.

4.2.4.2 QDs

As-prepared QDs usually have hydrophobic surface ligands. While hydrophilic versions can be prepared, they often suffer from a lower quality than their hydrophobic counterparts. QDs are now commercially available with a wide variety of surface functional groups present, *e.g.* amines, carboxyls, biotin, avidin and other proteins for subsequent functionalization. For QDs the most popular chemistry used for bioconjugation includes: (1) covalent modification of the QD surface with target functional groups on biomolecules, (2) electrostatic interactions between charged QDs and oppositely charged biomolecules, (3) dative bonding between the QD surface and biomolecules with available thiols or polyhistidines, and (4) QDs functionalized with streptavidin and biomolecules labelled with biotin or *vice versa*. As for AuNPs, there are well-developed standard chemistries available for all of these subclasses, with advantages and disadvantages to each as discussed previously.[34]

4.2.4.3 CNTs

CNT sample purification often starts by removing catalyst particles using oxidative methods, which introduce carbonyl and carboxylic acid groups on the open ends of the CNTs and at defect sites along the CNT sidewalls.[36] This has become one of the favoured routes of covalently attaching biomolecules to CNTs. The proliferation of amino functionalities on proteins, enzymes and antibodies, among other biomolecules, allows for facile amide functionalization with CNT carboxylates. A wide variety of biomolecules such as carbohydrate, oligonucleotides, proteins, enzymes and even DNA have been attached to CNTs in this fashion.[37] However, such functionalization can be difficult to reproduce since the extent of functionalization is dependent on the degree and type of nanotube carboxylation, which in turn varies according to CNT source. Covalent modification to the CNT sidewalls can disrupt the CNT structure and therefore properties.

Physical adsorption (non-covalent attachment) has been used to attach a variety of moieties to CNTs. Unfortunately, proteins and antibodies, in particular, may lose their biological activity when adsorbed on a CNT surface. This can be due to a change in conformation when binding with the CNT and/or unfavourable orientation of the active site. The interaction of biomolecules with CNTs has been of particular interest with a view to their use as biosensors or improving biocompatibility. Non-covalent binding of streptavidin to CNTs has been achieved *via* covalent attachment to linkers that are adsorbed along the CNT axis.[38] Pyrene-based molecules interact with CNT sidewalls *via* π–π stacking and can be covalently functionalized with biomolecules to afford conjugation. DNA has been shown to strongly interact with CNTs, forming uniform coatings.[39] The wrapping of CNTs has recently been extended to other biopolymers, including chitosan, chondroitin sulfate and hyaluronic acid.[40] Biomolecules of interest, including antibodies, may subsequently be anchored to these biopolymers wrapped around the CNTs.[41]

4.3 Nanoparticles Technologies in Detection Science

4.3.1 Optical Detection

4.3.1.1 AuNPs

One of the first uses of nanoparticles in biosensing was that of colloidal gold in lateral flow immunoassays, also known as lateral flow devices (LFD), which are one of the most important products of the diagnostics industry (Figure 4.6). The first commercial application of LFDs was Unipath's Clearview home pregnancy test in 1985. Most current LFDs are rapid membrane-based immunoassays that depend on the recognition properties of antibodies and that can be seen by eye. LFDs most commonly use colloidal gold and monodisperse latex as detection reagents. The majority of LFDs are based on antibodies attached to AuNPs by a method first reported in 1979,[42] while few reports exist of LFDs based on oligonucleotide-

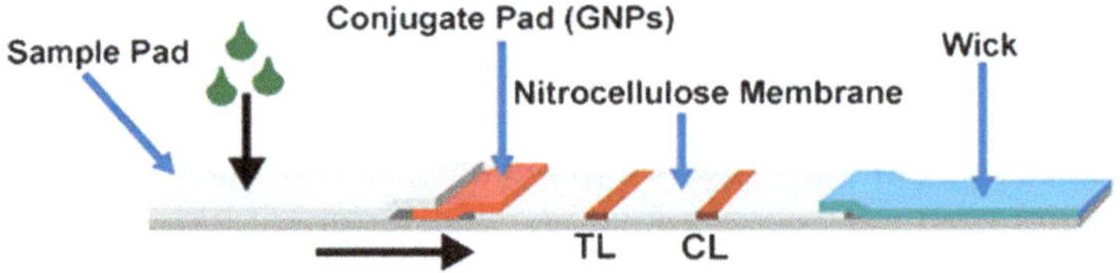

Figure 4.6 Schematic of a lateral flow device showing the main components: sample pad, conjugate pad, nitrocellulose membrane and wick; TL = test line and CL = control line.[46]

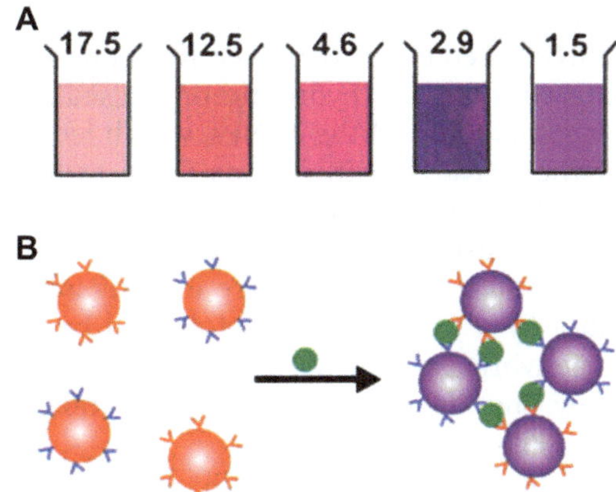

Figure 4.7 (A) Colour change of 15 nm AuNPs upon aggregation showing how the inter-particle distance decreases. (B) Schematic of distance-dependent sandwich assay for high molecular weight polyvalent antigen (*green circles*), leading to aggregation of AuNPs and a red shift in their extinction spectrum.[46]

functionalized AuNPs (functionalized by the recent method of mixing alka-nethiol-terminated oligonucleotides and citrate-capped AuNPs).[43–45]

Prostate-specific antigen,[47] ricin[48] and *Staphylococcus aureus*[49] are examples of high molecular weight molecules detectable by LFDs. The membranes in these immunoassays are striped with a test line of capture antibodies and a conjugate pad containing AuNP-labelled detection antibodies. When a LFD is inserted into a sample, AuNP-labelled antibodies migrate towards the test line antibodies, binding target molecules/antigens along the way. At the test line, target molecules bound to AuNPs are sandwiched between them and the capture antibodies, leading to a visible increase in colour (Figure 4.7). Typical run times are less than 10 min, with reported sensitivities between nanomolar and picomolar depending on the assay type.[46]

In 1996, oligonucleotide-modified nanoparticles were used to generate unusual optical properties. In that assay, 13 nm AuNPs were used, which aggregated in the presence of the DNA target analyte, changing the colour of the solution from red to blue.[50] This colour change was due to the particle surface plasmons interacting and scattering properties of the aggregates. This observation sparked immense interest in exploring the possibilities for nanomaterial use in bio-diagnostic applications. Such properties were not observed with microparticle probes of the same material, partly due to the lower loading of oligonucleotides. More recently, citrate-modified AuNPs were used with short single-strand DNA oligomers to detect sequences in PCR-amplified genomic DNA.[51,52] Again, complementary binding resulted in particle aggregation and a colour change, whereas if the oligomers were not complementary, no binding and therefore no colour change occurred, indicating the absence of target DNA.

Although many AuNP-based biodetection methods have been developed, the sensitivity of AuNP-based colorimetric assays is not impressive (~ 1 nM). To increase the detection signal, hydroquinone-based Ag(I) reduction was used to form a silver shell around the AuNPs, which led to a 105-fold signal amplification over the AuNP (sensitivity was down to 50 fM level, a nearly 100-fold improvement in sensitivity over traditional fluorescence-based methods). Mirkin *et al.* used magnetic and AuNPs to detect DNA with a detection limit of 10 fM.[53] DNA probes were initially hybridized with target DNA, then separated by a magnetic field, followed by washing to remove unbound AuNPs. Dehybridized AuNPs were silver stained, which allowed optical detection by the naked eye, absorbance measurement or light scattering measurement.

Surface-enhanced Raman scattering (SERS) greatly amplifies optical signals and therefore is one of the most sensitive diagnostic approaches available. Mirkin *et al.* describe detecting nucleic acids with SERS, where a Raman-active dye, oligonucleotide-modified AuNPs and a silver staining method were used. After hybridization of the target strand, silver staining on target-labelled AuNPs amplified the Raman signals compared to dyes, resulting in a detection limit as low as 20 fM. By using different Raman dyes for different targets, multiplexing was achieved.[54] Extending this system to allow protein detection enabled detection limits of 30 fM to be achieved in human serum.[55]

Surface plasmon resonance (SPR) is a surface-sensitive analytical technique that can detect dielectric constant changes induced by molecular adsorption to a thin metal film (Figure 4.8). Improvements in sensitivity were achieved when nanoparticles were introduced into established DNA assays using SPR. Real-time DNA hybridization detection limits were lowered

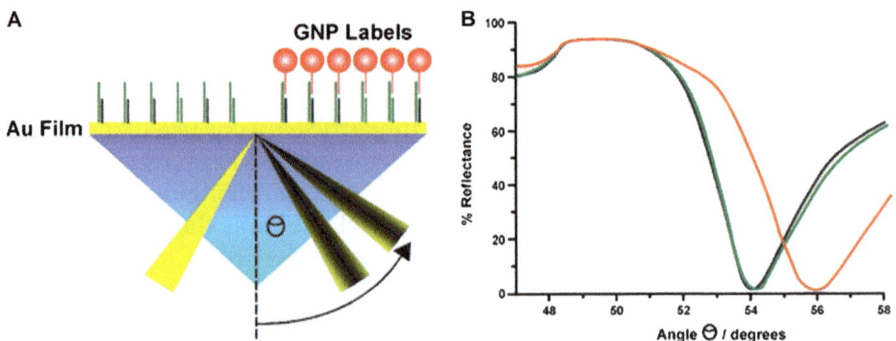

Figure 4.8 (A) Schematic of a SPR biosensor on a thin film of gold at the interface between a prism and the sample. When low molecular weight (MW) molecules bind to the surface, they produce very small changes in the deflection angle (difference between the black and green lines in B). Much larger changes in the deflection angle are produced when low MW target molecules are sandwiched between the surface and AuNPs (difference between the green and red lines in B).[46]

from ~150 nM to less than 10 pM target concentration between surface capture strands and oligonucleotide-functionalized AuNPs.[56] Further improvements were made to DNA detection by immobilizing captured oligonucleotide strands onto chemoresponsive diffraction gratings followed by capture of target DNA, which were detected with oligonucleotide-functionalized AuNPs. This resulted in femtomolar concentration detection limits.[57] By controlling the wavelength *via* planar surface plasmons on gratings, coupled to localized surface plasmons of nanoparticles, Wark *et al.* improved the sensitivity of diffraction grating systems to 10 fM target concentration.[58]

4.3.1.2   QD Bioassays

Highly fluorescent QDs can also lead to the development of sensitive bioassays. Fundamental drawbacks of traditional fluorophore-based assays include broad absorption and emission profiles, which reduce the possibility of multiplexing (which is discussed further later). Quantum dots, with their very narrow emission bands, may overcome this. Hahn *et al.* used streptavidin-conjugated QDs to detect a single *E. coli* O157:H7 bacterial pathogen, showing sensitivity improvements two orders-of-magnitude greater than conventional fluorescent dyes.[59] *E. coli* cells were labelled with biotinylated anti-*E. coli* O157:H7 antibodies which could complex to the streptavidin-conjugated CdSe/ZnS core/shell QDs. QDs were also used by Alivisatos and co-workers to detect single base-pair mutations in DNA in chip-based assays.[60] They could detect complementary DNA at 2 nM concentrations in the presence of background oligonucleotides with sequence mismatches. This study used QDs made from toxic CdSe, which will limit its use in the medical diagnostic arena until a safer alternative is found. Nanoparticle toxicity will be further discussed later in the chapter.

4.3.1.3   CNT Bioassays

Individual semiconducting SWNTs exhibit photoluminescence, with discrete bands in the near-IR region between 900 and 1600 nm. The sharp nanotube fluorescence spectra can be detected in a complex biological environment due to the low absorption of blood and tissue in this region. SWNTs were used to selectively probe cell-surface receptors as near-IR fluorescent tags for cell imaging.[61] The nanotubes were non-covalently functionalized with amine groups using the surfactant PL-PEG-NH$_2$, followed by conjugation with antibodies recognizing both the CD20 cell surface receptor (Rituxan) and the HER2/neu receptor on certain breast cancer cells (Herceptin). *In vitro* near-IR fluorescence imaging showed specific binding of the antibody-conjugated SWNTs to the host cells, with high specificity for the different antibodies (55:1 and 20:1 for host cells:non-host cells).

Barone *et al.* linked enzyme reactions to CNT fluorescence, creating a fluorescent enzymatic sensor.[62] The authors non-covalently functionalized

SWNTs with glucose oxidase (GOx) and potassium ferricyanide, with the latter quenching the SWNT fluorescence. When glucose was added to GOx–SWNTs, the ferricyanide ions left the surface of the CNT, yielding a recovery of the fluorescence. The authors related the CNT fluorescence to the glucose concentration and maintain that this type of sensor, enveloped in a small dialysis capillary, could be implanted in the body. The capillary could allow glucose to diffuse in, easily allowing sugar levels to be measured. This research demonstrates the feasibility of using CNT sensor systems in implantable biomedical sensors, assuming a lack of CNT toxicity.

4.3.2 Electrochemical Detection

The oldest and most commonly used transducers in biosensors are electrochemical. Electrical detection methods are appealing because of their low cost, low power consumption, ease of miniaturization and potential multiplexing capability.[63,64] Electrochemical sensors can be based on potentiometry, amperometry, voltammetry, coulometry, ac conductivity or capacitance measurements.[65] One of the advantages of using electrochemical detection is the portability and therefore variety of point-of-care environments available with these systems. A portable sandwich assay using handheld electrical detection was developed for DNA measurements using AuNPs combined with subsequent silver amplification.[66] This system used the gap between electrodes as the sensing surface. If oligonucleotides were immobilized in the gap, nanoparticle-labelled target strands would alter the electrical current or resistance between the electrodes. Un-optimized, this system had a detection limit of 500 fM and an impressive selectivity factor of 10 000 : 1. Such a simple device could potentially reduce the need for temperature control on chips, thereby greatly diminishing assay complexity.

A quartz crystal microbalance (QCM) is a mass-sensitive technique based on the piezoelectric response of materials. By measuring the change in frequency of a resonating quartz crystal, the mass per unit area of material deposited on the crystal surface can be attained.[67] The resonance is disturbed by the addition or removal of a small mass due to oxide growth/decay or film deposition at the surface of the acoustic resonator. AuNPs were used to amplify the signal from DNA on a QCM. In a direct assay, DNA capture strands were immobilized onto gold-coated piezoelectric crystals followed by hybridization to one half of the target strand. Subsequent complementary hybridization to an AuNP-labelled second half of the target strand greatly enhanced the output signal due to the weight of the AuNP on the quartz crystal sensing surface. To improve the signal further, $AuCl_4^-$ in the presence of hydroxylamine was electrochemically reduced onto the AuNP, resulting in a structural increase of the AuNPs and a detection limit of 1 fM target DNA concentrations.[68]

A popular approach in electrochemical assays uses electrochemical characteristics of nanomaterials and stripping voltammetry. A sandwich assay is carried out where target analytes are captured between two antibodies or

DNA. One target is generally labelled with signalling nanomaterials such as CNTs, AuNPs or QDs, enabling amplified detection with anodic stripping. Stripping voltammetry is a method for quantitative determination of specific metallic ion species and other electrochemical analytes. Large signal amplification can be achieved using stripping voltammetry detection due to the preconcentration step of the desired analytes on the surface of the electrode by electrodeposition or adsorption.

An electrochemical immunoassay for IgG using AuNP labels was reported where the target analyte was captured by an antibody immobilized on a microwell plate, followed by labelling with another antibody modified with AuNPs.[69] Under acidic conditions the AuNPs were dissolved by oxidation followed by anodic stripping voltammetry to determine the concentration of the Au(III) ion. A large number (1.7×10^5) of Au(III) ions were liberated, which amplified the electrochemical signal and lowered the detection limit for IgG to 3 pM. A similar approach used an anti-thrombin aptamers–adenine–AuNP complex where adenine nucleobases were released from the AuNPs after addition of sulfuric acid or nuclease. Differential pulse voltammetry in this case detected the purine nucleobases, with a detection limit of 2.8 pM due to the large numbers of aptamers and adenines per AuNP.[70]

In a different approach, the large surface area of CNTs was used to immobilize numerous enzymes and antibodies, resulting in greatly enhanced detection.[71,72] Protein targets and DNA target strands were captured by antibodies or DNA, respectively, which were immobilized on magnetic beads and CNTs, respectively. Approximately 9600 alkaline phosphatase molecules were immobilized on the CNTs, which were detected by a separate CNT-modified electrode. Since the products liberated by enzymatic reaction were detected at the CNT-modified electrode, the signals were greatly enhanced compared with those of a bare electrode, leading to detection limits of 3 fM for the IgG target and 54 aM for target DNA. A tetrapeptide (RGDS) was conjugated to SWNTs and used to capture BGC cells by specific recognition between the RGDS and integrin receptor on an electrode. Double amplification by SWNTs on the electrode and HRP-catalyzed H_2O_2 reduction allowed this electrochemical sensor to detect BGC cells with a sensitivity of 620 cells mL^{-1}.[73]

To boost the detection sensitivity of PSA (prostate-specific antigen, a biomarker for prostate cancer) in serum, an amplification step was incorporated by combining SWNT forest immunosensors with HRP–MWNT–Ab2 bioconjugates. The secondary antibody (Ab2) and HRP tag were covalently linked to MWNTs at high ratios of 1:200.[74] This amplification strategy improved the detection limit 100-fold to 4 pg mL^{-1} and the sensitivity by 800-fold, compared to conventional ELISA. These results highlight the excellent promise CNTs show in ultrasensitive immunoassay research in proteomics and systems biology.

Yang *et al.* reported an immunoassay for PSA and mouse IgG based on AuNPs.[75] Antibodies were immobilized on an ITO electrode, followed by antigen binding and finally detection with AuNP-labelled antibodies. The detection limit attained in buffer was 7 aM for IgG and 30 aM for PSA. Another sensitive electrochemical assay used the electrocatalytic reduction of hydrogen peroxide by hydrazine based on a conducting polymer and AuNPs. Dendrimers were attached to carboxylic acid-functionalized polythiophenes. Amines from the dendrimer facilitated interaction with multiple AuNPs (allowing amplification), which had been functionalized with either DNA capture probes or anti-IgG capture antibodies. Following incubation with the target DNA probe or antigen of interest (IgG), biotin-linked DNA detection probes (or anti-IgG detection antibodies) were attached to the system. Avidin-linked hydrazine was then used as the reporter system, which allowed the electrocatalytic reduction of hydrogen peroxide, detected by differential pulse voltammetry. The detection limits for human IgG and DNA were 1 fM and 50 aM, respectively.[76]

A thionine-doped magnetic AuNP-based immunoassay for carcinoembryonic antigen (CEA) was reported with electrochemical detection from a carbon fibre microelectrode.[77] The electrode was modified with AuNP, protein A and anti-CEA. The thionine-doped magnetic AuNPs were functionalized with anti-CEA as a secondary antibody and the HRP enzyme for signal detection. Without H_2O_2, the detection limit was 2.8 pM and improved to 28 fM with H_2O_2.

A low signal-to-noise ratio, low reproducibility, problems in target detection in serum and weak multiplexing capability still need to be addressed for wide applications of electrochemical detection platforms, despite research showing impressive detection limits and portability.

4.4   Multiplexed Immunoassays

Multiplexing strategies are involved in more recently developed techniques. An array of dispersible nanodisks prepared by on-wire lithography for a new encoding system is reported by Mirkin *et al.* These structures were functionalized with Raman-active chromophores and were used for DNA detection in a multiplexed format at a target concentration as low as 100 fM.[78] This study shows how, with further optimization, these finely controlled nanoparticle structures or isolated nanoparticle dimers could be promising candidates as highly sensitive, multiplexible SERS-active structures for biosensing applications.

A multiplexed detection system is required to detect different kinds of metal ions simultaneously. Redox-active nanoparticles are particularly interesting for electrochemical detection, especially when stripping analysis is used for quantitation. Assays have been developed whereby the NP is stripped from a DNA–NP conjugate to give quantitative DNA analysis. The format used was a sandwich assay, containing a magnetic bead at the centre with target capture strands attached (Figure 4.9). The target DNA is then hybridized with the magnetic bead, followed by labelling with NP-functionalized oligonucleotides. The NP labels, which are generally inorganic, are then dissolved, allowing electrochemical detection. Depending on their

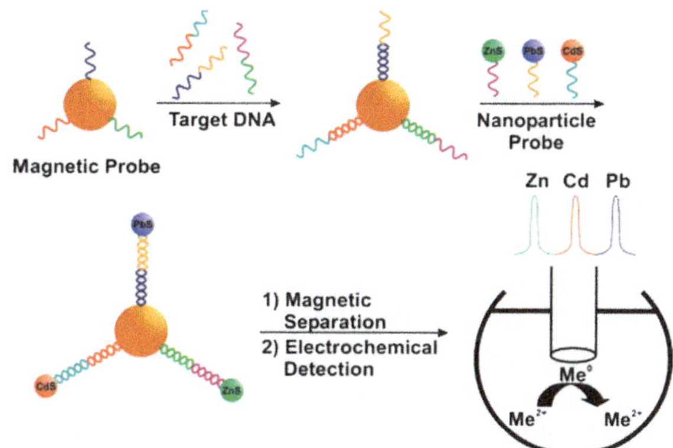

Figure 4.9 Multiplexing strategy where magnetic microparticles (*orange spheres*) are labelled with DNA capture strands that can bind target DNA. Subsequent incubation with oligonucleotide-functionalized QD labels (*small green, blue and red spheres*) with different electrochemical properties can be used to code for the specific target DNA of interest. (Reproduced from Rosi and Mirkin[19] with permission from the American Chemical Society.)

composition, different NPs yield different voltammetric detection, which means the NPs can code for the target strand of interest, allowing multiplexing to be carried out.[79,80] The optimized limit of detection of this system was 270 pM, which is not as high as other methods [namely electrical detection with nanowires and nanoparticles or resonant light scattering coupled with metal nanoparticles (a more comprehensive list can be found in Rosi and Mirkin[19])] and would require target amplification with PCR for enhanced detection.

The Wang group also developed a multianalyte electrochemical biosensor, which used different kinds of QD-labelled proteins and aptamers immobilized on a Au surface.[80] Surface-modified aptamers initially captured QD-labelled proteins. The latter were then displaced by unmodified target proteins because unmodified proteins have a higher binding affinity to aptamers than QD-tagged proteins. The liberated QDs were ultimately dissolved and formed ions, which were detected using stripping voltammetry and a glassy carbon electrode. Each displacing event released a large number of QD labels, which had the result of lowering the detection limit for the thrombin target to 0.5 pM.

4.5 Nanotoxicology

Currently there is increased interest in the potential use of nanomaterials for drug delivery, due to the potential of controlled drug release, increased targeting of cells, drug solubility and circulation time achievable using nanomaterials.[81] With this increased interest comes concerns with the unwanted toxic effects and the formulation of dedicated legislation for the regulation of nanoparticles is on-going within the EU, with final assessment on a review of occupational health and safety legislation with respect to nanomaterials expected in 2014.[2] Nanomaterials can enter the bloodstream following inhalation, ingestion, injection or dermal exposure. Once in the bloodstream, nanomaterials rapidly circulate within the body and are taken up in organs and tissues,[82] resulting in a number of potentially toxic interactions as shown in Figure 4.10.[83]

The major toxicity concern is the leaching of toxic ions, as some nanomaterials undergo a degree of solubility in biological fluids. The dissolution of ZnO nanoparticles, for example, releases toxic zinc ions, resulting in interference with mitrochondria, inflammation and induction of cytotoxicity.[84] Limbach and colleagues hypothesized a Trojan horse-type mechanism where the size and shape of the nanoparticles mean that they can easily pass into a cell and release toxic ions.[85] Cho *et al.* demonstrated that nanoparticles with high solubility release toxic ions during breakdown in the phagolysosome, resulting in inflammation.[86] However, the zeta potential of insoluble nanoparticles in the phagolysosome may also disrupt the membrane and cause inflammation.

In addition to toxic ion leaching, nanomaterials have oxidative potential and cause their effects *via* the final common pathway of oxidative stress. Natural antioxidant defences can neutralize the production of reactive oxygen stress at low levels; however, occupational levels of nanoparticle exposure can result in these systems being overwhelmed and could result in DNA damage and loss of metabolic function.[87–89] The effect on the lungs of

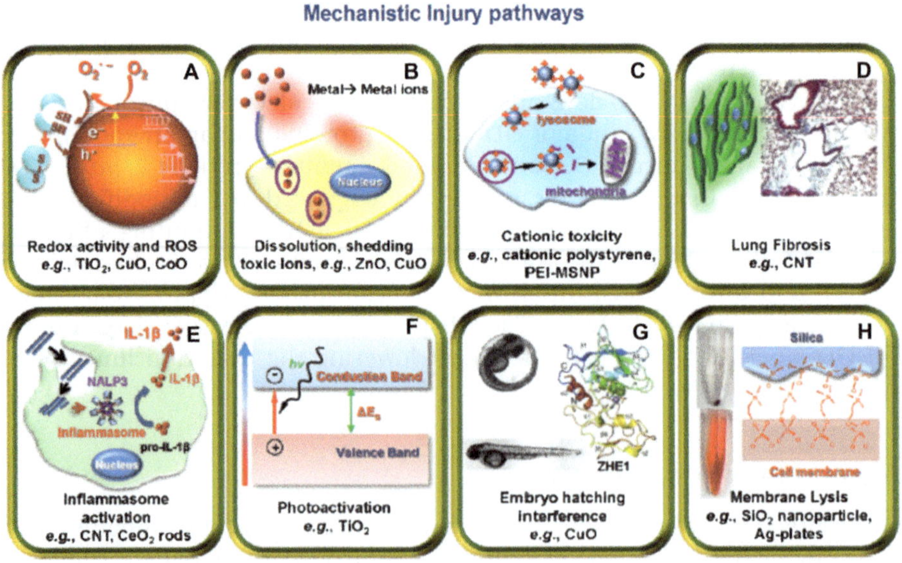

Figure 4.10 Possible biological tissue injury pathways due to nanomaterials. (Reproduced from Nel *et al.*[83] with permission from the American Chemical Society.)

high aspect ratio particles is well known as a result of asbestos. It is unsurprising therefore that concerns in relation to the high aspect ratio of nanomaterials, in particular CNTs, have been raised. Incomplete clearing of CNTs from lung tissues and incomplete phagocytosis due to the size of the fibres was demonstrated to be pro-inflammatory and may result in mesothelioma.[90,91] Recently, 4 μm in length was established as the threshold length beyond which fibres are pathogenic to lung tissue.[92]

Clearly, the nanotoxicology of a particular nanomaterial can vary and a clear understanding of the interactions at the nano/bio interface is critical to ensure their safe use. To enable the assessment of the nanomaterial hazard, high throughput screening approaches such as that developed by Nel *et al.*[83] are necessary. The high-throughput *in vitro* screening developed by Nel *et al.* makes predictions about the physicochemical properties of nanomaterials, thus informing pathology or disease outcomes *in vivo*, which helps to establish structure–activity relationships that allow hazard ranking. Increasing the information available on the hazards associated with nanomaterials is essential to inform the regulators to legislate, as the full potential of nanomaterials will not be realized without allaying the concerns in relation to toxicity.

4.5.1 Beneficial Effects of Nanotoxicology

By harnessing the toxic effects of nanomaterials to microbes, their potential as antimicrobial agents has received increasing attention in recent years.[93] Owing to the widely reported antimicrobial properties of silver, AgNPs were unsurprisingly one of the first nanomaterials tested for antimicrobial properties.[94] The toxic effects of silver increased when AgNPs were used, possibly due to the size and shape of the particle (the Trojan horse-type mechanism). These NPs not only demonstrated antiviral activity but also showed bacteriostatic effects against a number of pathogens, including *E. coli*, *S. aureus*, *Salmonella* spp. *Bacillus cereus*, and *L. monocytogenes*, and bactericidal activity against methicillin-resistant *S. aureus* (MRSA) and methicillin-resistant *S. epidermidis* (MRSE).[95–99] Although some reports suggest that AgNP antimicrobial activity is due to the silver rather than the size of the particles (nanoparticles made of other metals, *i.e.* gold, do not have the same effect on growth[95,98]), other nanoparticles made from zinc and sulfur did show antimicrobial activity and antifungal activity.[100,101] Results indicate that zinc nanoparticles may distort and damage the bacterial cell membrane, resulting in the leakage of intracellular contents and eventually the death of bacterial cells.[102] Strong antimicrobial activities were also exhibited by CNTs against *E. coli* and *B. subtilis*. CNT networks developed on the surface of the bacterial cell, which destroyed the bacterial envelopes, resulting in leakage of the intracellular contents and cell death.[103] The antimicrobial effects of nanomaterials have been exploited in the food industry as antimicrobial agents in different food packages and were shown to improve the shelf-life of the product.[104,105]

Apart from the antimicrobial effects shown by nanomaterials themselves, the improved efficacy of photo-thermal antimicrobial nanotherapy in the presence of nanomaterials has been reported. The high binding affinity of CNTs to bacteria dramatically improved the effectiveness of the inherent near-infrared (NIR) laser, as cell viability was affected neither by CNTs alone nor by NIR irradiations alone.[106] Nanoparticles that can target specific bacterial cells have numerous antimicrobial applications and capitalize on a global market for nanoparticles in biotechnology and pharmaceuticals, which was valued at US$17.5 billion in 2011.[107] The use of nanoparticles as antimicrobial agents shows promising results; however, the safety of nanoparticles themselves and their effect on human health is still unclear. Until a clear understanding of the processes by which nanoparticles kill bacteria and the potential health effects are determined, their widespread use remains unrealistic.

4.6   Conclusions and Outlook

Nanoparticle-based detection techniques have revolutionized detection science by reducing analysis times, increasing sensitivity and facilitating the multiplex detection of a wide range of molecules. The increased interaction between target molecules and detection nanoparticles facilitated by the large surface-to-volume ratios of nanoparticles has resulted in the development of biosensors with enhanced sensitivities and improved response times. While the benefits of nanoparticles in detection science have been demonstrated time and time again in controlled laboratory environments, the exploitation of these technologies with many biological samples can be problematic. This is very much in evidence when it comes to complex matrices like food and blood, where a high background signal, which limits sensitivity, is often caused by the presence of interferences in the sample matrix such as cells, fats, proteins, sugars and enzymes. In order to overcome this issue, functionalized magnetic nanoparticles show great promise. These magnetic nanoparticles capture target molecules from a sample matrix, thus eliminating the possibility of matrix effects. The quenching of QDs by magnetic particles can, however, lead to decreased sensitivity.

Paramount to widespread use of nanoparticles is robust functionalization techniques and simple cost-effective nanoparticle fabrication. As outlined above, many functionalization strategies exist and the challenge lies in choosing the strategy that is application appropriate while offering robust modification with minimal effect on the function of the target molecule. However, the most prominent concern in relation to nanoparticles is their potential toxicity. Aside from the toxic effects associated with the bulk materials, *e.g.* silver, the small size of nanomaterials, which is similar to asbestos, may be adverse to human health. Numerous studies on the toxicological effects of nanomaterials specifically are underway; however, a clear understanding of the possible health effects of nanoparticles is still unavailable.

Despite these possible limitations, the use of nanoparticles in detection science is increasing. This coupled with advances in miniaturization technologies (*e.g.*, lab-on-a-chip) will result in portable, easy-to-use, rapid and sensitive detection devices for many different target molecules, which will be used worldwide in many industries.

References

1. E. Lahiff, C. Lynam, N. Gilmartin, R. O'Kennedy and D. Diamond, *Anal. Bioanal. Chem.*, 2010, **398**, 1575.
2. European Commission (2013), *Second Regulatory Review on Nano-materials*, Communication from the Commission to the European Parliament, the Council and the European Economic and Social Committee, Brussels, 3.10.2012, COM(2012) 572 final.
3. M. R. J. Auffan, J. Bottero, G. Lowry, J. Jolivet and M. Wiesner, *Nat. Nanotechnol.*, 2009, **4**, 634.
4. V. L. Pushparaj, A. Kumar, S. Murugesan, L. Ci, R. Vajtai, R. J. Linhardt, O. Nalamasu and P. M. Ajayan, *Proc. Natl. Acad. Sci. U. S. A.*, 2007, **104**, 13574.
5. M. Cadek, K. P. Ryan, V. Nicolosi, G. Bister, A. Fonseca, J. B. Nagy, K. Szostak, F. Béguin and W. J. Blau, *Nano Lett.*, 2004, **4**, 353.
6. G. M. Spinks, M. Bahrami-Samani, P. G. Whitten and G. G. Wallace, *Adv. Mater.*, 2006, **18**, 637.
7. G. Fanchini, S. Miller, L. B. Parekh and M. Chhowalla, *Nano Lett.*, 2008, **8**, 2176.
8. N. Zhu, Z. Chang, P. He and Y. Fang, *Anal. Chim. Acta*, 2005, **545**, 21.
9. X. Yu, V. Patel, G. Jensen, A. Bhirde, J. D. Gong, S. N. Kim, J. Gillespie, J. S. Gutkind, F. Papadimitrakopoulos and J. F. Rusling, *J. Am. Chem. Soc.*, 2006, **128**, 11199.
10. Z. M. Liu, G. L. Shen and R. Q. Yu, *Anal. Lett.*, 2009, **42**, 3046.
11. J. Chen, C. Lynam, O. Ngamna, S. Moulton, W. Zhang and G. G. Wallace, *Electrochem. Solid-State Lett.*, 2006, **9**, H68.
12. Y. J. Ko, Y. Ahn, S. Y. Hwang, N. G. Cho and S. H. Lee, *Electrophoresis*, 2008, **29**, 3466.
13. D. A. Healy, P. Leonard, L. McKenna and R. O'Kennedy, *Trends Biotechnol.*, 2007, **25**, 125.
14. B. Byrne, N. Gilmartin and R. O'Kennedy, *Sensors*, 2009, **9**, 4407.
15. N. Sanvicens, N. Pascual and M. P. Marco, *Trends Anal. Chem.*, 2009, **28**, 1243.
16. S. K. Saha, S. S. Agasti, C. Kim, X. Li and V. M. Rotello, *Chem. Rev.*, 2012, **112**, 2739.
17. S. Link and M. A. El-Sayed, *J. Phys. Chem. B*, 1999, **103**, 4212.
18. M. C. Daniel and D. Astruc, *Chem. Rev.*, 2004, **104**, 293.
19. N. L. Rosi and C. A. Mirkin, *Chem. Rev.*, 2005, **105**, 1547.
20. S. Iijima, *Nature*, 1991, **354**, 56.
21. R. L. McCreery, *Chem. Rev.*, 2008, **108**, 2646.

22. M. S. Arnold, A. A. Green, J. F. Hulvat, S. I. Stupp and M. C. Hersam, *Nat. Nanotechnol.*, 2006, **1**, 60.

23. X. Tu, S. Manohar, A. Jagota and M. Zheng, *Nature*, 2009, **460**, 250.

24. H. Liu, D. Nishide, T. Tanaka and H. Kataura, *Nat. Commun.*, 2011, **2**, 309.

25. A. Nish, J.-Y. Hwang, J. Doig and R. J. Nicholas, *Nat. Nanotechnol.*, 2007, **2**, 640.

26. Y. Wang, R. Hu, G. Lin, I. Roy and K.-T. Yong, *ACS Appl. Mater. Interfaces*, 2013, **5**, 2786.

27. U. Resch-Genger, M. Grabolle, S. Cavaliere-Jaricot, R. Nitschke and T. Nann, *Nat. Methods*, 2008, **5**, 763.

28. P. Zrazhevskiy, M. Sena and X. H. Gao, *Chem. Soc. Rev.*, 2010, **39**, 4326.

29. *Bioconjugate Techniques*, ed. G. T. Hermanson, Academic Press, San Diego, 2nd edn, 2008.

30. W. R. Algar, D. E. Prasuhn, M. H. Stewart, T. L. Jennings, J. B. Blanco-Canosa, P. E. Dawson and I. L. Medintz, *Bioconjugate Chem.*, 2011, **22**, 825.

31. A. J. Khopade and F. Caruso, *Langmuir*, 2002, **18**, 7669.

32. H. Mattoussi, J. M. Mauro, E. R. Goldman, G. P. Anderson, V. C. Sundar, F. V. Mikulec and M. G. Bawendi, *J. Am. Chem. Soc.*, 2000, **122**, 12142.

33. G. Chen and M. Jiang, *Chem. Soc. Rev.*, 2011, **40**, 2254.

34. K. E. Sapsford, W. Russ Algar, L. Berti, K. Boeneman Gemmill, B. J. Casey, E. Oh, M. H. Stewart and I. L. Medintz, *Chem. Rev.*, 2013, **113**, 1904.

35. D. A. Giljohann, D. S. Seferos, W. L. Daniel, M. Massich, P. C. Patel and C. A. Mirkin, *Angew. Chem. Int. Ed.*, 2010, **49**, 3280.

36. D. Tasis, N. Tagmatarchis, A. Bianco and M. Prato, *Chem. Rev.*, 2006, **106**, 1105.

37. E. Katz and I. Willner, *ChemPhysChem*, 2004, **5**, 1084.

38. R. J. Chen, Y. G. Zhang, D. W. Wang and H. J. Dai, *J. Am. Chem. Soc.*, 2001, **123**, 3838.

39. Z. J. Guo, P. J. Sadler and S. C. Tsang, *Adv. Mater.*, 1998, **10**, 701.

40. S. E. Moulton, M. Maugey, P. Poulin and G. G. Wallace, *J. Am. Chem. Soc.*, 2007, **129**, 9452.

41. C. Lynam, N. Gilmartin, A. I. Minett, R. O'Kennedy and G. Wallace, *Carbon*, 2009, **47**, 2337.

42. W. P. Faulk and G. M. Taylor, *Immunochemistry*, 1971, **8**, 1081.

43. K. Glynou, P. C. Ioannou, T. K. Christopoulous and V. Syriopoulou, *Anal. Chem.*, 2003, **75**, 4155.

44. S. Deborggraeve, F. Claes, T. Laurent, P. Mertens, T. Leclipteux, J. C. Dujardin, P. Herdewijn and P. Buscher, *J. Clin. Microbiol.*, 2006, **44**, 2884.

45. J. Aveyard, M. Mehrabi, A. Cossins, H. Braven and R. Wilson, *Chem. Commun.*, 2007, 4251.

46. R. Wilson, *Chem. Soc. Rev.*, 2008, **37**, 2028.

47. C. Fernandez-Sanchez, C. J. McNeil, K. Rawson, O. Nilsson, H. Y. Leung and V. Gnanapragasam, *J. Immunol. Methods*, 2005, **307**, 1.

48. R. H. Shyu, H. F. Shyu, H. W. Liu and S. S. Tang, *Toxicon*, 2002, **40**, 255.

49. S. H. Huang, H. C. Wei and Y. C. Lee, *Food Control*, 2007, **18**, 893.

50. C. A. Mirkin, R. L. Letsinger, R. C. Mucic and J. J. Storhoff, *Nature*, 1996, **382**, 607.

51. H. Li and L. Rothberg, *Proc. Natl. Acad. Sci. U. S. A.*, 2004, **101**, 14036.

52. H. Li and L. J. Rothberg, *J. Am. Chem. Soc.*, 2004, **126**, 10958.

53. J. K. N. Mbindyo, B. D. Reiss, B. R. Martin, C. D. Keating, M. J. Natan and T. E. Mallouk, *Adv. Mater.*, 2001, **13**, 249.

54. Y. G. Sun and T. N. Xia, *Science*, 2002, **298**, 2176.

55. T. K. Sau and C. J. Murphy, *J. Am. Chem. Soc.*, 2004, **126**, 8648.

56. L. He, M. D. Musick, S. R. Nicewarner, F. G. Salinas, S. J. Benkovic, M. J. Natan and C. D. Keating, *J. Am. Chem. Soc.*, 2000, **122**, 9071.

57. R. C. Bailey, J.-M. Nam, C. A. Mirkin and J. T. Hupp, *J. Am. Chem. Soc.*, 2003, **125**, 13541.

58. A. W. Wark, H. J. Lee, A. J. Qavi and R. M. Corn, *Anal. Chem.*, 2007, **79**, 6697.

59. M. A. Hahn, J. S. Tabb and T. D. Krauss, *Anal. Chem.*, 2005, 77, 4861.

60. D. Gerion, F. Q. Chen, B. Kannan, A. H. Fu, W. J. Parak, D. J. Chen, A. Majumdar and A. P. Alivisatos, *Anal. Chem.*, 2003, 75, 4766.

61. K. Welsher, D. Daranciang and H. Dai, *Nano Lett.*, 2008, **8**, 586.

62. P. W. Barone, D. A. Heller and M. S. Strano, *Nat. Mater.*, 2005, **4**, 86.

63. F. Lisdat and D. Schafer, *Anal. Bioanal. Chem.*, 2008, **391**, 1555.

64. J. S. Daniels and N. Pourmand, *Electroanalysis*, 2007, **19**, 1239.

65. W. Goepel, K.-D. Schierbaum, in *Sensors: A Comprehensive Survey, Vol 2: Chemical and Biochemical Sensors, Part I*, ed. W. Goepel, T. A. Jones, M. Kleitz, J. Lundstroem and T. Seiyama, Verlag Chemie, Weinheim, 1991, p. 1.

66. S.-J. Park, T. A. Taton and C. A. Mirkin, *Science*, 2002, **295**, 1503.

67. M. D. Ward and D. A. Buttry, *Science*, 1990, **249**, 1000.

68. Y. Weizmann, F. Patolsky and I. Willner, *Analyst*, 2001, **126**, 1502.

69. M. Dequaire, C. Degrand and B. Limoges, *Anal. Chem.*, 2000, 72, 5521.

70. P. He, L. Shen, Y. Cao and D. Li, *Anal. Chem.*, 2007, **79**, 8024.

71. X. Yu, B. Munge, V. Patel, G. Jensen, A. Bhirde, J. D. Gong, S. N. Kim, J. Gillespie, J. S. Gutkind, F. Papadimitrakopoulos and J. F. Rusling, *J. Am. Chem. Soc.*, 2006, **128**, 11199.

72. J. Wang, G. Liu and M. R. Jan, *J. Am. Chem. Soc.*, 2004, **126**, 3010.

73. W. Cheng, L. Ding, J. Lei, S. Ding and H. Ju, *Anal. Chem.*, 2008, **80**, 3867.

74. J. X. Wang, Z. J. Shi, N. Q. Li and Z. N. Gu, *Anal. Chem.*, 2002, 74, 1993.

75. J. Das, M. Abdul Aziz and H. Yang, *J. Am. Chem. Soc.*, 2006, **128**, 16022.

76. M. J. A. Shiddiky, M. Aminur Rahman and Y.-B. Shim, *Anal. Chem.*, 2007, **79**, 6886.

77. D. Tang, R. Yuan and Y. Chai, *Anal. Chem.*, 2008, **80**, 1582.

78. L. Qin, M. J. Banholzer, J. E. Millstone and C. A. Mirkin, *Nano Lett.*, 2007, 7, 3849.

79. J. Wang, D. Xu, A.-N. Kawde and R. Polsky, *Anal. Chem.*, 2001, **73**, 5576.

80. J. Wang, G. Liu and A. Merkoci, *J. Am. Chem. Soc.*, 2003, **125**, 3214.

81. D. M. Webster, P. Sundaram and M. E. Byrne, *Eur. J. Pharm. Biopharm.*, 2013, **84**, 1.

82. B. Wang, X. He, Z. Zhang, Y. Zhao and W. Feng, *Acc. Chem. Res.*, 2013, **46**, 761.

83. A. Nel, T. Xia, H. Meng, X. Wang, S. Lin, Z. Ji and H. Zhang, *Acc. Chem. Res.*, 2013, **46**, 607.

84. T. Xia, M. Kovochich, M. Liong, L. Madler, B. Gilbert, H. Shi, J. I. Yeh, J. I. Zink and A. E. Nel, *ACS Nano*, 2008, **2**, 2121.

85. L. K. Limbach, P. Wick, P. Manser, R. N. Grass, A. Bruinink and W. J. Stark, *Environ. Sci. Technol.*, 2007, **11**, 4158.

86. W. S. Cho, R. Duffin, F. Thielbeer, M. Bradley, I. L. Megson, W. MacNee, C. A. Poland, C. L. Tran and K. Donaldson, *Toxicol. Sci.*, 2012, 477.

87. E. J. Petersen and B. C. Nelson, *Anal. Bioanal. Chem.*, 2010, **398**, 613.

88. A. Nel, T. Xia, L. Mädler and N. Li, *Science*, 2006, **311**, 622.

89. S. J. Cho, D. Maysinger, M. Jain, B. Röder, S. Hackbarth and F. M. Winnik, *Langmuir*, 2007, **23**, 1974.

90. K. Donaldson, F. Murphy, R. Duffin and C. Poland, *Part. Fibre Toxicol.*, 2010, 7, 5.

91. C. A. Poland, R. Duffin, I. Kinloch, A. Maynard, W. A. Wallace, A. Seaton, V. Stone, S. Brown, W. Macnee and K. Donaldson, *Nat. Nanotechnol.*, 2008, **3**, 423.

92. A. Schinwald, F. Murphy, A. Prina-Mello, C. Poland, F. Byrne, D. Movia, J. R. Glass, J. C. Dickerson, D. Schultz, C. E. Jeffree, W. Macnee and K. Donaldson, *Toxicol. Sci.*, 2012, **128**, 461.

93. P. Dallas, V. K. Sharma and R. Zboril, *Adv. Colloid Interface Sci.*, 2011, **1–2**, 119.

94. T. V. Duncan, *J. Colloid Interface Sci.*, 2011, **363**(1), 1.

95. H. H. Lara, E. N. Garza-Trevino, L. Ixtepan-Turrent and D. K. Singh, *J. Nanobiotechnol.*, 2011, **9**, 30.

96. Q. Feng, J. Wu, G. Chen, F. Cui, T. Kim and J. Kim, *J. Biomed. Mater. Res.*, 2000, **4**, 662.

97. B. De Gusseme, L. Sintubin, L. Baert, E. Thibo, T. Hennebel, G. Vermeulen, M. Uyttendaele, W. Verstraete and N. Boon, *Appl. Environ. Microbiol.*, 2010, **4**, 1082.

98. S. S. Khan, A. Mukherjee and N. Chandrasekaran, *Colloids Surf. B*, 2011, **1**, 129.

99. J. S. Kim, E. Kuk, K. N. Yu, J. Kim, S. J. Park, H. J. Lee, S. H. Kim, Y. K. Park, Y. H. Park, C. Y. Hwang, Y. K. Kim, Y. S. Lee, D. H. Jeong and M. H Cho, *Nanomedicine*, 2007, **1**, 95.

100. T. Jin, D. Sun, J. Y. Su, H. Zhang and H. J. Sue, *J. Food Sci.*, 2009, **74**, M46.

101. S. R. Choudhury, M. Ghosh, A. Mandal, D. Chakravorty, M. Pal, S. Pradhan and A. Goswami, *Appl. Microbiol. Biotechnol.*, 2011, **2**, 733.

102. Y. Liu, L. He, A. Mustapha, H. Li, Z. Q. Hu and M. Lin, *J. Appl. Microbiol.*, 2009, **4**, 1193.

103. S. Liu, A. K. Ng, R. Xu, J. Wei, C. M. Tan, Y. Yang and Y. Chen, *Nanoscale*, 2010, **12**, 2744.
104. C. Costa, A. Conte, G. G. Buonocore and M. A. Del Nobile, *Int. J. Food Microbiol.*, 2011, **3**, 164.
105. A. L. Incoronato, A. Conte, G. G. Buonocore and M. A. Del Nobile, *J. Dairy Sci.*, 2011, **4**, 1697.
106. J. W. Kim, E. V. Shashkov, E. I. Galanzha, N. Kotagiri and V. P. Zharov, *Lasers Surg. Med.*, 2007, **39**, 622.
107. J. A. Lemire, J. J. Harrison and R. J. Turner, *Nat. Rev. Microbiol.*, 2013, **11**, 371.

Smart Indicator Technologies for Chemical and Biochemical Detection

SUBRAYAL M. REDDY

Department of Chemistry, University of Surrey, Guildford, Surrey,
GU2 7XH, UK
Email: s.reddy@surrey.ac.uk

5.1 Chemical Indicators

For over 40 years, biosensor research has offered the promise of inexpensive, portable and easy-to-use analytical devices for the rapid determination of a range of analytes of biomedical and environmental significance. However, to date, besides the glucose biosensor and the pregnancy testing kit, other biosensor success stories are few and far between. The success of the latter two sensors is based on a global demand by the layperson for convenient self-regulation of glucose in diabetes patients and the convenience of an early indicator of pregnancy, respectively. It could be argued that other biomarkers of significance do not require day-to-day monitoring. For example, cholesterol levels can be monitored once a month. An equally valid argument is the cost of the detector in the biosensor leading to a digital read-out. For example, is it necessary to have a digital read-out, when a semi-quantitative visual indicator would be sufficient? This chapter looks at selected chromophore chemistries which have been used as smart colour-changing indicators for diagnostic testing of environmental markers. Chemicals including azo compounds (as used in fabric dyes), pH indicators

RSC Detection Science Series No. 3
Advanced Synthetic Materials in Detection Science
Edited by Subrayal Reddy

Published by the Royal Society of Chemistry, www.rsc.org

Table 5.1 Absorbed and observed colours of visible light.

Wavelength of Maximum Absorption (nm)	Colour Absorbed	Colour Observed
380–440	Violet/violet-blue	Green-yellow/yellow
440–470	Blue	Orange
470–500	Blue-green	Red
500–520	Green	Purple
520–580	Yellow-green/yellow	Violet/violet-blue
580–620	Orange	Blue
620–680	Red	Blue-green
680–780	Red	Green

(also known as halochromophores) and coordination complexes will be discussed. The application of nanoparticle arrays, which can split light and change colour with change in lattice spacing within their array, is also considered.

The portion of a chemical which is responsible for it to exhibit colour is known as the chromophore. The colour arises when the chemical in solution absorbs certain wavelengths of light compatible with causing specific electronic (HOMO–LUMO) transitions within its molecular orbitals. The light that is not absorbed by the molecule is transmitted or reflected and corresponds to the observed colour of the compound. Various factors in the molecule affect the colour observed, including the presence of π-bonds (typically found in food colourings) and their extent of conjugation (alternating single and double bonds) and metal complexation (as in, for example, chlorophyll or hemoglobin). A spectrophotometer can be used to determine the wavelength of maximum light absorption for a chemical in solution. It should be noted that the wavelength of absorbed light is not directly indicative of the colour of the compound. For example, a compound that absorbs photons from the orange part of the light spectrum will appear blue. A useful method of predicting the observed colour of a chemical in solution from its absorption wavelength is summarized in Table 5.1; the ability to predict molecular colour was recently reviewed and discussed by Williams *et al.*[1]

5.1.1Glucose Tests

Glucose is an important metabolic marker used in the within-day management of diabetes. Normal glucose levels are in the range 60–95 mg dL^{-1}. The frequent self-monitoring diabetic blood glucose concentration is essential to prevent both short- and long-term complications from hypo- and hyperglycemia. For example, people with diabetes who self-administer insulin could be required to test their blood glucose up to 10 times in a day, in order to self-manage their blood glucose. The most convenient method for diabetic patients is to use a glucose biosensor. The device uses a disposable, single-use, plastic substrate-based test strip with in-built capillary (wicking) mechanism for the biological sample (typically blood or urine). The strip also has integrated electrodes and the enzyme glucose oxidase to specifically recognize glucose and convert it into a product that can be detected by optical or electrochemical means. The device offers the height of convenience for the layperson and returns a precise and accurate digital read-out of glucose within seconds. For example, on the market today are glucometers such as Accu-Check® Aviva, One Touch® and Freestyle® Freedom.[2] In contrast to this, the low-tech approach, which preceded the glucose biosensor revolution and is still used today for a simple low-cost and rapid measure of glucose (mainly in urine), is the colorimetric dip-stick test strip.[3] The test-strip methods are paper-based (with no need for plastic substrate, electrodes and a reader) and therefore offer the key advantages of being a low-cost visual indicator based on recyclable materials.

Colorimetric (dry-reagent) strip methods of determining glucose in urine rely on a two-enzyme reaction process (Eqns 5.1 and 5.2). Glucose oxidase, a suitable reducing agent (generically termed RH_2) and peroxidase enzyme are immobilized on the strip. The biological sample (containing glucose) is then simply added to the strip. In the first reaction, glucose in the presence of oxygen is enzymatically converted to gluconic acid and hydrogen peroxide. In a second enzyme reaction (using peroxidase), and in the presence of RH_2, the hydrogen peroxide is catalytically reduced to water and the RH_2 compound is oxidized to give a coloured product. Typically, benzidines such as *o*-dianisidine or tetramethylbenzidine (TMB) have been used, resulting in a colour change from colourless to orange/red. The colour intensity can then be compared with a colour chart to estimate the glucose level.

$$\text{Glucose} + O_2 \xrightarrow{\text{Oxidase}} H_2O_2 + \text{Product} \qquad (5.1)$$

$$H_2O_2 + RH_2 \xrightarrow{\text{Peroxidase}} 2H_2O + R \qquad (5.2)$$

Cha *et al.*[4] have demonstrated the use of cellulose paper for the successful preparation of glucose test strips, using 2,4,6-tribromo-3-hydroxybenzoic acid (TBHBA) as the precursor colour agent. In the presence of glucose and glucose oxidase, the enzymatically produced hydrogen peroxide is used in a second enzyme reaction (with HRP) to oxidize TBHBA. The oxidized product readily couples with 4-aminoantipyrine (4-APP) to give a red azo compound, as shown in Figure 5.1

The intensity of the red colour produced due to the presence of glucose is reliable in the range 0.18–9.91 mg mL^{-1}, as confirmed from spectrometric measurements at 540 nm.[4]

5.1.2 Chlorine Test Strips

Chlorine in the form of freely available chlorine (diatomic chlorine, hypochlorous acid and the hypochlorite ion) as well as combined available chlorine (such as organic chloramines and ammonium chloramines) has been

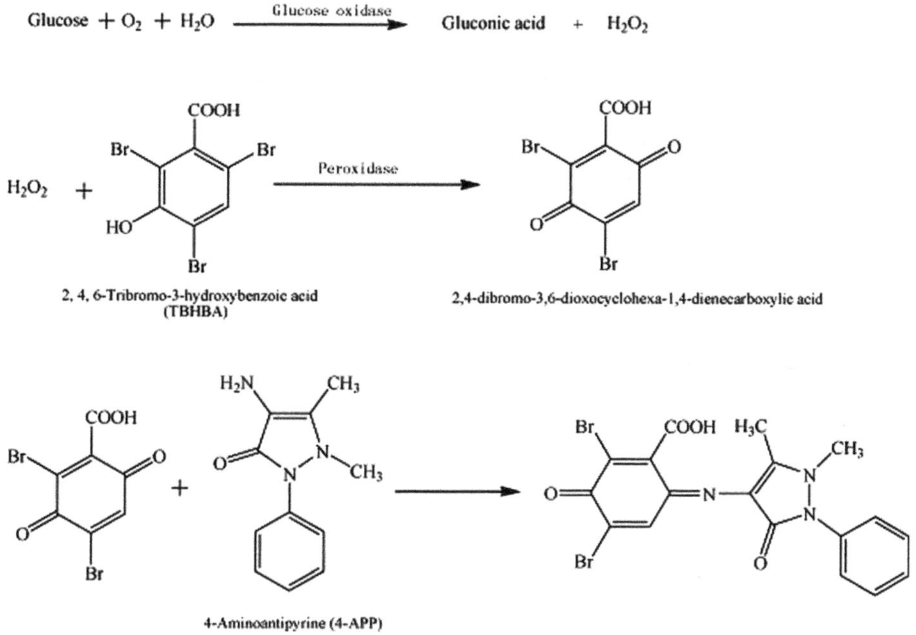

Glucose + O_2 + H_2O —— Glucose oxidase ——→ Gluconic acid + H_2O_2

2, 4, 6-Tribromo-3-hydroxybenzoic acid
(TBHBA)

2,4-dibromo-3,6-dioxocyclohexa-1,4-dienecarboxylic acid

4-Aminoantipyrine (4-APP)

Figure 5.1 The reaction mechanism of glucose with TBHBA and 4-APP. (Reproduced from Cha *et al.*[4] with permission from Elsevier.)

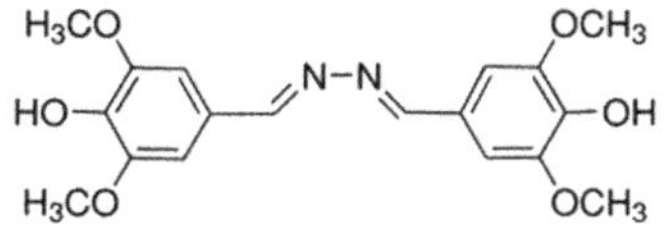

Figure 5.2 Structure of syringaldazine.

commonly used to disinfect swimming pools, public drinking water and waste water. Chlorine levels in swimming pools are an important measure of bacterial contamination from typical species such as *Streptococcus faecalis*, *E. coli* and *Pseudomonas aeruginosa*. The last of these organisms can be a cause of infection in the ear and in some external wounds and is thus of considerable interest in determining the safety of, for example, pool water. A simple colorimetric test for the determination of free chlorine is based on a plastic patch containing absorbent paper treated with a mixture of syringaldazine (Figure 5.2; 4-hydroxy-3,5-dimethoxybenzaldehyde azine) and vallinazine, which turns from yellow to purple in the presence of free chlorine.

Syringaldazine has also been used on its own. The reagent is sensitive to hypochlorite but insensitive to bound chlorine such as chloramines, unlike *o*-tolidine when used as the chromogenic reagent which is unable to discriminate between free and bound chlorine.[5–7] Upon oxidation the syringaldazine is converted into a red azo complex (azo-bis-quinonemethide) which has an absorption maximum at 530 nm and can reliably measure chlorine in the range 0.1–1 ppm.

Chlorine content in washing powder and bleach formulations is important for supplier authentication and detection of adulteration. The standard test for chlorine content in washing powders and bleaches involves two key reactions (Eqns 5.3 and 5.4). The OCl^- (hypochlorite; bleaching compound) is the component being measured. The first reaction uses hypochlorite to oxidize iodide (I^-) to iodine (I_2). The iodine level is a measure of original hypochlorite level. The second reaction requires a colour indicator such as starch (starch forms a dark blue colour complex with iodine). The iodine is reacted with a known amount of thiosulfate ($S_2O_3^{2-}$) in the second reaction (by titrating the thiosulfate dropwise) until the blue colour just disappears. The exact point at which the blue colour permanently disappears is the end-point where all the iodine has been converted back to iodide. By working backwards, first calculating exactly how much (as a concentration) of thiosulfate was used, half this value is the concentration of bleach.

$$OCl^- + 2I^- + 2H_3O^+ \rightarrow I_2 + Cl^- + 3H_2O \qquad (5.3)$$

$$I_2 + 2S_2O_3^{2-} \rightarrow 2I^- + S_4O_6^{2-} \qquad (5.4)$$

Claver *et al.*[8] reported a dry-reagent chemistry method for the chemiluminescent determination of hypochlorite. The $HClO_3^-$-sensitive test strip comprises a 10 mm×9 mm piece of a range of materials (including polyurethane, PVC and cellulose acetate) cellulose anion exchange paper all treated with fluoresceinate anions (uranine). The optimum test strip was with polyester. The test strip responded linearly to hypochlorite in the range 2–51.4 mg L^{-1}, with a lower detection limit of 0.4 mg L^{-1}.

5.1.3   Drugs of Abuse Tests

Law enforcers require reliable methods to be able to determine the presence of contraband at a crime scene or on a criminal suspect. The presumptive test used for cocaine, also known as the Scott's test, relies on a three-stage process.[9] First, the sample (typically 1 mg) is introduced into 0.02 mL of an aqueous solution of cobalt dithiocyanate. The first indication that cocaine may be present is the formation of a blue precipitate within the solution, due to the formation of a water-insoluble cobalt complex with the amido group within cocaine. However, the test is not confirmatory at this stage as the cobalt reaction also occurs with secondary and tertiary amines and other alkaloids including, for example, lidocaine (an anesthetic) and diphenhydramine (an antihistamine and also the sleep-inducing element in common over-the-counter sleeping pills). In a second stage, a couple of drops of concentrated hydrochloric acid is added, resulting in the blue precipitate disappearing. The third stage of the test is to add 0.2 mL of chloroform. The latter results in phase separation, with the organic layer being at the bottom and the aqueous layer on top, but with the organic layer assuming the blue colouration. The blue complex is thought to be an octahedral complex between the cobalt dithiocyanate and two molecules of cocaine, where each cocaine molecule forms two dative bonds with the central cobalt.[10] Recent publications have focused on the voracity of this test in relation to avoiding false positives.[11] The test has recently been developed into a simple swab, but the major drawback is that the swab only focuses on the first of the three stages and therefore is not an entirely reliable presumptive test.

A modification of the Scott's test has also been developed for the determination of ketamine.[12] Ketamine hydrochloride, or 2-(2-chlorophenyl)-2-(methylamino)cyclohexanone hydrochloride, is widely used as an anesthetic drug, stimulating NMDA (*N*-methyl-D-aspartate) receptors on neuronal cells. Although primarily used as an anesthetic, more recently it has been used with criminal intent as a club drug, strongly implicated with date rape. The modified Scott's test requires the sample (typically 1 mg) or one or two droplets of the sample in solution to be treated with alkali by adding a drop of 1 M KOH followed by addition of cobalt tetrathiocyanate. A resultant purple/blue precipitate is indicative of a sample containing ketamine hydrochloride. A negative result is indicated by no change in original colour or formation of a blue solution. The positive result is thought to be due to an octahedral complex being formed between the cobalt tetrathiocyanate and two molecules of ketamine.

An exhaustive list of rapid testing methods of drugs of abuse by law enforcement and narcotics laboratory personnel has been published by the United Nations as part of the UN International Drug Control Program and readers are directed to this document for more methodologies for individual drugs of abuse, including cannabis, heroin and club drugs.[13]

5.2 pH Indicators

pH indicators can serve a useful purpose in the indirect measurement of CO_2. There are other conditions besides altered CO_2 levels in which the pH within food packaging can change. For example, lactic acid bacteria production during ground beef spoilage[14,15] and production of ammonia-based compounds during fish spoilage.[16–19] There are also conditions in which the pH of wearable patch sensors can change, for example for pollution monitoring or providing an extra safety function in professional garments, thereby improving workers' safety in toxic acidic textile industries.[20]

The pH indicators above are an integration of well-established indicator dye systems such as methyl red[20] and bromocresol green[21] with a suitable polymer matrix, such as a sol–gel system to encapsulate the dye but allow relatively unimpeded diffusion of protons and hydroxide ions.

Jeronimo *et al.*[22] used the ability of acetazolamide (an anti-glaucoma agent) to inhibit carbonic anhydrase activity as a method to measure acetazolamide. The enzyme and a pH indicator dye (cresol red) were immobilized in a sol–gel thin film sandwich. The enzyme catalyzes the dehydration of carbonic acid to give CO_2 and H_2O, thereby increasing the pH. In the presence of acetazolamide, the enzyme reaction is inhibited. The authors demonstrated an acetazolamide concentration-dependent change in the spectrometric absorbance at 570 nm, with linearity in the range 1–10 mM.

5.2.1 Volatile Amine Sensors and Indicators

Spoilage of food products can be mainly attributed to microbes residing upon and within the product. Under appropriate temperature and environmental conditions, the microbes can metabolize the food material, resulting in direct and indirect degradation due to interaction of the food with waste products from metabolism.[23] Volatile amines, also known as total volatile basic nitrogen (TVB-N) content due to microbial growth, can be used as a fish freshness indicator in both vacuum-packaged and air-packaged fish.[24] The indicators generally rely on pH colour indicators which change colour in the presence of the alkaline TVB-N compounds. Crowley *et al.* immobilized several pH indicator dyes within a gas permeable polymer film, allowing the measurement of spoilage products in the headspace of packaged fish.[16] As the fish degrades due to microbial spoilage, the colour change within the pH patch (integrated into the food packaging) is indicative of a shift to higher pH values. The visual change due to the presence of TVB-N was monitored quantitatively using a custom-built LED probe,[25] but it was also possible to monitor the change with the human eye using a colour code.[26]

Kuswandi *et al.* developed a colorimetric method using the ability of polyaniline (PANI) to change colour as a function of pH.[17] Under acidic conditions, the PANI undergoes a blue shift and exhibits maximum absorbance at approximately 750 nm. Under alkaline conditions, the PANI undergoes a red shift with a new maximum absorbance at approximately 550 nm. The change in absorbance at either peak is indicative of a change in pH. In the presence of TVB-N, the expected increase in pH was observed as a corresponding change in these two peaks. The total colour difference correlated well with TVB-N levels in fish. The responses allowed the real-time tracking of fish spoilage either at constant temperature or within fluctuating temperature environments. The PANI film could be used as a low-cost sensor suitable for smart packaging applications. Kuswandi *et al.* also showed that TVB-N from microbial degradation of shrimp could be monitored using a natural dye, curcumin. Curcumin is the major yellow pigment extracted from turmeric, a commonly used spice. The curcumin was immobilized within a biodegradable cellulose film. With the production of the basic volatile amines within the headspace of the food packaging due to food spoilage, the ensuing pH increase results in a colour change from yellow to orange and reddish orange with increasing TVB-N.[27]

5.2.2 CO$_2$ Sensors and Indicators

CO$_2$ in food packaging can modulate microbial degradation of many food types in controlled environment food packaging. Little short of sterilizing the food prior to packaging or even after using aseptic techniques during packaging, it is difficult to avoid the presence of live microbes indigenous to fruit and fresh meat. The levels of CO_2 present can either promote or delay microbial degradation. For instance, near 100% levels of CO_2 in the food packaging atmosphere will minimize the oxygen content, thereby inhibiting aerobic spoilage microbes. A modified atmosphere package including CO_2 and oxygen could promote both anaerobic and aerobic microorganisms to thrive, resulting in food spoilage. CO_2 is also produced by food during natural respiration, which could then lead to biodegradation of the food due to activation of anaerobic microbes. Therefore, controlling the level of CO_2 in food packaging using modified atmosphere packaging (MAP) is import-ant. Measuring the change in CO_2 concentration as a result of, for example, respiration, and more importantly damage to the food packaging itself, is a useful indicator of contamination and degradation.

Conventional CO_2 sensors can be divided into two types: electrochemical and optical. The commercially available optical technique is based on non-distributed infrared (NDIR). This method is used to measure gaseous CO_2. However, the NDIR instrument is bulky and expensive and does not lend itself to a facile CO_2 measurement in the field. The sampling technique is an invasive method which requires the destruction of the packaging, typically using a needle and syringe in order to draw out a sample of the headspace gas. The commercially available electrochemical device, also known as the Severinghaus sensor, can measure CO_2 in an aqueous solution sample and is based on a modified pH electrode consisting of a bicarbonate solution filled glass electrode covered by a thin CO_2 permeable membrane. The CO_2 in the aqueous sample is in equilibrium with carbonic acid, which can dissociate into a bicarbonate anion and a proton. Only the CO_2 gas can permeate the selective membrane. The gas exiting the membrane on the pH electrode side dissolves in water to form carbonic acid, which subsequently dissociates into the bicarbonate anion and the proton. The proton-induced pH change can then be measured by the pH probe, the change being proportional to the concentration of CO_2 in the aqueous sample. Neither the optical nor elec-trochemical techniques above are suitable for use in the food packaging integrity-testing industry, given that they both require access to the en-vironment inside the package.

There is therefore a desire to develop smart indicator patch-type tech-nologies that can be integrated into the packaging, which then respond in real-time to the changing conditions of the food contained therein.[28] The response is typically a change in colour of the patch, and therefore requires any colour changing reagent sealed and secured within a gas permeable polymer matrix (such as polypropylene), and the reagent then only responds if the analyte of interest is present. In the case of a CO_2 indicator, given that dissolved CO_2 can induce a change in pH, then pH sensitive dyes are obvious candidates to be used in the smart patch. Diffusion of CO_2 from the test medium (gaseous or aqueous) through a 10 μm thick gas permeable mem-brane-based patch containing the pH sensitive dye will result in a change in colour of the dye due to the rapid establishment of an equilibrium where the gaseous CO_2 dissolves in the dye solution to produce carbonic acid, thereby releasing protons. The protons are associated with the basic form of the indicator dye (In$^-$) to produce the protonated form (HIn), which in turn produces a colour change of the indicator solution. Typical pH indicators include bromocresol purple and methyl red [2-(4-dimethylaminophenyla-zo)benzoic acid]. The methyl red pH indicator, for example, undergoes a colour change on protonation in the range 5.4–6.0, in which the changes in molecular and electronic structure result in a change in colour from yellow in basic or neutral solutions to red in acidic solutions (Figure 5.3).[20]

Table 5.2 gives a list of dyes *versus* the critical pH at which colour change is observed.

The sensitivity of this colour change is a function of the acid dissociation constant of the pH indicator and the buffer conditions of the indicator solution. It should be noted that because this method of measuring CO_2 is based on measuring pH, sources of error or interference would be if acidic or basic volatile compounds were produced during food spoilage within the packaging. Water-based indicators integrated with a base such as sodium hydrogen carbonate and a plasticizer such as glycerol within a patch format

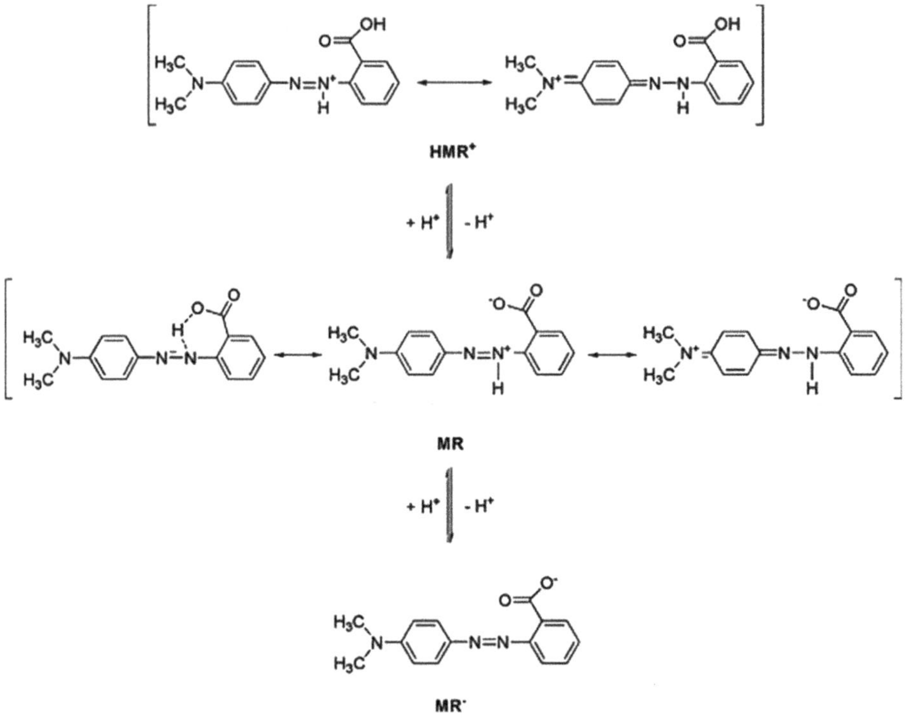

Figure 5.3   Acid (HMR$^+$), neutral (MR) and basic (MR$^-$) forms of the methyl red molecule.
(Reproduced from Caldara *et al.*[20] with permission from Elsevier.)

Table 5.2   Some commonly used pH indicators.

Indicator	Acid Colour	Base Colour	pH Transition Range
Thymol blue	Red	Yellow	1.2–2.8
Methyl orange	Red	Yellow	3.1–4.4
Bromocresol green	Yellow	Blue	3.8–5.4
Bromocresol purple	Yellow	Purple	5.2–6.8
p-Nitrophenol	Colourless	Yellow	5.6–7.6
Cresol red	Yellow	Red	7.2–8.8
Phenolphthalein	Colourless	Pink	8.0–9.6
Alizarin yellow	Yellow	Orange-red	10.1–12.0

have improved operational lifetimes as well as rapid response and recovery times (<2 min).[29] Non-aqueous solvent-based indicators have also been explored, but require a phase transfer agent (PTA) to solubilize the hydrophobic dye. The dye indicator molecule in the anionic form is ion paired with a quaternary ammonium ion (Q^+In^-) and immobilized typically in a hydrated polymer layer on a polyester or glass support. In the presence of CO_2 the ion pair reacts according to the following equation, resulting in the protonation of the indicator dye and thereby changing its absorption maximum (Eqn 5.5):

$$\{Q^+In^- \cdot xH_2O\} + CO_2 \rightleftharpoons \{Q^+HCO_3^- \cdot (x-1)H_2O\} + HIn \qquad (5.5)$$

A major drawback is that the indicator dyes have a poor shelf-life due to issues with membrane ageing.

Nopwinyuwong *et al.* demonstrated a colorimetric indicator label for monitoring spoilage of desserts due to moisture build-up.[30] A mixed pH dye indicator, designed as a chemical barcode, was integrated into the dessert packaging. The dyes included bromothymol blue and methyl red, which respond to microbial-produced CO_2 during food spoilage due to moisture. The total colour difference of the mixed pH dye-based indicator correlated well with CO_2 levels of intermediate moisture levels in the dessert. Spoilage could be monitored at various constant temperatures and even correlated with temperature fluctuation, potentially allowing the label to also act as a temperature indicator.

Simple methods to indicate levels of CO_2 can therefore be used to determine the quality and degree of deterioration of the packed food during the various stages of production and transport to outlets, as well as for the consumer to make an informed choice prior to and post-purchase. Food traceability throughout the food supply chain is increasingly becoming a necessary task, mandatory in the European Union (EU) since 2005. In light of the above, there is an increasing need to develop low-cost, accurate, rapid, robust, non-invasive and non-destructive indicators. The indicators also need to be non-toxic, and insoluble in water.[31] They should also exhibit an irreversible response towards the analyte and be tamper proof. While the latter is a little more difficult to police, the former is important in order to confirm traceability and history of the handling of the product.

5.3   Oxygen Indicators

Often taken for granted, packaging is important for much of the quality we expect from food products. Traditionally, packaging fulfils four basic roles: containment, protection, communication and convenience. On the simplest scale the packaging acts as a barrier, protecting the food from outside contamination such as microbes and chemicals reaching sterilized or cleaned food. Such contamination would degrade the quality of the food, or possibly make it unsafe to eat. Especially in the information driven culture of today, important safety and nutritional information about the food is often conveyed through the packaging medium, such as printed labels and branding.

When the integrity of the package is relied upon to preserve the food, there is a higher expectation of the packaging materials to ensure that integrity is maintained, *i.e.* diffusion of gases through the materials, ruptures of the packaging and failures during the packaging process. Possibly more importantly, if the packaging fails to maintain the MAP environment, then this should be easily discernable and the food can be removed from circulation or repackaged.

Current methods of analyzing the atmosphere within MAP are often destructive. Sampling techniques are used which break the seal of the package and analyze a sample of the gas for its composition. These tests are called headspace analyses. Often carbon dioxide contribution is analyzed by FT-IR and oxygen levels determined electrochemically,[32] which has resulted in, for example, research into intelligent packaging.

Mills *et al.* have worked extensively on developing colorimetric oxygen indicators and intelligent inks for food packaging.[33,34] Colorimetric indicators use redox-active dyes, also referred to as electrochromes, which show a distinct colour change from the reduced to oxidized state. A reducing agent or process converts the electrochrome to its reduced form. In its simplest form a colorimetric indicator would be a mixture of reducing agent and electrochrome. Two ways of formulating a simple indicator would be: (a) limited reducing agent and (b) excess reducing agent.

In the case of (a), once oxygen has re-oxidized the electrochrome there is insufficient reducing agent or process to re-initiate reduction. This effectively makes the colour change irreversible so if the indicator shows positive for oxygen, oxygen could have been present at any stage in the lifetime of the package. An ideal colorimetric oxygen indicator would need to be kept in the reduced state while this process occurs; thus excess reducing agent (method b) is required. Exposure to ambient oxygen during the storage of these indicators also represents a problem, as premature oxidation further depletes the reducing agent. Two colorimetric oxygen indicators widely studied are methylene blue and viologen.

5.3.1   Methylene Blue

The Ageless Eye™ colorimetric indicator uses methylene blue as the dye. Other redox dyes that can be used in a colorimetric indicator are thiazines, oxazines, azines, indophenols, indigo dyestuffs and polyaniline. In the Ageless Eye™ indicator the dye is reduced using a strong reducing agent such as glucose under basic conditions according to Eqn (5.6):[33,34]

$$D_{Ox} + \text{Reductant} \rightarrow D_{Red} + \text{Oxidized reductant} \qquad (5.6)$$

$$D_{Red} + O_2 \rightarrow D_{Ox} + H_2O \qquad (5.7)$$

D represents the dye in either its reduced D_{Red} or oxidized D_{Ox} forms. In the case of methylene blue, the dye is blue in the oxidized form and colourless in the reduced leuco form.

A semiconductor can be used to reduce the electrochrome after exposure to UV light. When the semiconductor absorbs light energy, an electron is excited from the valance band into the conduction band. This generates a free electron in the conduction band and an electron hole in the valence band. The excited electron can then reduce the electrochrome, which is now in the active state ready to be exposed to atmospheric oxygen and be oxidized. After the UV excitation of the semiconductor, many of the electrons fall back to the holes they left in the valance band, which limits the amount of electrochrome the semiconductor can reduce. A sacrificial electron donor such as triethanolamine[35,36] is used in the formulation to donate an electron to the hole in the semiconductor valence band and prevent the recombination of the excited electrons and valance band holes.[37,38] The redox cycles within the so-called Strathclyde system are shown diagrammatically in Figure 5.4.[39] Titanium(IV) oxide is widely used as a photocatalyst. There are two principle phases for the structure of titanium oxide: rutile and anatase. The anatase phase is more efficient for UV photocatalysis as its band gap of 3.2 eV corresponds to the absorption of light with wavelengths less than 385 nm, which falls in the UV region.[39,40]

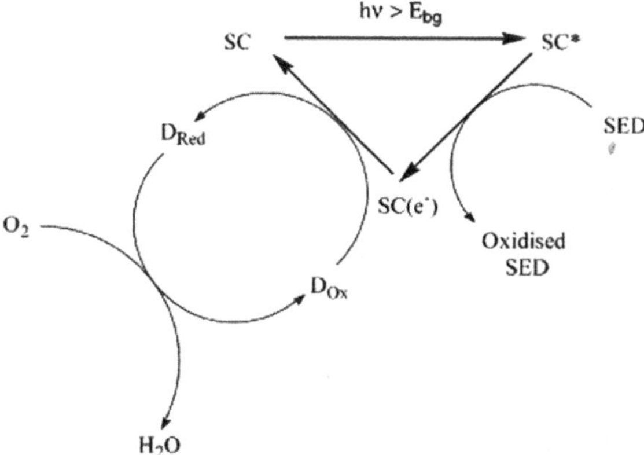

Figure 5.4   Redox cycle between semiconductor (SC), electrochrome (D), oxygen and sacrificial electron donor (SED).[39]

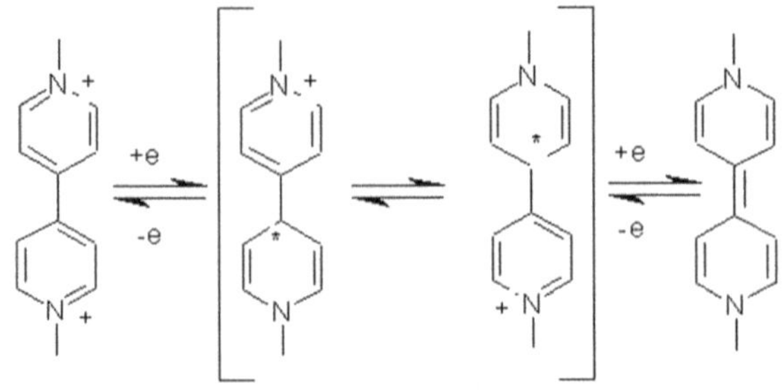

Figure 5.5 Redox steps of the viologens, and resonance stabilization in the monocation radical state.

5.3.2 Viologens

A viologen-based oxygen sensor has also been described, which replaces methylene blue as the redox dye component. The viologen is entrapped within a poly(vinyl alcohol) (PVA) film.[41] More recently, polyviologens have been exploited as potential colour indicators of oxygen levels. The viologen monomers have three redox forms. The oxidized dication species (V^{2+}) is reduced to the monocation radical species ($V^{+}*$), which can be reduced further to a neutral quinoid species (V^{0}). In 4,4′-dimethylviologen (MV), the dication is colourless, the monocation radical species is a dark violet and the quinoid species a slight brown colour; the structures are shown in Figure 5.5. Roberts *et al.* demonstrated that the intense blue colour of the reduced form of polyviologens reversibly faded to grey in the presence of oxygen.[42] The methylene blue and polyviologen systems demonstrate great potential to be used in food packaging where oxygen tension is crucial to the freshness of the product. The patch could therefore be an indicator of the food packaging seal being compromised.

5.4 Hydrogels and Colloids as Indicators

Recently, nanotechnology and bionanotechnology has seen the use of long-established nanochemistry such as colloids, quantum dots and gold nanoparticles to develop a range of smart colour-changing indicators for determining the presence of biomarkers. Gold nanoparticles bioconjugated with analyte-specific antibodies have been used to develop lateral flow assays such as the blue positive result in, for example, the pregnancy test kit. This technology has already been reviewed in Chapter 4 and is discussed in Section 5.5 below. There are opportunities also to extend the application of such bioconjugated nanoparticle systems for the determination of protein markers indicative of early stage cancer.[43–45]

Novel sensor devices composed of crystalline colloidal arrays (CCAs) immobilized in a hydrogel polymer matrix have been used in the development of smart colour changing patches to indicate the presence of metal ions and biomarkers.[46,47] The hydrogels are able to shrink and swell in response to the specific presence of a chemical compound. The lattice structure spacing of the embedded CCA changes as a function of an analyte-specific triggered gel volume change, thereby altering the wavelength of light diffracted by the CCA. This results in an indicative visible colour change in the gel (Figure 5.6).[48] A blue shift in absorbance (and Bragg diffraction) is due to gel shrinking causing the inter-particle spacing of the CCA to decrease. A red shift in the diffraction spectrum will occur upon gel swelling due to an increase in inter-particle spacing. Lack of control in ionic strength when measuring a biologically relevant target could cause major interference due to non-specific shrinking and swelling.

The sensor devices are made specific for a given target chemical through immobilizing the appropriate recognition element (*e.g.* an enzyme) within the CCA–gel system. For example, to detect Pb^{2+} ions, a crown ether was used as the ion-selective recognition agent within the gel. Pb^{2+} between 0.1 μM to 20 mM was detected visually using this technique. Researchers also employed the CCA–gel systems to measure glucose and galactose by using glucose oxidase and galactosidase, respectively, as the gel-immobilized recognition elements. The enzymes undergo conformational changes during catalytic conversion of their respective substrate. It is thought that the latter has an impact on the gel swelling, which in turn modulates the lattice spacing of the CCA.[49–51] The sensors have various applications in areas

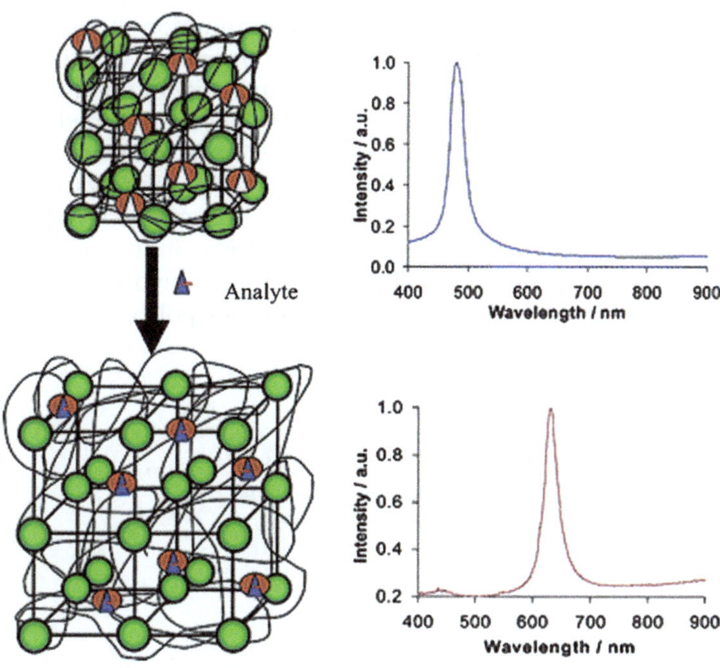

Figure 5.6 Schematic of analyte-specific gel swelling resulting in red-shift CCA Bragg diffraction.[48]

including environmental and medical diagnostics. Recently, this technique was used with gel-based molecularly imprinted polymers (MIP), selective for club drugs such as ketamine.[52] The colour changing patch was sensitive to ketamine levels in the range 1 pg mL^{-1} to 1 µg mL^{-1}, corresponding to a shift in absorption wavelength between 550 and 650 nm, respectively. This level of discrimination in concentration depended on the cross-linking density of the polymer. Additionally, the MIP only responded to ketamine when tested against a range of drugs, including morphine, PCP and amphetamine.

Stimuli-responsive polymers have attracted much research interest over the past several decades. They are effective in converting environmental stimuli such as pH, ionic strength, temperature and biomolecule concentration into a signal manifest as a change in physical property of the polymer. Such stimuli-responsive, largely reversible self-regulating systems come under the category of intelligent or smart materials. Polymer gels which are responsive to glucose concentration show much promise for the development of novel glucose measurement sensors and potential self-regulation (*e.g.* insulin infusion systems) for diabetic patients.[53]

Glucose is able to specifically bind with phenylboronic acids (PBAs). PBAs and their derivatives are another class of hydrogels known to selectively bind diols, including sugars, to form anionic cyclic boronate esters. A variety of chromophores and fluorophores (including anthracene, pyranine, fluorescein and rhodamine dyes) have been chemically coupled with PBA, allowing the colorimetirc or fluorimetric determination of sugars.[54,55] Azobenzene dyes have been used for the development of colorimetric sugar sensors;[56–58] the absorption wavelength and intensity of the dye changed as a function of the nature and concentration of the sugars added (Figure 5.7).

The incorporation of the glucose oxidizing enzyme, glucose oxidase (GOx), into a pH-sensitive hydrogel material can result in solute-gating polymers which swell and shrink as a function of glucose concentration (*cf.* Eqn. 5.1), where the gluconic acid specifically produced can induce size and shape change in pH-sensitive gels. Typical polymers investigated include poly(methacrylic acid-*g*-ethylene glycol) hydrogels and porous poly(vinylidene fluoride)

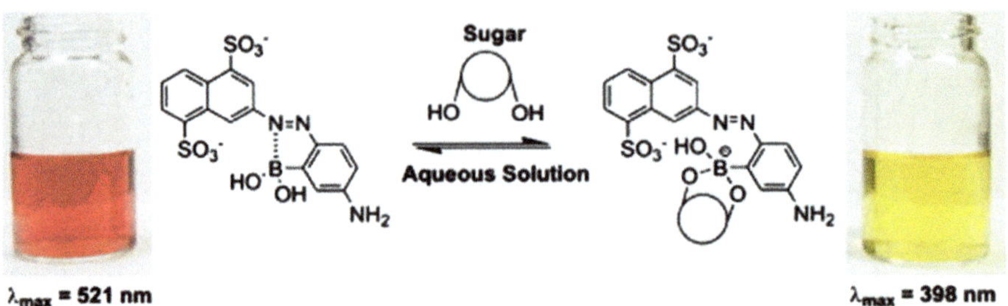

Figure 5.7   Colour change of *ortho*-azo-substituted PBA solution induced by sugar addition.
(Reproduced from Egawa *et al.*[54] with permission from Elsevier.)

(PVDF) and cellulose membranes grafted with poly(acrylic acid) (PAAC) within the pores.[53] With the GOx covalently linked to the PAAC in the pores, glucose diffusion would produce acid that protonated the carboxylate groups of the PAAC, causing a shrinkage of PAAC chains due to reduced electrostatic repulsion, which in turn would open the pores in the membrane and allow the release of insulin. The latter system has major implications for the development of a self-regulating artificial pancreas (for diabetic patients) for auto-insulin release in conditions of high glucose concentration.

The dyes have been used in solution, but have also been integrated into hydrogel polymer films, nanoparticles and cadmium selenide quantum dots (CdS QDs) to develop smart and highly sensitive colour changing materials for measuring glucose.[59–61] Holographic glucose sensors have also been developed by integrating PBA-coupled hydrogels with photonic CCAs.[62,63]

Holographic sensor systems again work on the swelling/shrinking characteristics of hydrogels. The basic mechanism of the holographic sensor relies in measuring changes in interference patterns between the incident beam through a grating and a reflected beam. A change in the distance of fringes would bring about a significant change in the wavelength (colour) of the reflected hologram, according to Bragg's law. Using this principle, Kabilan *et al.*[64] developed a colour changing glucose-responsive polymer matrix. In the presence of glucose, the matrix swelled and caused a red-shift of the hologram due to increased spacing between the fringes. A blue shift occurred with decreasing exposure to glucose, due to gel shrinkage. The same workers integrated PBAs (namely, 3-acrylamiodphenylboronic acid or 2-acrylamido-5-fluorophenylboronic acid) for better selective recognition of glucose over lactate. The sensor worked at physiological pH and ionic strength. Besides lactate interference, another drawback was that other saccharides with *cis*-diol groups present in blood were also able to react with boronic acid. Application of derivatized versions of PBA allowed for improved selectivity for glucose over other such sugars.

A disadvantage of the PBA/CCA colorimetric system is that it only responded in low ionic strength conditions and so would not work in physiological conditions. The ionic strength problem was addressed by introducing a polyacrylamide–poly(ethylene glycol) (PEG) hydrogel and a polyacrylamide-15-crown-5 hydrogel suspending phenylboronic acid groups. In the presence of glucose, the polymers self-assembled into a supramolecular structure causing blue shifts of the photonic crystal diffraction (gel shrinkage). These systems have been applied to ocular inserts, giving colour changing diagnostic contact lenses for patients with diabetes. Asher and co-workers improved this system for longer-term use by controlling the elasticity and the hydrophilic–lipophilic balance of the hydrogel system by co-polymerizing *n*-hexyl acrylate into an acrylamide–bisacrylamide hydrogel. They were able to measure glucose in blood in approximately 90 s and glucose concentrations in tear fluid in a significantly longer period of 300 s. They also demonstrated that the sensor was reversibly stable for up to one week.

Toxic metal ion sensing is possible with a range of chromophores. The chromophore is also an ionophore (ion complexing agent) and is chosen to be capable of selectively chelating with target metal ions. For example, diphenylcarbazide (DPC), 4-*n*-dodecyl-6-(2-thiazolylazo)resorcinol (DTAR) and 4-*n*-dodecyl-6-(2-pyridylazo)phenol (DPAP) chromophore molecules can selectively bind with cadmium ions, resulting in a visual colour change at sub-nanomolar concentrations.[65] Recently developed materials comprising ionophoric components are discussed in depth in Chapter 6.

Smart patches integrated into food packaging as a measure of food freshness and ripeness are commonly found in supermarkets in New Zealand and Australia, but have yet to be commonplace in food packaging in the northern hemisphere. The patches rely on the changing composition of certain room temperature gases (volatiles) in the sealed food packaging. Ethylene gas is a hormone produced and released by ripening fruit. Bananas produce significant amounts of the hormone during ripening and it is commonly known to keep ripe bananas away from other fruit to reduce cross-ripening, or actually close to other fruit that one wishes to ripen faster since the ethylene released from the banana can stimulate the ripening of neighbouring fruit. Molybdenum chemistry has been used to develop a smart indicator patch for the measurement of this ripeness indicator.[66] The smart patch is made of a thin film ceramic (combination of filter paper and colloidal silicon dioxide using a layer-by-layer self-assembly method) with ammonium molybdate and palladium sulfate catalyst immobilized within the ceramic matrix. In the presence of ethylene, the gas is catalytically oxidized to ethanal (acetaldehyde) and the molybdenum salt forms a dark blue mixed oxidation state (V and VI) oxide complex (Figure 5.8). The intensity of the blue colour is used as a semi-quantitative and qualitative measure of fruit ripening within the package. As an integrated component of food packaging, this patch is informative to all users in the food chain, including producers, retailers and consumers.

Such patches have appeared on the front display label of packaged apples,[67] with a three-label colour change indicating, for example, crispness, firmness and juiciness,[68] thereby giving the consumer information as to when to open the package based on their preference.

Oil rancidity technologies have been developed which rely on an optical cap 'sniffing' the vapours of extra virgin olive oil. The smart cap is made of an array of metalloporphyrin-based optochemical sensors, the colours of which are altered in the presence of aldehyde, the key component indicative of rancid off-flavours. The cap allowed for non-invasive testing of bottled oil

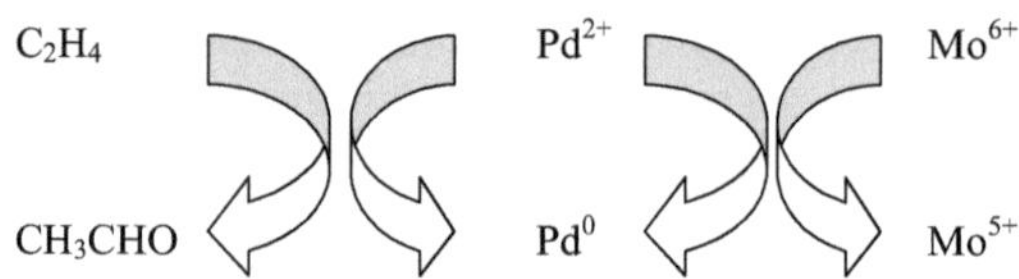

Figure 5.8   Schematic of molybdenum blue reaction.

quality. The spectral analysis of the chromophore array could be analyzed using multivariate data analysis to determine parameters such as oil ageing, rancidity and aroma.[69]

The development of reliable and sophisticated techniques for the authentication of food products continues to be a challenge in public food safety. The authenticity of food oils, for example, has become very important in order to retain consumer confidence in brand-name products with the ever increasing issues of counterfeit oil marketing. The conventional methods to authenticate products include chromatography and mass spectrometry and, more recently, microarray DNA-based methods have received attention.[70] Colorimetric methods offer cheaper and less labour-intensive methods to determine oil authenticity.

5.5 Biochemical Indicator Technologies

Jenison *et al.* have developed a thin film technology to allow the naked eye to detect nucleic acid sequences.[71] The technique relies on ultra-thin films being able to interfere with white light and split it into the different colours in the visible spectrum, depending on film thickness. For example, a change in thickness of 20 Å can be visualized. Such thickness changes are of the same order as protein–protein biomolecular interactions or hybridization of complementary nucleic acid sequences. This thin film technology has been successfully used on clinically relevant samples for the visual determination of pathogenic bacterial and viral nucleic acid sequences, as well as the discrimination of sequence variations in human genes tentatively related to disease susceptibility.

5.5.1 Lateral Flow Assays and Dipstick Assays

Lateral flow immunoassays (LFIAs) come under the category of dipstick tests and there is a rather high degree of materials engineering which has gone into making them efficient. Applications of LFIAs include measurement of pathogens, drugs, hormones and metabolites in biomedical and environmental settings, and they are specifically designed for single use and then disposal. Such tests offer a level of convenience for use by the layperson for the qualitative (yes/no) and semi-quantitative measurement of biomarkers in resource-poor and non-laboratory environments. The main components include a wicking surface to take up a suitable biological sample (such as blood or urine), a polymer membrane on a plastic backing (to transport the analyte and indicator component) and test and control lines some distance down from where the sample is introduced which change colour as a result of a target molecule being present. The latter lines typically have a biological antibody specific for the target molecule immobilized. Each aspect needs to be finely tuned to allow the assay to work efficiently. The indicator component is key for target molecule visualization on the test and control lines. The most prevalent LFIA is of course the pregnancy test kit (which has been specifically reviewed in Chapter 4). We will focus here on the recent advances in the indicators (labels) used to develop a colour change on the test and control lines. Whereas the test line will indicate a positive result, the control line is there to guarantee that the test is working properly. As the liquid front of the biological sample traverses laterally, the time taken to reach the test and control lines is crucial for the optimum deposition and recognition of the target molecule. The liquid sample moves under capillary action through the polymer matrix, typically nitrocellulose[72] although other polymer materials have been investigated.[73] The porosity and pore density of the membrane is an important parameter in determining diffusion of the sample and can range from 0.05 to 12 μm pores distributed throughout the membrane. The pore size needs to be compatible with the movement of not only the target molecule but also the indicating label. The target molecules can be quite large. For example, in the pregnancy test, the hormone being measured (human chorionic gonadotrophin; hCG) has a molecular mass of 36 000 Da.

The indicating labels are made of coloured or fluorescent nanoparticles ranging in size from 15 to 800 nm. These are also required to flow unimpeded through the membrane in order to allow the efficient development of colour in the test and control lines. The nanoparticles used include colloidal gold (see Chapter 4) or coloured latex particles. The particles are pre-modified with a monoclonal antibody specific to the target molecule and are retained at the beginning of the polymer membrane, near the junction between the wicking layer (for a bio-sample) and the polymer membrane. With the introduction of the liquid sample the target molecule (if present in the sample) is transported through into the membrane on the traversing liquid front and specifically picks up the antibody-coated nanoparticle following immunobinding. The bioconjugate then continues to traverse laterally through the membrane, reaching first the test line, followed by the control line. The test line is immobilized with a cheaper polyclonal antibody for the target and a colour change produced here is due to aggregation of a complex comprising the target analyte sandwiched between antibody-modified nanoparticles and the immobilized antibodies on the test strip. The control line contains an analyte–protein mixture and will interact with excess antibody-modified nanoparticles which will specifically bind to this line. The control line will always indicate a colour change if the test is working properly (*i.e.* the liquid sample traverses across the polymer membrane and releases and transports antibody-modified nanoparticles).

In addition to the pregnancy test kit, the lateral flow immunoassay is now being exploited for other bio-applications, including detection of plague-causing pathogens as well as cancer markers. Anish *et al.*[74] recently demonstrated a dipstick approach to measuring *Yersinia pestis*, the bacterium responsible for the black death plague (septicemic, pneumonic and bubonic). The technique again exploits the use of an antibody to detect the bacteria. This relies on identifying a key antigen produced by the bacteria, so that an antibody can be raised against it. The latter workers identified a heptose trisaccharide (a hapten) as a candidate antigen. In order to raise the antibodies, a synthetic form of the trisaccharide was conjugated with a carrier protein in order to induce a T-cell-dependent immune response in mice. The anti-hapten monoclonal antibodies were then tested for *Y. pestis* bacteria recognition and analyzed by immunofluorescence and confocal laser microscopy. Specific bacterial labelling was visualized using a secondary polyclonal antibody (anti-mouse IgG FITC). The antibodies were shown to be highly specific for *Y. pestis* as they did not recognize other bacteria, including *E. coli*, *Salmonella typhi* or *Neisseria mengingitidis*. Although prevalence of the disease is severely diminished compared with its dominance in the mid-14th century, there are still up to 2000 cases a year in sub-Saharan Africa. In such remote areas with a lack of laboratory technologies to perform complex analyses, such as phenotyping and the polymerase chain reaction, a simple diagnostic tool offers many benefits. Efforts to incorporate *Y. pestis* antibodies to point-of-care diagnostics platforms are currently under way.

While lateral flow immunoassays offer advantages of being simple to use, low-cost, specific and convenient with a prolonged shelf-life, there are also disadvantages. For example, they only provide semi-quantitative results; imprecise sample volumes could reduce precision; pores in the membrane may become occluded due to biological sample variation and matrix effects; and good antibody preparation is required. The latter need for good antibody preparation could be a long, involved and expensive process. Molecularly imprinted polymers (also known as plastic antibodies) may offer a cheaper alternative to replace antibodies in lateral flow assays. Reddy's group is actively researching in this area to offer such cheaper alternatives (see Chapter 3).

A disposable dip-stick biosensor has been developed enabling the visual determination of DNA by eye.[75] This has been extended to visual genotyping of single nucleotide polymorphisms (indicators of DNA alteration or damage). The dip-stick dry reagent format eliminates the need for multiple incubation and washing steps, which are often necessary in DNA assays. The presence of the correct complement DNA target in the sample solution elicits the formation of a biotinylated hybrid along the dip-stick, which is captured at the test zone of the sensor by immobilized streptavidin, which is in turn linked to strong colour-producing gold nanoparticles in the test zone. Such techniques lend themselves to small laboratory and field testing applications. Such dip-stick approaches have also been applied to detection of pathogens in biological fluids[76,77] and genetically modified organisms in food.[78–80] Dye-doped latex polymer beads have also been used in place of gold nanoparticles to induce the desired colour change for DNA detection in the test zone.[81] Although primarily a qualitative test for DNA, the test can be made quantitative by using a strip reader.

Bai *et al.* have developed an optical thin film biosensor chip for the rapid detection and discrimination of vegetable oils.[82] The method uses an array of aldehyde-labelled DNA probes covalently attached to a hydrazine-derivatized optical chip surface. In the presence of biotinylated complementary strands (PCR amplicons), hybridization occurs between the probe and complement. The hybrid authenticates the oil brand and is visualized with an anti-biotin IgG–horseradish peroxidase conjugate and a precipitatable substrate of the latter enzyme, resulting in a colour change on the chip surface from gold to blue/purple. The use of the enzyme horseradish peroxidase (HRP) to label antibodies for visual recognition of an antibody–antigen binding event (as in the latter example) is not new and has been used widely in enzyme-linked immunosorbent assays (ELISA) for the determination of biomarkers. More recently, this ELISA technique has been conjugated with plasmonics to develop optical (colour changing) array-based sensors for multiple protein biomarkers.[83] Arrays of gold nanoparticles were deposited using electron beam lithography, making multi-protein discriminations possible. Plasmonics refers to the laser excitation of a metal surface, where the surface electron plasma absorbs the laser energy and resonates, making the surface sensitive to mass adsorption.[84] When single nanoparticles are used, it is possible to reach single-molecule detection sensitivities. Chen *et al.*[83] coupled the protein-selective HRP-based ELISA system to gold nanoparticles, making the gold surface sensitive to the precipitation product from the HRP–substrate reaction, which only occurs if the target protein is present.

The use of a fluorophore or chromophore as the visual indicator in an optical sensor strategy adds an extra step for the optical transduction in order to indicate the presence of a target analyte. Galban *et al.* have suggested the use of reagentless optical sensing for organic compounds based on auto-indicating proteins such as flavoenzymes and heme proteins.[85] The proteins themselves are used as transducers by monitoring changes in their spectroscopic properties, thereby obviating the need for a separate colour-changing label.

A range of fluorescent proteins such as green, yellow and cyan fluorescent proteins are commercially available and are being used in a range of biological tests and for imaging protein movement and dynamic protein interactions in biological systems.[86,87] Biosensors have subsequently been developed to track a wide spectrum of intracellular proteins and employ the FRET effect (Förster resonance energy transfer) between two fluorescent proteins as a means of transduction and indication of the analyte of interest, such as change in pH, calcium ion activity, enzyme activity and membrane potential. Typically, an active peptide sensitive to one of the above markers is sandwiched between two fluorescent proteins, the emission wavelength of one protein being able to excite the other. The two proteins need to be very close together to allow this FRET energy transfer. The appropriate distance for energy transfer is achieved by the selective interaction of the sandwiched peptide with the analyte of interest.[88]

5.6 Metal Complexes as Indicators

Metal complexes are formed when electron-rich moieties of organic molecules are able to form dative bonds with electron-deficient metal ions.

Metal complexes are essential to all living systems. Chlorophyll *a* and *b* comprise a macrocyclic organic molecule with an Mg centre. Hemoglobin is a tetrameric quaternary metalloprotein with four porphyrin rings, each with an Fe centre. Owing to the electronic transitions that occur during light absorbance, metal complexes are generally coloured and the colour relies on the HOMO–LUMO band gap absorption energy of the complex. This band gap is not only a function of the conjugated organic ligands bound to the metal centre, but it is also a function of the d–d transitions in the metal itself from partially occupied to partially unoccupied orbitals. Thus, depending on the metal centre and depending on the bound ligands, a spectrum of coloured compounds is possible.

Such complexes have been used to identify metal ions, by a simple colour response. The level of selectivity for a particular metal ion depends on the binding constant (or stability constant) for that metal–ligand binding event, and in some cases the macrocyclic ligand not only forms dative bonds but also offers a cavity size fit for a given ionic radius. The ligands can also be referred to as ionophores, and have been used as the ion-selective component in ion-selective electrodes. For example, the antibiotic valinomycin is highly selective for K^+ binding. The following focuses on some developments where the chromophoric ability of the ionophore is exploited for recognition of important biomarkers.

Porphyrins and phthalocyanines (PCs) are able to complex with a range of metals to produce metalloporphyrins (MPs) and metallophthalocyanines (MPCs), resulting in an array of different coloured products. The latter compounds are able to change their colour in the presence of a range of volatile organic compounds (VOCs) and have led to the development of novel optoelectronic nose technologies. Indeed, it has been recognized that the mammalian olfactory receptors are based on metalloproteins.[89] Porphyrins have many desirable properties compared with the organic chromophores typically used as pH indicator dyes. They have greater thermal and photo-chemical stabilities; their extended π-conjugated macrocyclic rings lead to large non-linear optical effects; and subtle changes in their physical properties can be produced through chemical modification of their peripheral structure.[90]

For MPs and MPCs to be used as colorimetric chemoresponsive sensors, they require a chromophoric centre to strongly interact with the analyte. MPs exhibit excellent capability for the detection of metal-ligating VOCs, because of their open coordination sites for axial ligation (above and below the porphyrin plane) to the metal centre. The compounds are intensely coloured and can exhibit large spectral shifts upon ligand binding, which is an advantage. The key interactions and their relative strengths for MP and MPC dyes and the target molecule decrease in significance in the order: bond formation and coordination (>100 kJ mol^{-1})$>$Lewis acid–base interactions$>$hydrogen bonding$>$charge transfer and π–π delocalization$>$dipole–dipole interactions$>$van der Waals (physisorption/adsorption; <5 kJ mol^{-1}). While pH indicator dyes respond to change in Brønsted acidity and basicity, the MPs and MPCs exhibit solvatochromic properties and change colour in response to changes in general polarity in their chemical environment. Typically, these chemoresponsive dye systems can be immobilized on a variety of inert solid supports, including porous polymeric membranes (typically, nylon or PVDF), acid-free paper or silica gel plates to produce arrays of coloured spots. Suslick *et al.* produced a 24-dye array for a series of different VOCs covering the most common organic functional groups, namely amine, arenes, alcohols, aldehydes, carboxylic acids, esters, halocarbons, ketones, phosphines, sulfides and thiols. By using digital scans, they could subtract the RGB (red–green–blue) components from the before and after images, resulting in a difference map with a distinctive pattern.[89] The resultant colour values are the absolute values of the colour change in each dye spot. Excellent discrimination between these analytes is observed even without performing any statistical analyses. For complex mixture analyses, each analyte response can be represented as a change in red, green and blue values across the 24 dyes, which effectively gives a 72-dimensional vector. Hierarchical cluster analysis can be performed on digitally recorded data to examine the multivariate distances between the analyte responses of a 72-dimensional RGB colour space. A principal component analysis (PCA) can find the independent linear combinations of the changes in RGB values. This allows the analytes to be discriminated in 2-D or 3-D space. The reader is directed to Chapter 7 if they wish to develop a further understanding of multivariate and PCA techniques for multi-analyte discrimination. The major drawback of electronic nose technologies has been interference from changes in relative humidity. Water vapour can range in the environment from <2000 ppm to >20 000 ppm. Therefore, even very low interference from water can cause significant issues if we are interested in determining volatiles in the low ppm or ppb range. However, the colorimetric sensor arrays using MPs and MPCs are based on water-insoluble and hydrophobic dyes which have been printed on highly hydrophobic surfaces, making the arrays essentially immune to changes in humidity.

Sarin has been reported as being used as a chemical weapon during civil unrest and also in terrorist attacks. It is highly toxic, affecting muscle function and in high doses will cause death by asphyxiation. Therefore, a method of rapid detection of such a chemical attack would be crucial both for civilians and the armed forces alike. A metal complex that can be used to develop simpler and more sensitive detection devices for the nerve agent sarin has been described by Ordronneau *et al.*[91] The technique is elegant in that it relies upon the discolouration/decomplexation of a chromophore from its metal centre. In this example, the chromophore was based on a bipyridine ligand coordinated to iron [namely, a tris(bipyridine)iron(II) coordination complex], the electronic transitions giving rise to a red colour. However, in the presence of the organophosphate sarin, the ligand's structure is irreversibly altered causing decomplexation of the metal, thereby preventing electronic transitions, and the colour vanishes. This technique can be the basis of a quick visual sensor, but may also be integrated with a hand-held reflectometer or a flat bed scanner with dedicated software for rapid measurement and semi-quantitative read-out.[92]

5.7 Time–Temperature Indicator Labels

In the interest of convenience for consumers, but also as a health and safety check on the processing history of fresh foods, time–temperature indicators (TTIs) have been developed. The TTIs, in the form of a colorimetric strip, can offer an irreversible timeline indicator of the temperature history of a food item, from packaging production and transit to retailer receipt and consumer purchase. This test strip can be informative throughout the latter supply chain. For example, a perishable food batch that should be stored below 4 °C, or frozen[93–95] throughout the supply chain, can be flagged-up as potentially spoiled food if it has spent any significant time above the storage temperature (as indicated by the TTI).[96] Yoon *et al.*[97] demonstrated the integration of a phospholipase enzyme system into frozen food packaging. Although an enzyme was used, they were able to show that it could still demonstrate sufficient catalytic function to determine frozen pork meat quality. The TTI contained glycerol and sorbitol to stop the enzyme from freezing in the test strip. The strip also contained a mixture of pH indicators, namely bromothymol blue, neutral red and methyl red; the total colour change was indicative of microbial food degradation due to CO_2 metabolite production. Such enzymatic reactions have also been used in the colorimetric evaluation of quality changes in frozen vegetables.[97]

The latter are examples of dyes that respond indirectly to temperature rises resulting from meat degradation. There is the possibility to use thermochromic dyes, which directly respond to temperature change.[98] For example, cresol red and the betaine dye 2,6-diphenyl-4-(2,4,6-triphenylpyridino)phenolate (DTPP) exhibit thermochromism when embedded in a hydrogel network. In the case of DTPP at pH 8.5, the colour changes from colourless at 10 °C to a deep violet at 80 °C. In the case of cresol red, the colour change is from yellow to red. The temperature-induced colour change is thought to be due to a shift in the proton-transfer equilibrium between the phenolate and phenol forms of the dye molecules in the gel microenvironment. The current drawback with these dye systems is that the colour change is reversible, when the temperature is reversed. Therefore, a temperature indicator based on such dyes, while useful for real-time monitoring of temperature fluctuations, will not give a history of such fluctuations.

With the best intentions, all such authentication methodologies can be subject to abuse through tampering and therefore adding the extra element of making them tamper-proof is required. For example, by integrating the TTI with a radiofrequency identification (RFID) protocol[99] for instant reporting of temperature fluctuations to the key stake-holders (producers, transporters and retailers) would be beneficial in circumventing any tampering. In a typical RFID system, a receiver can transmit radio waves to capture data from an RFID tag. The data can then be passed on to a host computer, which can then relay the information through the supply chain management.[100] In a typical RFID, there is a microchip, smaller than a grain of rice, connected to a tiny antenna. RFIDs can be passive or active, the former being powered by the energy supplied by the reader and the latter with its own battery for powering the microchip's circuitry and transmitting signals to the reader. Their range depends on the rf band used, but typically ranges from a few cm to 1 or 2 m. The RFIDs need to be cheap for global uptake of this technology, and with major users, such as Wal-Mart and the U.S. Department of Defense, mandating the use of RFID tracking technology, this has driven down the price per RFID to a fraction of a US dollar. RFIDs have pervaded all aspects of modern society, including uses in commerce (social media, inventory systems and payment for goods by mobile phones), transportation logistics, passports, animal and human identification and institutions including libraries, healthcare and universities.

Mainetti *et al.* demonstrated that RFID technology could be used for tracing and tracking fresh vegetables through the supply chain, allowing the consumer to know the complete history of the purchased product.[100] Steinberg and Steinberg demonstrated the integration of a battery-free RFID tag with an optoelectronic interface comprising a silicon photodiode with two LED sources. The detector could indicate any pH change by the colour change in bromocresol green entrapped within a sol–gel system. The device could therefore be used for microbial and temperature related degradation of food. The RFID wireless communication platform offers advantages of informing stakeholders of real-time changes in a remote fashion.[101]

5.8   On-going Developments and Future Perspectives

Colloidal crystals, such as monodisperse silica particles (of approximately 100 nm diameter), have been used to develop colour-sensitive and chemically-selective polymers and hydrogels and are undergoing rapid development.[46,47] The polymers and gels can be designed to be highly selective for a target analyte of interest of medical or environmental significance. In the presence of the analyte, a corresponding shift in the gel volume results in a shift of the Bragg diffraction due to a change of the periodic lattice spacing. This results in a visually perceptible change in colour of the gel.

Electrospinning involves a technique of applying a dc electric field while a polymer solution is drawn through a narrow nozzle. This results in polymer nano-scale fibres being produced. Integration of electrospun polymer fibres with photochromic materials offers promise for developing sensitive optical devices and/or biosensors.[102] This technique holds much promise for the future of highly selective and simple-to-use visual indicators.

DNAzymes are biocatalysts with a promising capacity to selectively identify charged organic and inorganic compounds at ultra-trace levels and can be applied as nano-biological recognition probes in sensing in conjunction with laser-based optical detection. Promising areas of application include *in situ* monitoring of contamination.[103]

Shape memory materials (SMMs) and shape memory alloys (SMAs) are a class of active smart materials which change their shape reversibly due to environmental stimuli such as temperature and pH. Shahinpoor and Martinez reported a shape memory alloy which was able to detect a one-time temperature excursion into non-allowable temperatures for non-frozen food as well as pharmaceutical or other medical products.[104] A range of polymeric materials including hydrogels and cellulosic- and polyurethane-based systems have been employed as SMMs.[93,105–107] Integration of SMMs with markers of biological change could lead to a visually appreciable physical effect. Such indicator technologies have been little studied in this regard, but hold much promise in developing simple visual indicators. Indeed, the CCA integration with hydrogels as described in Section 5.4 can be regarded as a shape memory effect, although it is not generally described in this manner. Further exploration of bio-responsive polymeric materials which can induce colour change or shape change is warranted.[84,108–110]

References

1. D. L. Williams, T. J. Flaherty and B. K. Alnasleh, *J. Chem. Educ.*, 2009, **86**, 333.
2. C. Tack, H. Pohlmeier, T. Behnke, V. Schmid, M. Grenningloh, T. Forst and A. Pfuetzner, *Diabetes Technol. Ther.*, 2012, **14**, 330.
3. G. P. James and D. E. Bee, *Clin. Chem.*, 1979, **25**, 996.
4. R. Cha, D. Wang, Z. He and Y. Ni, *Carbohydr. Polym.*, 2012, **88**, 1414.
5. T. B. Savage and L. P. Stratton, *Appl. Microbiol.*, 1971, **22**, 809.
6. B. Saad, W. T. Wai, S. Jab, W. S. W. Ngah, M. I. Saleh and J. M. Slater, *Anal. Chim. Acta*, 2005, **537**, 197.
7. W. J. Cooper, C. A. Sorber and E. P. Meier, *J. Am. Water Works Assoc.*, 1975, **67**, 34.
8. J. B. Claver, M. C. V. Miron and L. F. Capitan-Vallvey, *Anal. Chim. Acta*, 2004, **522**, 267.
9. R. Haddoub, D. Ferry, P. Marsal and O. Siri, *New J. Chem.*, 2011, **35**, 1351.
10. K. Oguri, S. Wada, S. Eto and H. Yamada, *Jpn. J. Toxicol. Environ. Health*, 1995, **41**, 274.
11. Y. Tsumura, T. Mitome and S. Kimoto, *Forensic Sci. Int.*, 2005, **155**, 158.
12. P. Dubey, S. K. Shukla and K. C. Gupta, *Aust. J. Forensic Sci.*, 2013, **45**, 165.
13. United Nations Drug Control Programme, *Rapid Testing Methods of Drugs of Abuse, Manual for Use by National Law Enforcement and Narcotics Laboratory Personnel*, 1-3-1994, United Nations, New York, 1994.
14. A. Parks, M. Brashears, W. Woerner, J. Martin, L. Thompson and J. Brooks, *J. Anim. Sci.*, 2012, **90**, 2054.

15. A. Parks, M. Brashears, W. Woerner, J. Martin, L. Thompson and J. Brooks, *J. Anim. Sci.*, 2012, **90**, 642.

16. K. Crowley, A. Pacquit, J. Hayes, K. T. Lau and D. Diamond, *Proc. IEEE Sens.*, 2005, 754.

17. B. Kuswandi, Jayus, A. Restyana, A. Abdullah, L. Y. Heng and M. Ahmad, *Food Control*, 2012, **25**, 184.

18. A. Pacquit, K. T. Lau and D. Diamond, *Proc. IEEE Sens.*, 2004, 365.

19. A. Seephueng, H. Manuspiya, R. Magaraphan and M. Nithitanakul, *Abstr. Pap. Am. Chem. Soc.*, 2008, **235**, 337-PMSE.

20. M. Caldara, C. Colleoni, E. Guido, V. Re and G. Rosace, *Sens. Actuators B*, 2012, **171**, 1013.

21. S. Jurmanovic, S. Kordic, M. D. Steinberg and I. M. Steinberg, *Thin Solid Films*, 2010, **518**, 2234.

22. P. C. A. Jeronimo, A. N. Araujo, M. Conceicao, B. S. M. Montenegro, D. Satinsky and P. Solich, *Analyst*, 2005, **130**, 1190.

23. N. Ibrisimovic, M. Ibrisimovic, M. Barth and U. Bohrn, *Monatsh. Chem.*, 2010, **141**, 125.

24. A. Azizishirazi and S. S. Shekarforoush, *Adv. Food Sci.*, 2010, **32**, 41.

25. A. Pacquit, K. T. Lau and D. Diamond, *Proc. Soc. Photo-optical Instrum. Eng.*, **5826**, 545.

26. A. Pacquit, J. Frisby, D. Diamond, K. T. Lau, A. Farrell, B. Quilty and D. Diamond, *Food Chem.*, 2007, **102**, 466.

27. B. Kuswandi, Jayus, T. S. Larasati, A. Abdullah and L. Y. Heng, *Food Anal. Methods*, 2012, **5**, 881.

28. A. Mills, *Sens. Environ. Health Security*, 2009, 347.

29. A. Mills, G. A. Skinner and P. Grosshans, *J. Mater. Chem.*, 2010, **20**, 5008.

30. A. Nopwinyuwong, S. Trevanich and P. Suppakul, *Talanta*, 2010, **81**, 1126.

31. A. Mills and K. Eaton, *Quim. Anal.*, 2000, **19**, 75.

32. F. C. O'Mahony, T. C. O'Riordan, N. Papkovskaia, J. P. Kerry and D. B. Papkovsky, *Food Control*, 2006, **17**, 286.

33. A. Mills, *Chem. Soc. Rev.*, 2005, **34**, 1003.

34. A. Mills and S. K. Lee (University of Strathclyde), *U.S. Pat.*, 8 114 673, 2012.

35. S. K. Lee, A. Mills and A. Lepre, *Chem. Commun.*, 2004, 1912.

36. S. K. Lee, M. Sheridan and A. Mills, *Chem. Mater.*, 2005, **17**, 2744.

37. A. Mills, D. Hazafy and K. Lawrie, *Catal. Today*, 2011, **161**, 59.

38. A. Mills, S. K. Lee and M. Sheridan, *Analyst*, 2005, **130**, 1046.

39. A. Mills, K. Lawrie, J. Bardin, A. Apedaile, G. A. Skinner and C. O'Rourke, *Analyst*, 2012, **137**, 106.

40. T. Murakata, R. Yamamoto, Y. Yoshida, M. Hinohara, T. Ogata and S. Sato, *J. Chem. Eng. Jpn.*, 1998, **31**, 21.

41. A. Mills, M. McFarlane and S. Schneider, *Anal. Bioanal. Chem.*, 2006, **386**, 299.

42. L. Roberts, R. Lines, S. Reddy and J. Hay, *Sens. Actuators B*, 2011, **159**, 342.

43. C. Fang, Z. Chen, L. Li and J. Xia, *J. Pharm. Biomed. Anal.*, 2011, **56**, 1035.
44. Z. Li, Y. Wang, J. Wang, Z. Tang, J. G. Pounds and Y. Lin, *Anal. Chem.*, 2010, **82**, 7008.
45. D. Tang, Y. Cui and G. Chen, *Analyst*, 2013, **138**, 981.
46. J. H. Holtz and S. A. Asher, *Nature*, 1997, **389**, 829.
47. J. H. Holtz, J. S. W. Holtz, C. H. Munro and S. A. Asher, *Anal. Chem.*, 1998, **70**, 780.
48. G. R. Hendrickson and L. Lyon, *Soft Matter*, 2009, **5**, 29.
49. M. Ben-Moshe, V. L. Alexeev and S. A. Asher, *Anal. Chem.*, 2006, **78**, 5149.
50. V. L. Alexeev, S. Das, D. N. Finegold and S. A. Asher, *Clin. Chem.*, 2004, **50**, 2353.
51. S. A. Asher, V. L. Alexeev, A. V. Goponenko, A. C. Sharma, I. K. Lednev, C. S. Wilcox and D. N. Finegold, *J. Am. Chem. Soc.*, 2003, **125**, 3322.
52. L. Meng, P. Meng, Q. Zhang and Y. Wang, *Anal. Chim. Acta*, 2013, **771**, 86.
53. L. Y. Chu, Y. Li, J. H. Zhu, H. D. Wang and Y. J. Liang, *J. Controlled Release*, 2004, **97**, 43.
54. Y. Egawa, T. Seki, S. Takahashi and J. Anzai, *Mater. Sci. Eng. C*, 2011, **31**, 1257.
55. Y. Ma, N. Li, C. Yang and X. R. Yang, *Colloids Surf. A*, 2005, **269**, 1.
56. Y. Egawa, R. Gotoh, S. Niina and J. Anzal, *Bioorg. Med. Chem. Lett.*, 2007, **17**, 3789.
57. Y. Egawa, R. Gotoh, T. Seki and J. Anzai, *Mater. Sci. Eng. C*, 2009, **29**, 115.
58. N. DiCesare and J. R. Lakowicz, *Org. Lett.*, 2001, **3**, 3891.
59. Q. Cui, M. M. Muscatello and S. A. Asher, *Analyst*, 2009, **134**, 875.
60. M. M. Muscatello, L. E. Stunja and S. A. Asher, *Anal. Chem.*, 2009, **81**, 4978.
61. M. M. Muscatello, L. E. Stunja and S. A. Asher, *Anal. Chem.*, 2009, **81**, 4978.
62. K. Gawel, D. Barriet, M. Sletmoen and B. T. Stokke, *Sensors*, 2010, **10**, 4381.
63. Q. Wu, L. Wang, H. Yu, J. Wang and Z. Chen, *Chem. Rev.*, 2011, **111**, 7855.
64. S. Kabilan, A. J. Marshall, F. K. Sartain, M. C. Lee, A. Hussain, X. P. Yang, J. Blyth, N. Karangu, K. James, J. Zeng, D. Smith, A. Domschke and C. R. Lowe, *Biosens. Bioelectron.*, 2005, **20**, 1602.
65. S. A. El Safty, D. Prabhakaran, A. A. Ismail, H. Matsunaga and F. Mizukami, *Adv. Funct. Mater.*, 2007, **17**, 3731.
66. J. H. Kim and S. Shiratori, *Jpn. J. Appl. Phys., Part 1*, 2006, **45**, 4274.
67. C. Lang and T. Huebert, *Food Bioprocess Technol.*, 2012, **5**, 3244.
68. K. Sharrock, in *Emerging Food Packaging Technologies*, ed. K. L. Yam and D. S. Lee, Woodhead, Cambridge, 2012, p. 175.
69. A. G. Mignani, L. Ciaccheri, H. Ottevaere, H. Thienpont, L. Conte, M. Marega, A. Cichelli, C. Attilio and A. Cimato, *Anal. Bioanal. Chem.*, 2011, **399**, 1315.

70. L. J. Ou, P. Y. Jin, X. Chu, J. H. Jiang and R. Q. Yu, *Anal. Chem.*, 2010, **82**, 6015.
71. R. D. Jenison, R. Bucala, D. Maul and D. C. Ward, *Expert Rev. Mol. Diagn.*, 2006, **6**, 89.
72. G. E. Fridley, C. A. Holstein, S. B. Oza and P. Yager, *MRS Bull.*, 2013, **38**, 326.
73. M. Kehrel, N. Dassinger, S. Merkl, D. Vornicescu and M. Keusgen, *Phys. Status Solidi A*, 2012, **209**, 917.
74. C. Anish, X. Guo, A. Wahlbrink and P. H. Seeberger, *Angew. Chem. Int. Ed.*, 2013, **52**, 9524.
75. I. K. Litos, P. C. Ioannou, T. K. Christopoulos, J. Traeger-Synodinos and E. Kanavakis, *Biosens. Bioelectron.*, 2009, **24**, 3135.
76. C. Duran, F. Nato, S. Dartevelle, T. P. Lan Nguyen, N. Taneja, M. N. Ungeheuer, G. Soza, L. Anderson, D. Benadof, A. Zamorano, Tai The Diep, Q. N. Truong, H. N. Vu, C. Ottone, E. Begaud, S. Pahil, V. Prado, P. Sansonetti and Y. Germani, *PLoS One*, 2013, **8**, e80267.
77. L. A. W. Smith, R. Hillman, J. Ward, D. M. Whiley, L. Causer, S. Skov, B. Donovan, J. Kaldor and R. Guy, *Sexually Transmitted Infections*, 2013, **89**, 320.
78. D. S. Elenis, D. P. Kalogianni, K. Glynou, P. C. Ioannou and T. K. Christopoulos, *Anal. Bioanal. Chem.*, 2008, **392**, 347.
79. D. P. Kalogianni, T. Koraki, T. K. Christopoulos and P. C. Ioannou, *Biosens. Bioelectron.*, 2006, **21**, 1069.
80. E. Michelini, P. Simoni, L. Cevenini, L. Mezzanotte and A. Roda, *Anal. Bioanal. Chem.*, 2008, **392**, 355.
81. X. Mao, W. Wang and T. E. Du, *Talanta*, 2013, **114**, 248.
82. S. Bai, S. Li, T. Yao, Y. Hu, F. Bao, J. Zhang, Y. Zhang, S. Zhu and Y. He, *Food Control*, 2011, **22**, 1624.
83. S. Chen, M. Svedendahl, T. J. Antosiewicz and M. Kall, *ACS Nano*, 2013, **7**, 8824.
84. T. Endo, A. Shibata, Y. Yanagida, Y. Higo and T. Hatsuzawa, *Mater. Lett.*, 2010, **64**, 2105.
85. J. Galban, V. Sanz, E. Mateos, I. Sanz-Vicente, A. Delgado-Camon and S. de Marcos, *Protein Pept. Lett.*, 2008, **15**, 772.
86. M. A. Hink, T. Bisseling and A. J. W. G. Visser, *Plant Mol. Biol.*, 2002, **50**, 871.
87. M. Wolff, S. Kredel, J. Wiedenmann, G. Nienhaus and R. Heilker, *Comb. Chem. High Throughput Screening*, 2008, **11**, 602.
88. H. Ai, K. L. Hazelwood, M. W. Davidson and R. E. Campbell, *Nat. Methods*, 2008, **5**, 401.
89. K. S. Suslick, *MRS Bull.*, 2004, **29**, 720.
90. K. S. Suslick, N. A. Rakow, M. E. Kosal and J. H. Chou, *J. Porphyrins Phthalocyanines*, 2000, **4**, 407.
91. L. Ordronneau, A. Carella, M. Pohanka and J. P. Simonato, *Chem. Commun.*, 2013, **49**, 8946.
92. B. Jayawardane, I. D. McKelvie and S. D. Kolev, *Talanta*, 2012, **100**, 454.

93. J. W. Akers, G. A. Coles and J. M. Zerkus, *U.S. Pat.*, 7 604 398-B1.
94. R. P. Singh and J. H. Wells, *Food Technol.*, 1985, **39**, 42.
95. J. H. Wells, R. P. Singh and A. C. Noble, *J. Food Sci.*, 1987, **52**, 436.
96. M. Nuin, B. Alfaro, Z. Cruz, N. Argarate, S. George, Y. Le Marc, J. Olley and C. Pin, *Int. J. Food Microbiol.*, 2008, **127**, 193.
97. S. H. Yoon, C. H. Lee, D. Y. Kim, J. W. Kim and K. H. Park, *J. Food Sci.*, 1994, **59**, 490.
98. A. Seeboth, J. Kriwanek and R. Vetter, *J. Mat. Chem.*, 1999, **9**, 2277.
99. I. M. Steinberg and M. D. Steinberg, *Sens. Actuators B*, 2009, **138**, 120.
100. L. Mainetti, F. Mele, L. Patrono, F. Simone, M. L. Stefanizzi and R. Vergallo, *Int. J. Antennas Propagation*, 2013, 531364.
101. K. L. Yam, P. T. Takhistov and J. Miltz, *J. Food Sci.*, 2005, **70**, R1.
102. S. Liu, L. Tan, W. Hu, X. Li and Y. Chen, *Mater. Lett.*, 2010, **64**, 2427.
103. R. Vannela and P. Adriaens, *Crit. Rev. Environ. Sci. Technol.*, 2006, **36**, 375.
104. M. Shahinpoor and D. R. Martinez, in *Smart Temperature Sensors*, ed. J. H. Jacobs, SPIE-Int Soc Optical Engineering, Bellingham, WA, USA, 1999, 3674, 127.
105. E. B. Ozkutuk, S. E. Diltemiz, E. Ozalp, T. Gedikbey and A. Ersoz, *Mater. Chem. Phys.*, 2013, **139**, 107.
106. X. Qiu and S. Hu, *Materials*, 2013, **6**, 738.
107. E. I. Saavedra Flores, M. I. Friswell and Y. Xia, *J. Intell. Mater. Syst. Struct.*, 2013, **24**, 529.
108. M. Shahinpoor, Y. Bar-Cohen, T. Xue, J. O. Simpson and J. Smith, *Proc. SPIE*, 1998, 251.
109. X. Li, Y. Wang, Z. Zheng, X. Ding and Y. Peng, *Prog. Chem.*, 2013, **25**, 1726.
110. S. Song, R. Dong, D. Wang, A. Song and J. Hao, *Soft Matter*, 2013, **9**, 4209.

Calixpyrrole: From Fundamental Studies to the Development of Ion Selective Electrodes

ANGELA F. DANIL DE NAMOR,* OLIVER A. WEBB, ABDELAZIZ EL GAMOUZ, WEAM ABOU HAMDAN AND MAAN AL-NUAIM

Thermochemistry Laboratory, Department of Chemistry, Faculty of Engineering and Physical Sciences, University of Surrey, Stag Hill, Guildford, GU2 7XH, UK
*Email: A.Danil-De-Namor@surrey.ac.uk

6.1 Calixpyrrole and Derivatives: Structural Features

Calix[n]pyrroles ($n = 4$, 5, 6) belong to the family of hetero-calixarenes.[1] These hetero-calixarenes have been the subject of numerous investigations, ranging from fundamental studies to a wide variety of applications. A simple calix[4]pyrrole,[2–4] namely *meso*-octamethylcalix[4]pyrrole, can be synthesized in a single-step procedure in high yield by the condensation reaction between pyrrole and acetone in acid medium, as shown in Scheme 6.1.

It was for more than a century that no attention was paid to this calix[4]pyrrole (CP), first reported by Baeyer in 1886.[5] The first report for the

---

RSC Detection Science Series No. 3
Advanced Synthetic Materials in Detection Science
Edited by Subrayal Reddy
© The Royal Society of Chemistry 2014
Published by the Royal Society of Chemistry, www.rsc.org

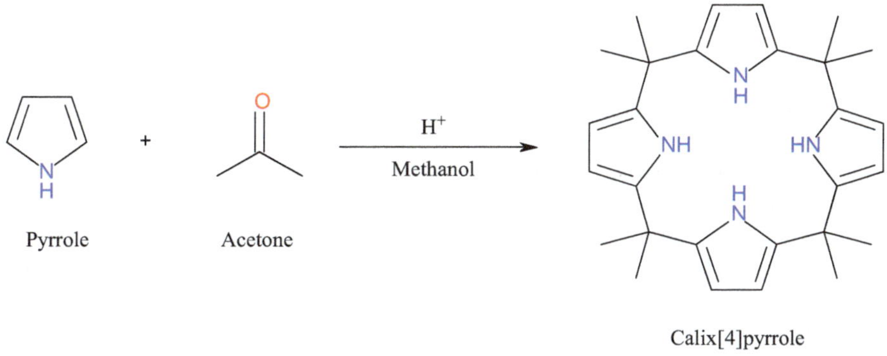

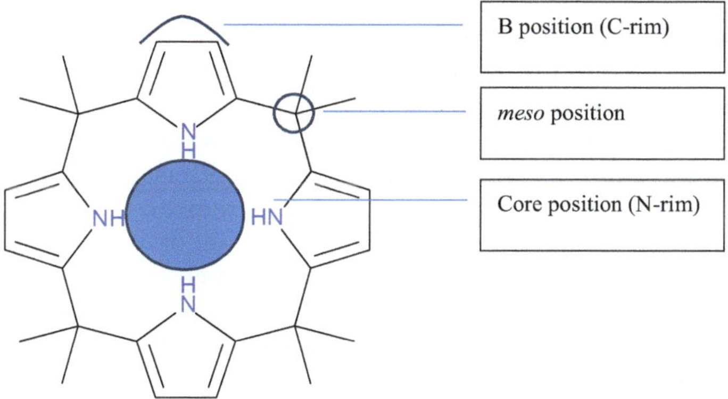

Scheme 6.1 Synthesis of calix[4]pyrrole from pyrrole and acetone in methanol.

Figure 6.1 Anatomy of calix[4]pyrrole.

synthesis of modified calixpyrroles appeared in 1958 with the contribution of Brown and co-workers.[6] The metallation of deprotonated calixpyrroles in 1996 by Fioriani and co-workers[7] renewed interest in these macrocycles. Research on calixpyrroles has increased significantly in recent years following the discovery in the mid-twentieth century that the presence of the NH array in the calixpyrrole structure can provide a binding site for complexation with anionic and neutral species. Particular emphasis has been placed on anions due to their role in biological and environmental systems. This is clearly reflected in the contributions made by Sessler and co-workers.[8–16] In recent years, considerable attention has been paid to this receptor and particularly to its wide number of derivatives. Thus research has been directed towards the design and synthesis of calix[4]pyrrole-based receptors[17–25] to the extent that these constitute a major area within the field of supramolecular chemistry. Subsequently, many transformations and modifications have been carried out on this macrocycle (Figure 6.1), including *meso*-substitution(s), β-substitution(s), core modification and core expansion.

Calix[4]pyrroles and derivatives have found a wide range of applications. Among them it is worth mentioning their use as (i) anchor groups in polymeric materials for the removal of polluting species from water, which can lead to new technological approaches for water decontamination purposes;[21,22,25–27] and (ii) as monitoring systems such as ion selective electrodes and sensors. Therefore the production of a single calixpyrrole receptor may lead to important developments in the environmental field. Several aspects need to be considered for the construction of calixpyrrole-based ion selective electrodes (ISEs) or any other macrocycle based electrodes,[28] such as (i) the ability of a receptor to interact selectively with the target species relative to another species present in solution; (ii) the structural characteristics of the receptor, given that it should contain hydrophilic regions to provide the active sites of complexation with the target species, but also sufficient hydrophobic character to be retained in the ion selective membrane; (iii) the ion complex must diffuse freely through the membrane in accord with the direction of the potential gradient; (iv) processes such as ion transfer at the solution/membrane interface as well as ion–receptor complexation must be kinetically fast; (v) an ion–receptor complex of moderate stability is required. Therefore to fulfil the above criteria for applications as ISEs, fundamental research is required to determine the parameters outlined above. Within requirement (i), thermodynamics plays a fundamental role in the determination of the selectivity of the receptor, in particular the stability constant which provides a quantitative measure of the strength of ion–receptor interactions. Stability constants can then be used to determine the selectivity factor, which is calculated from the stability constant ratio involving the same receptor and two ions in a given solvent at a given temperature.[29,30] At this stage, it is relevant to make a distinction between selectivity as defined above and the hosting capacity of the receptor. The latter requires the determination of the composition or stoichiometry of the complex formed.[31] Essentially, in the detection of analytes by ISEs, the receptor is expected to play a major role in differentiating between analytes with similar properties so as to allow the electrode to exhibit selectivity for the target analyte relative to others present in solution. Selectivity was defined by IUPAC as "refer[ing] to the extent to which the method can be used to determine particular analytes in mixtures or matrices without interferences from other components of similar behaviour".[32] Specificity therefore is the ultimate of selectivity and defines a host–guest system in which the host interacts only with a single guest; the terms should not be confused. The methods and techniques for performing these experiments and interpreting the results have been well defined.[29–31]

Other important parameters to be considered in assessing the suitability of a receptor are the transfer Gibbs energies of ions from water to membrane-like solvents. These data provide information regarding the difference in solvation of an ion between two solvents.[33–37] It is relevant to mention that in any complexation process there is a competition between the ligand and the solvent for the ion and this issue has been widely addressed in the literature.[29,30,38] The solubility of the complex in membrane-like solvents and in water is an important parameter to consider in an attempt to fulfil requirement (ii). Knowledge regarding the mechanism of charge transfer is needed in order to assess whether or not requirement (iii) is fulfilled. Kinetic investigations are required to ensure the fast response of the membrane in the case of sensing devices as required in (iv).

In this chapter, some aspects of calixpyrrole chemistry are discussed with illustrative examples with a view to highlighting the scope of these receptors for the development of sensing devices.

6.1.1　Calix[4]pyrrole

Calix[4]pyrrole (**1**; see Figure 6.3 below) is the most widely investigated macrocycle in this series and most of the derivatives are based on this receptor. Within the structure of calix[4]pyrrole, the pyrrole moieties through the secondary amine NH groups provide the sites of complexation with anions through hydrogen bond formation. Several crystal structures of calix[4]pyrrole macrocycles have been elucidated. These reveal that the macrocycle adopts a 1,3-alternate conformation in the solid state, with adjacent pyrrole moieties oriented in opposite directions (Figure 6.2). Interestingly, in contrast to what is known for calix[4]arenes,[29,30,39–41] in calix[4]pyrrole there is no possibility for the formation of a hydrogen bonded array between the various pyrrolic NH groups. Thus, in the absence of an added substrate there is no propensity for the free macrocycles to adopt the cone conformation, the predominant motif in calix[4]arene chemistry.

Calix[4]pyrrole is the smallest and simplest calixpyrrole; much larger calix[*n*]pyrroles have been reported in the literature[42] (*n* denotes the number of repeating pyrrole units).[43] The latter possess larger inner pockets than calix[4]pyrrole, leading to altered anion selectivities. The syntheses of larger structures require more sophisticated techniques in comparison with that

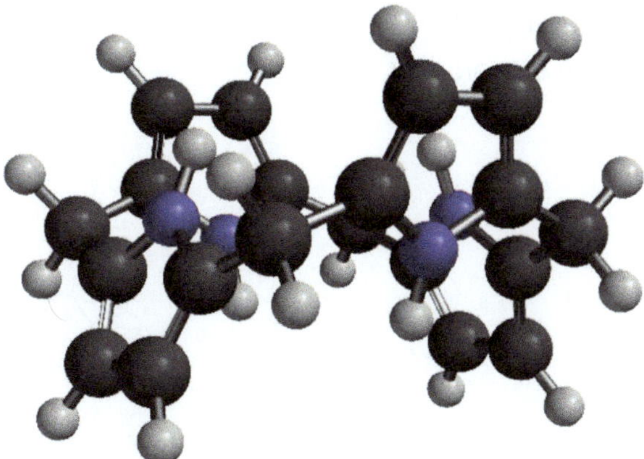

Figure 6.2　A molecular model displaying the 1,3-conformation of calix[4]pyrrole.

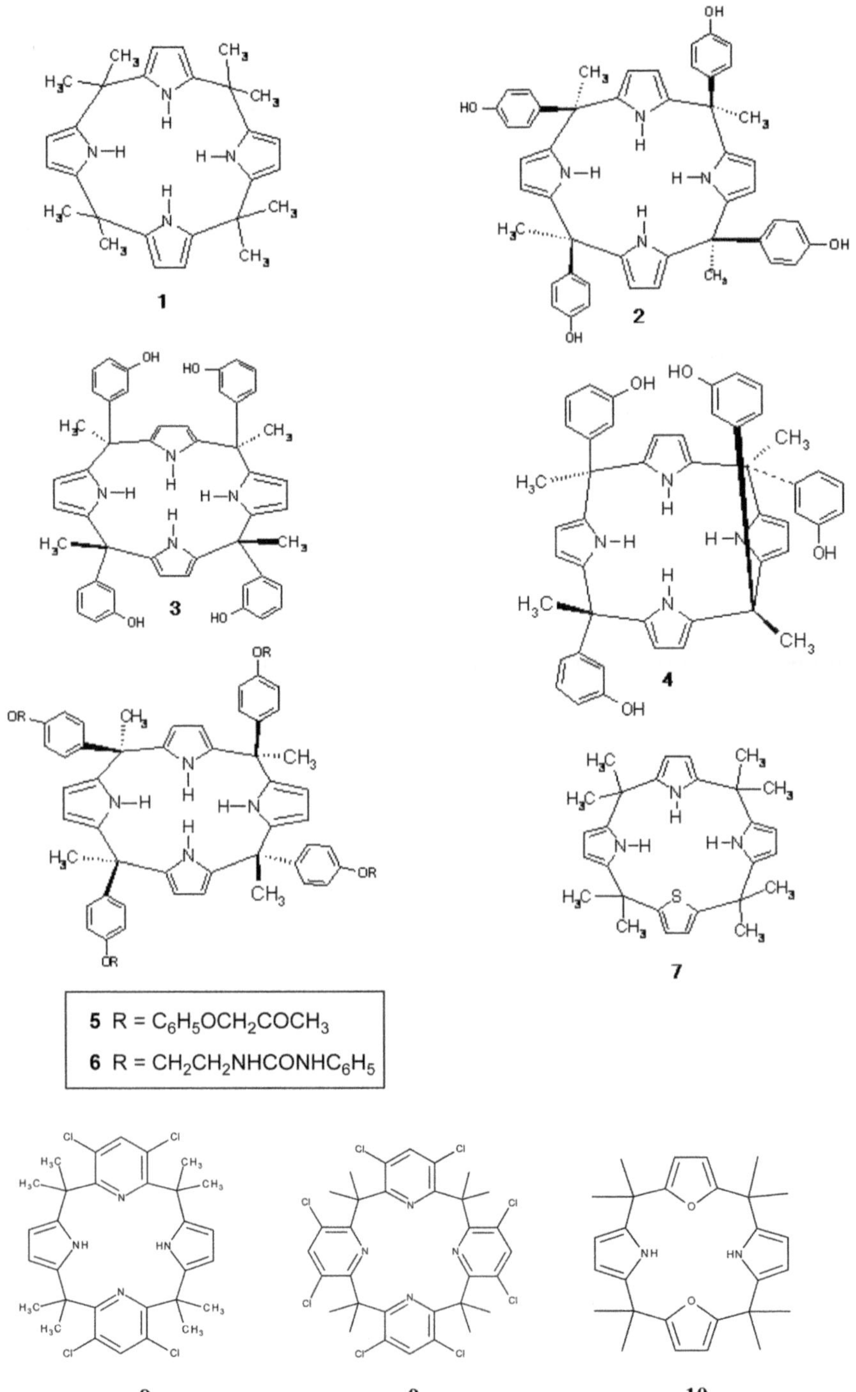

Figure 6.3 The structures of the calixpyrroles discussed.

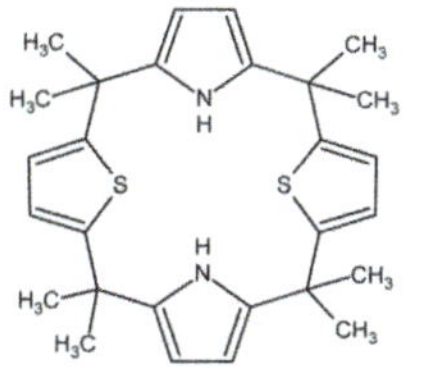

Figure 6.3 (Continued)

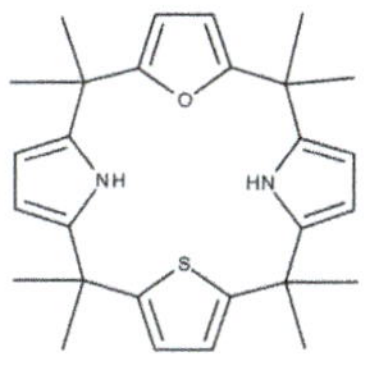

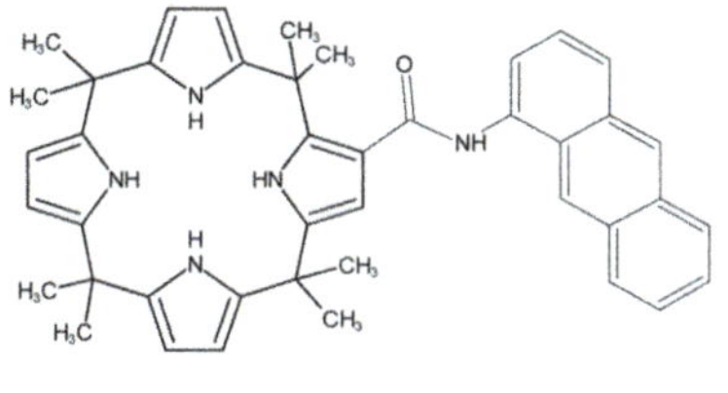

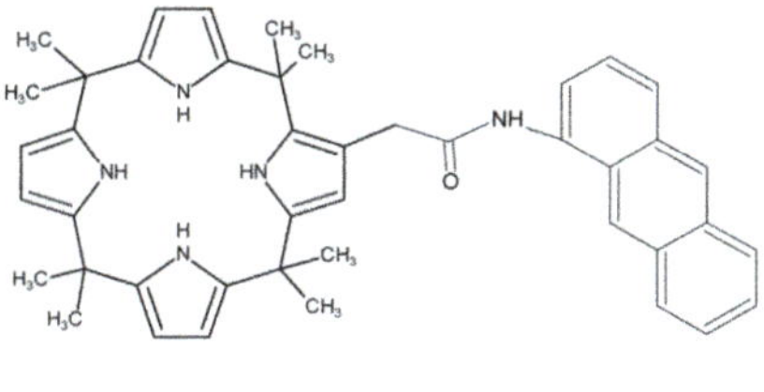

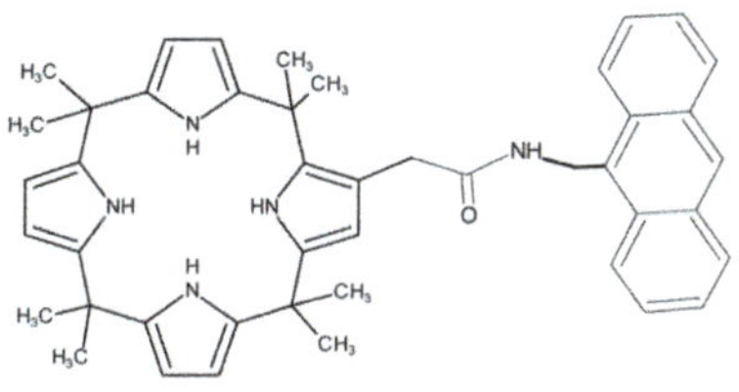

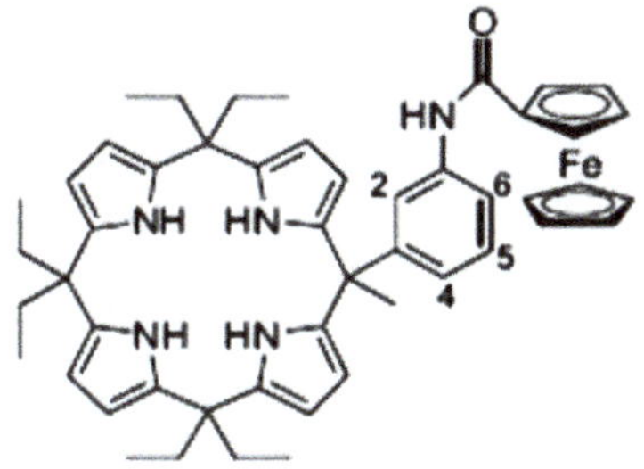

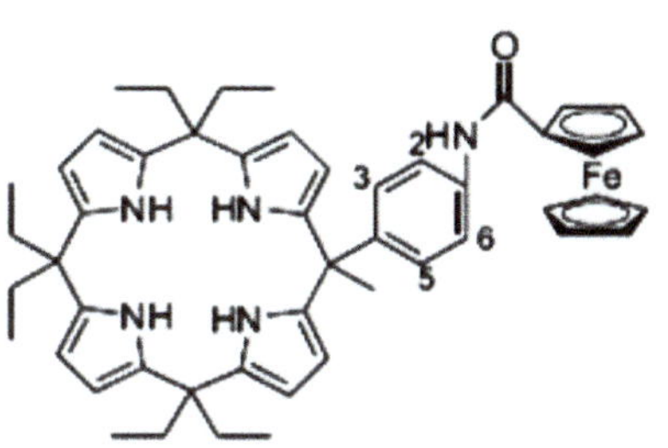

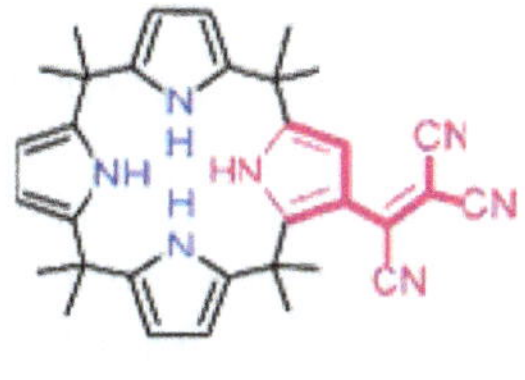

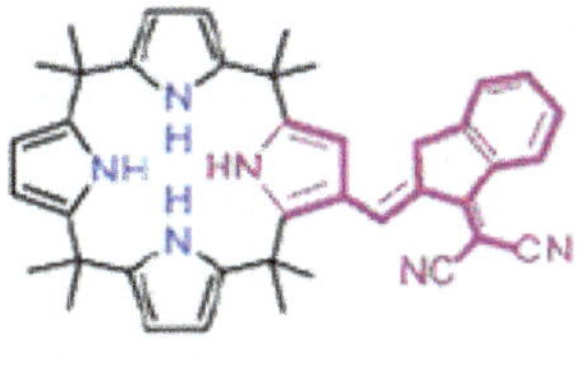

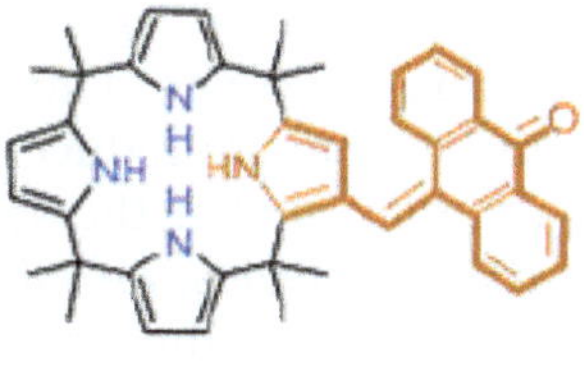

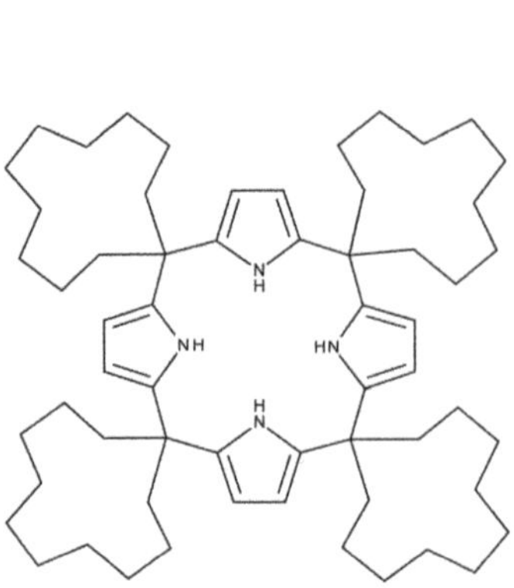

21

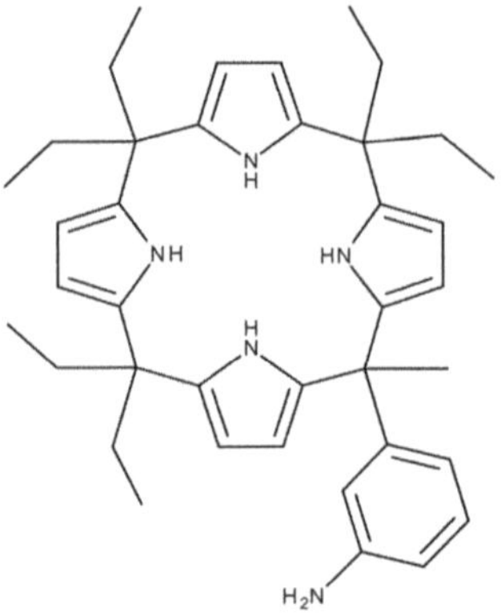

22

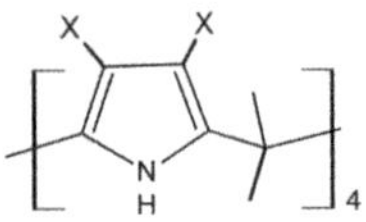

23

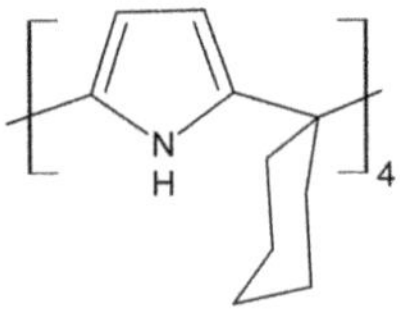

24

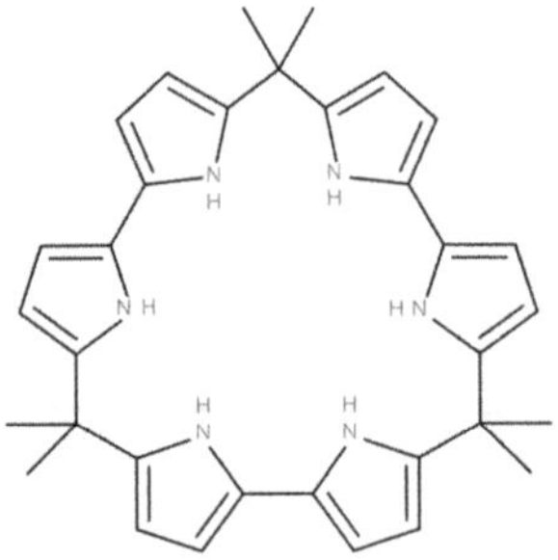

25

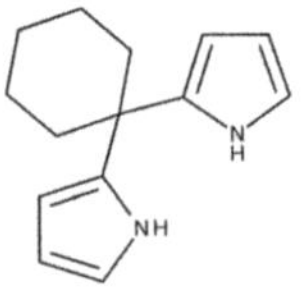

26

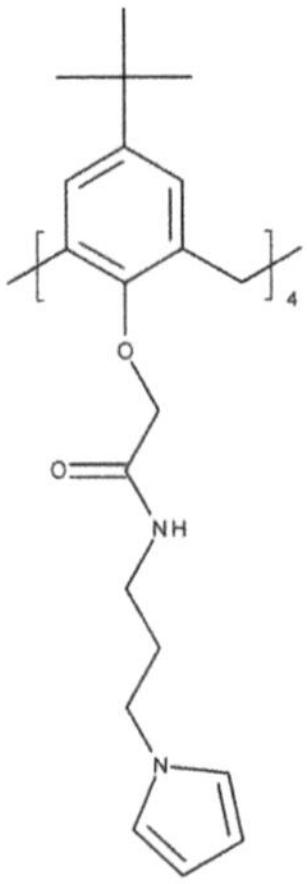

27

Figure 6.3 (Continued)

for calix[4]pyrroles. Thus the involvement of additional steps in their preparation leads to reduced yields.

In order to alter the selectivity of calix[4]pyrrole towards other molecules and ionic species, modifications to the structure were required such as the substitution of the bridging group for another, which allows for further functionalization or substitution of the pyrrolic ring for thiophene or furan,[44,45] as described below.

6.1.2 Modified Calixpyrroles

The great advantage resulting from the development of ISEs based on calixpyrroles is that the selectivity of the macrocycle for a given guest can be altered and subsequently the complex stability may be studied in solution by thermodynamics in order to select the ionophore before it is inserted into the membrane. Thus the calixpyrrole can be regarded as a platform with three addressable sites which can be separately modified by the introduction of different functional groups (the C-rim, the N-rim and the *meso*-position). Within this context, many modifications have been attempted and their ion recognition properties have been studied. Representative examples of calix[4]pyrrole derivatives for which detailed thermodynamic studies have been reported are shown in Figure 6.3. Other interesting examples are those involving heterocyclic receptors in which one or more of the pyrrole rings have been replaced by pyridine (**8, 9**),[46] furan (**10**) or thiophene (**11**), or those in which two pyrrole units have been replaced by one thiophene and one furan unit (**12**), leading to dramatic changes in their selectivity. A number of calix[4]pyrrole derivatives with chromogenic properties have been synthesized (**13–20**) and their binding and sensing properties have been investigated,[47–51] as well as other calix[4]pyrroles functionalized at the *meso*-position (**21, 22, 24**) or at the C-rim (**23**). The next section is concerned with the thermodynamics associated with some ion–receptor interactions.

6.2 Thermodynamic Aspects of Calixpyrrole and Functionalized Calixpyrroles: The Selectivity Issue

Calix[4]pyrrole is mainly known as an anion receptor[2–4] and the most detailed thermodynamic study on complexation has been carried out by Danil de Namor and Shehab.[17,18] From the viewpoint of the use of calix[4]pyrroles in the design of ISEs, the most relevant parameter to consider is the stability constant and hence the standard Gibbs energy of complexation. The stability constants (expressed as $\log K_s$) and derived standard Gibbs energies, $\Delta_c G^\circ$, for calix[4]pyrrole and derivatives with anions in a variety of solvents at 298.15 K are reported in Table 6.1. The remarkable feature of the data in Table 6.1 is the selective behaviour of the calix[4]pyrrole ligands towards the halide anions in acetonitrile and *N,N*-dimethylformamide (DMF). Thus these receptors are able to recognize the anions in the following order:

$$F^- > Cl^- > Br^- > I^-$$

It is therefore concluded that these receptors are more selective for fluoride relative to other halide anions in acetonitrile and DMF. The selectivity series down the group of the halide ions by calix[4]pyrrole was rationalized as the ligand competing successfully with the solvent to an extent that the binding energy is greatest for the smallest anion.[17] The stability

Table 6.1 Thermodynamic parameters for complexation of calix[4]pyrrole and derivatives with anions as tetra-*n*-butylammonium salts in acetonitrile and DMF at 298.15 K.

Receptor	Anion	$L:X^-$	$\log K_s$	$\Delta_c G^\circ$	$\Delta_c H^\circ$	$\Delta_c S^\circ$
Acetonitrile						
1	F^-	1 : 1	6.21^a	-35.4^a	-43.5^a	-27^a
	Cl^-	1 : 1	4.70^a	-26.8^a	-44.7^a	-60^a
	Br^-	1 : 1	3.65^a	-20.8^a	-30.7^a	-33^a
	$H_2PO_4^-$	1 : 1	5.00^a	-28.5^a	-48.1^a	-66^a
Acetonitrile						
2	F^-	1 : 1	5.44^c	-31.1^c	-32.4^c	-5^c
	Cl^-	1 : 1	3.82^c	-21.8^c	-20.5^c	5^c
	Br^-	1 : 1	3.20^c	-18.2^c	-15.4^c	9^c
	$H_2PO_4^-$	1 : 1	4.82^c	-27.5^c	-32.1^c	-15^c
Acetonitrile						
3	F^-	1 : 1	5.00^d	-28.5^d	-31.4^d	-10^d
		1 : 2	4.72^d	-27.0^d	-61.5^d	-116^d
	Cl^-	1 : 1	2.36^d	-13.5^d	-86.3^d	-244^d
	$H_2PO_4^-$	1 : 1	4.80^d	-27.4^d	-20.2^d	25^d
		1 : 2	4.66^d	-15.2^d	-29.9^d	-50^d
Acetonitrile						
4	F^-	1 : 1	5.00^d	-28.5^d	-31.4^d	-10^d
		1 : 2	4.72^d	-27.0^d	-61.5^d	-116^d
	Cl^-	1 : 1	2.36^d	-13.5^d	-86.3^d	-244^d
	$H_2PO_4^-$	1 : 1	4.80^d	-27.4^d	-20.2^d	25^d
		1 : 2	4.66^d	-15.2^d	-29.9^d	-50^d
***N,N*-Dimethylformamide**						
5	F^-	1 : 1	4.3^b	-24.6^b	-21.1^b	12^b
		1 : 2	2.8^b	-15.7^b	-6.4^b	32^b
	$H_2PO_4^-$	1 : 1	4.11^b	-23.4^b	-22.74^b	3^b
Acetonitrile						
6	F^-	1 : 1	5.0^f	-28.7^f	-20.2^f	29^f
	Cl^-	1 : 1	3.3^f	-18.7^f	-3.1^f	52^f
	Br^-	1 : 1	2.2^f	-12.7^f	-1.8^f	36^f
Acetonitrile						
7	F^-	1 : 1	5.25^e	-29.9^e	-31.5^e	-5^e
	Cl^-	1 : 1	4.15^e	-23.7^e	-51.7^e	-95^e
	Br^-	1 : 1	3.46^e	-19.7^e	-34.6^e	-49^e
	$H_2PO_4^-$	1 : 1	3.97^e	-22.7^e	-43.7^e	-70^e
		2 : 1	3.72^e	-21.2^e	-17.8^e	11^e

[a]Reference 17.
[b]Reference 19.
[c]Reference 22.
[d]Reference 21.
[e]Reference 20.
[f]Reference 78.

constants follow the pattern observed for the transfer Gibbs energies of halide anions from dipolar aprotic to protic solvents, with the strongest hydrogen bond formation with the anion of highest charge density (F^-).[17] As far as non-spherical anions are concerned, the dihydrogen phosphate is taken here as a representative example of the type of anion interaction with calix[4]pyrrole derivatives. All the receptors represented in Table 6.1 were found to interact with dihydrogen phosphate anions in acetonitrile and DMF, but not all of them are able to discriminate selectively between these anions and the fluoride ion in these solvents. The exception is **3**, which shows a slightly higher selectivity for dihydrogen phosphate relative to the fluoride anion $[(K_s(F^-)/K_s(H_2PO_4^-)=0.02]$ in acetonitrile; the selectivity factor of other receptors for fluoride is higher than that for dihydrogen phosphate ion.

The stability constant values for the complexation process between **2–5** and anions in acetonitrile are lower than those of **1** in this solvent. This decrease is due to the introduction of the phenol groups at the bridge between the pyrrole rings. Two main factors can affect the complexation in this case. Firstly, the steric effect where the phenol groups form a rigid wall, restricting easy access of the anions to the pyrrole protons. The second factor could be attributed to the repulsive effect of the aromatic rings that act as negative charges.

Examining the thermodynamic parameters for the complexation of **3** and anions in acetonitrile, it can be seen that **3** is able to interact with anions in the following order:

$$H_2PO_4^- > F^- > Cl^- > Br^- > I^-$$

Receptor **4** shows a similar trend in terms of selectivity in acetonitrile except that the affinity for the fluoride anion is similar to that of dihydrogen phosphate, with a slight preference for the former anion. A $1:2$ (ligand: anion) stoichiometry was observed with F^- and $H_2PO_4^-$, while a $1:1$ complex is formed in the case of Cl^-, with a modest stability constant.

Studies on the interaction of **5** with anions in DMF show that the thermodynamic data fit into a $1:2$ (ligand : anion) stoichiometry for the fluoride anion, while for the dihydrogen phosphate anion the data correspond to a model involving a $1:1$ process. Thermodynamic parameters for complexation of **7** with halide anions in acetonitrile show that the ligand selectivity follows the order:

$$F^- > Cl^- > Br^-$$

The selectivity factor, defined as the ratio between the thermodynamic stability constant of the receptor and two anions in a given solvent and temperature, as shown in Eqn (6.1), gives a quantitative assessment of the affinity of the receptor for one ion relative to another.

$$S_{F^-,X^-} = \frac{K_s^\theta(F^-)(s_1)}{K_s^\theta(X^-)(s_1)} \tag{6.1}$$

As far as receptor **7** is concerned, the selectivity factors taking fluoride as the reference ion show that **7** is more selective for fluoride relative to chloride, bromide and dihydrogen phosphate in acetonitrile by factors of ~ 13, 62 and 19, respectively. Thermodynamic data for complexation of this receptor with anions were compared with those of **1** and the same anions in the same solvent. The most distinctive feature of the data is that stability of anion–**1** complexes is greater than that of anion–7 and this must be due to the replacement of one pyrrole ring in **1** by a thiophene ring in 7, which leads to a decrease in the anion complex stability of the latter with respect to the former. The thermodynamic characterization of the above receptors with these anions provides a clear indication of the scope of these receptors for the design of ISEs. By incorporation of calix[4]pyrrole into the membrane of an ISE it is possible for the macrocycle to infer its selectivity onto the electrode. In solution-based isothermal titration calorimetry (ITC), acetonitrile (anhydrous) is commonly used as the model solvent to allow for complete dissociation of all components. Consequently the effect of transition between solvents from ITC analysis to the usual aqueous environment for electrode measurements must be taken into account. Also, immobilization of the macrocyclic ionophore into the membrane may affect the selectivity of the electrode and careful calibration of the equipment must be performed by exposure to competing ions in solution. Subsequently, it would appear that a fluoride ISE could be produced from membrane functionalization with calix[4]pyrrole; however, its selectivity for fluoride over the competing ions mentioned would be required. The advantage of a calixpyrrole-based electrode relative to the well-established and commercially available fluoride ISE (without calixpyrrole) will need to be assessed.

It is of interest to give a brief account on the selectivity of some calixpyrrole derivatives which have been used for the production of ISEs and sensors to highlight the role of thermodynamics in the choice of receptors for sensing purposes. A few representative examples are given. Anthracene-calix[4]pyrroles (**13**, **14**, **15**) were used for the complexation studies of these receptors with fluoride, chloride, bromide, dihydrogen phosphate and hydrogen sulfate in acetonitrile and are shown in Table 6.2. Moderate to weak complexes of 1 : 1 ligand : anion stoichiometry were found for the halide and $H_2PO_4^-$ anions as determined by fluorescence quenching analysis at 298 K.

Ferrocene as a redox-active organometallic group has been incorporated into a number of macrocycles and acyclic ligands including calix[4]pyrroles, leading to the production of receptors able to interact with anionic species as detected by the electrochemical response of the redox couple found by cyclic voltammetry.[52]

The stability constants of **16** and **17** with various anions in acetonitrile at 298.15 K are summarized in Table 6.3. The results show the selective

Table 6.2 Stability constants for receptors **13**, **14**, **15** with various anions (as tetrabutylammonium salts) in acetonitrile and dichloromethane at 298 K.

Anion	$\log K_s$ in dichloromethane			$\log K_s$ in acetonitrile		
	13	**14**	**15**	**13**	**14**	**15**
F^-	4.94	4.52	4.49	5.17	4.69	4.69
Cl^-	3.69	2.96	2.79	4.87	3.81	3.71
Br^-	3.01	–[a]	–[a]	–[a]	–[a]	–[a]
$H_2PO_4^-$	4.20	3.56	–[a]	4.96	3.90	–[a]

[a]Quenching insufficient to provide an accurate stability constant value.

Table 6.3 Binding properties of compounds **16** and **17** in acetonitrile at 298.15 K.

Anion	16 log K_s	17 K_s
No anion	$-^a$	$-^a$
F^-	>5	>5
Cl^-	3.55	3.72
Br^-	1.90	2.17
$H_2PO_4^-$	2.88	3.27
CH_3COO^-	>5	>5
HSO_4^-	n.d.^{b}	n.d.^{b}

^{a}Not available.
^{b}Not determined.

Table 6.4 Stability constants for receptors **18–20** (as log K_s) calculated for anionic substrates in DMSO (0.5% water) at 295 K.

Anion	18	19	20
F^-	>6	>6	5.71
Cl^-	3.14	2.88	2.97
CH_3COO^-	5.38	4.34	4.02
$HP_2O_7^{3-}$	5.76	4.68	4.00

recognition of these receptors for fluoride and acetate anions relative to other halides and $H_2PO_4^-$ but unable to interact with the hydrogen sulfate anion in this solvent. However, the stability constants of these ligands with either the fluoride or the ethanoate (acetate) anion has not been quantitatively assessed (log $K_s > 5$ for both anions) and therefore it is not possible from these data to conclude the selectivity sequence for these two anions.

A great deal of colorimetric chromophores have been reported, the signal units being nitrophenyl,[53,54] anthraquinone,[55,56] as well as nitrobenzene azo groups.[57,58] Other moieties (electron withdrawing)[47,59] bonded covalently to the anion receptor have been also investigated.

Nishiyabu and Anzenbacher[47] prepared three calixpyrrole-based chromogenic receptors *via* an electrophilic aromatic substitution using octamethylcalix[4]pyrrole with tetracyanoethylene for the production of **18**, while **19** and **20** were obtained by the condensation of formyl-octamethylcalix-[4]pyrrole with 1-indanylidenemalononitrile and anthrone, respectively. Thermodynamic data (Table 6.4) obtained with these receptors and anions in dimethyl sulfoxide (DMSO; 0.5% water content) clearly indicate that these receptors are selective for the fluoride anion. In fact the selectivity pattern is $F^- > HP_2O_7^{3-} > CH_3COO^- > Cl^-$ in the case of receptors **18** and **19**, while **20** does not distinguish between $HP_2O_7^{3-}$ and CH_3COO^- since the stability constant values are within the experimental error and approximately the same.

A distinction should be made between ISEs, chemically modified electrodes[52] and electrochemically active receptors.[60] An ISE has a membrane which separates an internal solution from that being measured and is accompanied by a reference electrode.[61] The membrane contains the receptor, and upon detection (host–guest complexation) of the change in membrane potential, the two half-potentials (ISE and reference) enable the membrane potential to be calculated as a function of analyte concentration. The formation of a chemically modified electrode requires a layer of material to be applied directly onto the surface of the electrode. The favoured way commonly discussed is the electropolymerization of pyrrole[62,63] directly onto the surface; additional molecules can be linked into the polymer to improve or tailor the response.[52] An electrochemically active reporter is a solid state device which has the membrane attached to the gate of a field effect transistor.[60] All of these competing yet similar technologies are mentioned; however, their individual advantages and disadvantages are not considered here.

6.3 Calixpyrrole Applications in ISEs and Chemically Modified Electrodes

The application of calixpyrroles for ISEs can be assigned to two distinct groups: either the calixpyrrole is mixed with PVC to form the ion selective membrane, or a matrix is formed from the electropolymerization of pyrrole or N-functionalized pyrrole through the 2- and 5-positions[12] with the inclusion of calixpyrrole. Each method may be employed to produce a desired ISE device; indeed, it is possible for a combination of the two methods to be employed. As outlined above, the formation and study of calixpyrroles has advanced greatly during recent years and the possibilities for producing selective molecules which may be characterized thermodynamically in solution are wide-reaching. Their subsequent application as ISEs is a natural progression and an excellent example of how low-cost fundamental research may lead to real-world applications.

The study of the selectivity of an ISE device by competitive techniques is required to determine the ultimate selectivity of the device *in situ*. This is necessary to understand any and all alterations caused by the membrane/matrix and solvent variations.

6.4 Calix[4]pyrrole ISEs

In a paper in this area, Král *et al.*[64] described a calixpyrrole–PVC ISE (based on structure **1**) along with two pyridine-containing analogues based on structures **8** and **9**, respectively. Their findings for the calixpyrrole-based ISE employed a membrane containing tridodecylmethylammonium chloride (TDDMA) as the lipophilic additive, which increased the magnitude of the electrode response with a detrimental effect to selectivity, as discovered by control experiments performed in the absence and presence of TDDMA in the membrane. This is not surprising, given that the interaction between **1** and the chloride anion in non-aqueous media is well established. Within this context, a relevant issue to address is that related to the selection of an additive in the composition of an ISE. Indeed, this research would be greatly benefited by preliminary studies on the possibility of ion–additive interactions which, as shown in this case, may impact negatively on the selectivity of the receptor for the target ion. As far as the ISE based on receptor **1** is concerned, the selectivity sequence at low pH appears to be largely influenced by the strength of the ion–receptor interaction, as reflected in the stability constant (hence standard Gibbs energy of complexation) of **1** and these anions in non-aqueous media as shown in Table 6.1. At this stage it seems appropriate to emphasize that it is the Gibbs energy which is the most relevant thermodynamic parameter to consider for assessing either the role of the receptor in the membrane or the transfer of the ion from water to the membrane phase. It is interesting and somehow expected that through a pH switching mechanism at high pH and as a result of the interaction between the pyrrole units of the receptor and the OH^-, the anionic complex rather than the free ligand was found (at least in part) responsible for the response of the electrode to the counter-ion (cation). The findings at high pH displayed a selectivity sequence as follows: $Br^- < Cl^- < OH^- \approx F^- < HPO_4^-$ from the calixpyrrole functionalized membrane.[2,8–16] Following this work, a detailed thermodynamic study of the interaction between the ligand and the halide ions in acetonitrile confirmed the sequence $I^- < Br^- < Cl^- < F^-$ in acetonitrile-d_3 at 298 K.[17] Selectivity was observed by the replacement of pyrrole units by pyridine in **8** and **9**. The authors attributed the behaviour of this electrode to hydrogen bond formation between a donor (pyrrole) and an acceptor (pyridine) and this depends upon the conformation of the receptor in the membrane, which will dictate the proximity between donor and acceptor as to whether it will enter into hydrogen bond formation. The fact that some cation interference is observed at this pH may be an indication that the pyridine ring is free to interact with cations. Whether or not hydrogen bonding takes place between the donor and acceptor functionalities in the ionophore component of the membrane, the results obtained with **8** at high pH follow the pattern observed in the transfer Gibbs energies of these anions from water to membrane-like solvents which seem to control the selective behavior of the membrane. Indeed, it is expected that an anion, like fluoride, with a high charge density sitting very comfortably in a protic solvent like water will be more reluctant to be transferred to the membrane than others along the series. In fact, as the polarizability of the anion increases the transfer process will be more favoured and this is indeed the pattern observed for the ion selective membrane at high pH. At low pH, on the assumption that the pyridine ring is protonated, the ability of the receptor to attract anions (depending on the counter-ion used to modify the pH) will increase not only due to the availability of the free NH moieties of the pyrrole units but even more to the protonation of the pyridine ring, given that ion–ion interactions are stronger than hydrogen bond formation, as shown by the results obtained with the fluoride anion. As far as phosphates are concerned, pH changes lead to different speciation in solution. It is interesting to observe that under the appropriate experimental conditions the membrane containing **8** can respond selectively to fluoride and phosphate. This is an important finding given the problems associated with fluoride in the fertilizer industry. In fact this receptor, which proved to be useful as a membrane component for monitoring purposes, has the potential to serve as a decontaminating agent in the fertilizer industry. With some slight variation the introduction of ionophore **9** in the membrane showed similar behaviour to that observed for **8**. Piotrowski *et al.*[65] explored the use of calix[4]pyrrole derivatives (**24** and **21**) using PVC as the polymeric matrix and dioctyl phthalate (DOP) as a plasticizer for producing liquid membrane electrodes for neutral nitrophenols. In doing so, the response of the electrode to the proton was assessed in the 3.5–9 pH range. Potential changes were observed in the presence of the receptors, with a decrease in the potential by increasing the pH of the aqueous solution. The reversibility of the process led the authors to assume that the structure of the receptors is not altered during the protonation/deprotonation processes. Based on previous work on the protonation of polypyrroles, the authors put forward the suggestion about the possibility of protonation through the high electron density of the α and β carbon atoms of the calix[4]pyrroles. It was found that these electrodes respond selectively to neutral *p*-nitrophenol in the presence of other nitrophenols and dihydroxybenzene isomers.

It would be interesting to see the selectivity series produced from an electrode based on calix[4]pyrrole as a comparison along with the control employed (PVC membrane with and without calixpyrrole). This work could be extended with the implementation of solution studies of the calixpyrrole receptor with the target molecules, which would produce answers for the hypothesis presented (that the interactions are based on proton transfer). Conductance measurements carried out in tetrahydrofuran and in water-saturated tetrahydrofuran would conclude this.

In a subsequent paper, Piotrowski and co-workers[66] continued the aforementioned work and introduced into their research two additional structures: **23** (X = Br) and dipyrrocyclohexane (**26**). By comparison of the results produced, with no selectivity or pH sensitivity displayed for the ISE based on **23** (X = Br), they were able to conclude that the mechanism by which a potentiometric signal was generated was due to the formation of a calix[4]-pyrrole–nitrophenol complex at the water/organic solvent interface as a result of hydrogen bond formation between the NH functionality of the pyrrole and the OH moiety of the nitrophenol, with the subsequent dissociation of the complex and the release of the proton to the aqueous phase adjacent to the non-aqueous phase boundary. The selectivity series of the other ligands remained in the same order (*para*-nitrophenol > *meta*-nitrophenol ≫ *ortho*-nitrophenol). The variances in selectivities between the ISEs were likely caused by steric factors or conformational stabilization afforded by the incorporation into the host molecules of the cyclic alkane chains present at the bridging sections; it may also be that the calix structure is reinforced by these cyclic chains, and their findings support this hypothesis.

The work carried out on calixpyrrole without introducing additional selective functional groups is fascinating. It shows the true versatility of a molecule first created in 1886,[5] able to accept neutral and anionic species. However, the ability of the molecule to be tuned with 'inert' moieties, unable to introduce new supramolecular bonds, does not allow for full exploitation of the capabilities of this useful molecule.

A recent publication containing a huge wealth of information on how to produce and physically and electrically characterize a chemically modified electrode was published by Taner *et al.*[67] using calix[4]pyrrole **22** synthesized previously[10] and attached to a glassy carbon (GC) electrode.[68] The work was performed by an electrochemical oxidation process in the presence of a solution of tetrabutylammonium tetrafluoroborate in acetonitrile. In the original publication the calix[4]pyrrole was bound to a fluorescent moiety for use as a fluorescent anion sensor[10] (*vide infra*) and includes some information regarding the selectivity of the molecule, namely $F^- > H_2PO_4^- > HP_2O_7^{3-} \gg Cl^-$ [as determined by fluorescence emission intensity with tetrabutylammonium (TBA^+) as the counter-ion]. Although it must be mentioned that the linker molecules for the fluorescent moieties were specifically chosen to improve the response and increase the binding affinities,[10] their role in the selectivity series must be incorporated. Their publication sets the standard of how a chemically modified electrode should be characterized, but does not offer an application for the technology. The authors set out to determine the structure and film stability of their GC–C4P electrode with electrochemical, optical and spectroscopic studies; the surface was studied by AFM (atomic force microscopy), surface roughness analysis, RAIRS (reflection–absorption infrared spectroscopy), XPS (X-ray photoelectron spectroscopy) and the surface stability was also analyzed.[67] This great array of analytical techniques was required in this situation as the authors assert the novelty of 3-aminophenylcalix[4]pyrrole grafted onto a carbon surface. However, these characterization techniques may not be equally applied to a true ISE due to the delicacy of membranes. In their work they display the calixpyrrole as being bound to the GC through the etheric bond resulting from the functionalities (–C–OH, –C=O, –COOH) generated on the electrode (carbon) surface and the β-carbon of the pyrrole rings As the intention is for the resulting modified electrode to be employed for anion sensing, binding to the solid support in this way should not interfere with the selectivity of the attached molecule (see Figure 6.4). Indeed, it is the NH groups of the pyrrole moiety which are of interest as hydrogen bond donors to the anion and binding in this way may aid the orientation of the NH groups outward from the surface and towards the solution when immersed, improving the sensitivity through organization and perhaps leading to an instrument able to operate over a greater activity range.

As mentioned earlier, an alternative to calixpyrrole functionalization is the prospect of introducing groups into the macrocyclic structure of the calixpyrrole. As shown in structure **11**, a calix[4]bipyrrole-bithiophene structure was found to possess remarkable selectivity for mercury(II).[24] Later work incorporated the calix[2]thieno[2]pyrrole into the membrane of an ISE.[69,70]

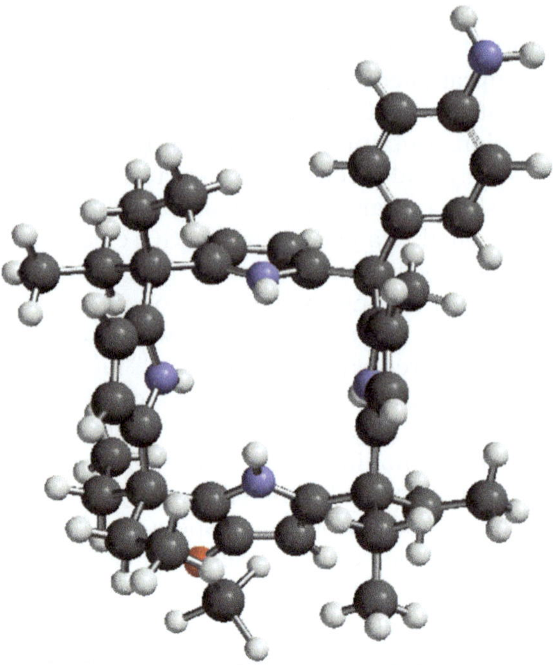

Figure 6.4 A 'top-down' rendering of how 3-aminophenylcalix[4]pyrrole may appear when attached through the oxygen (*red sphere*) onto the surface of the electrode.

Using the method outlined by Gholivand *et al.*[71] for the preparation of the membrane, an ISE (DOS as the plasticizer and $NaBPh_4$ as the additive) was produced which retained the mercury selectivity. In the electrochemical cell, Ag/AgCl//KCl was used as the reference electrode. The electrode was found to exhibit a Nernstian response of 27.8 mV decade^{-1} with a response time of 20 s over a Hg^{2+} concentration range of 1×10^{-6} to 1×10^{-2} M in aqueous solutions. The electrode response was found to be pH independent in the 1.0–6.0 range. At higher pH the decrease in potential was attributed to the formation of other mercury speciation in alkaline aqueous media. Selectivity coefficients were determined using the fixed interference method (FIM). These data reflect some interference with the silver cation, as expected from the thermodynamics of these systems. Indeed, metal cation complexes of these receptors have been previously investigated.[24] Thus the mercury(II) complex was found to exhibit a 1:1 Hg^{2+}:calix[2]thieno[2]pyrrole stoichiometry, with moderate complex stability in acetonitrile at 298.15 K ($\log K_s = 4.00$). Recent work carried out by Danil de Namor and co-workers[72] on the preparation of a Hg(II) ISE based on a calix[4]pyrrole derivative showed that $NaBPh_4$ is not the optimum additive to be used in the preparation of ion selective membranes for Hg(II) due to its interaction with this cation. Replacement of $NaBPh_4$ by a neutral additive significantly improved the electrode's response to Hg(II).

These studies need not be constrained to the limitations of utilizing pyrrole in the formation of the host, but should be considered from almost any host–guest system which may be studied in this way.[30,31]

Lim *et al.*[73] produced a calixpyrrole ISE from calix[2]furano[2]pyrrole **10**[72] and compared it analytically with calix[2]thiopheno[2]pyrrole **11** and calix[2]furanthiopheno[2]pyrrole **12** incorporated into membranes of respective ISEs for the selective recognition of Ag^+ in solution. Their findings showed a distinct advantage of the ISE containing calix[2]furano[2]pyrrole **10** over the other two (**11** and **12**) for the recognition of the Ag^+ cation. These results are rather surprising, given the nature of the donor atoms present in these ligands. Indeed, it is a well-known fact that a soft donor atom such as S is likely to interact strongly with a soft metal cation such as Ag^+. On the other hand, the ISE based on receptor **10** containing hard donor atoms is expected to interact with hard metal cations. However, the selectivity co-efficients obtained by the membrane electrode based on receptor **10** shows a good selectivity for silver, relative to hard metal cations. Here the authors conclude that the selectivity and recognition is due to ion–dipole interactions, but the total interaction is likely to be more complex. These notions bring us to the key concept, as enzymes recognize substrates by the 'lock and key mechanism'. Subsequently, the production of ISEs can be considered part of the human interest in biomimicry and the future for these technologies is bright as the study of natural complex supramolecular interactions is at the forefront of biology and biochemistry research. Much is known regarding the complexity in biological systems and the technologies remain in their infancies in comparison. As such, the field can continue to grow along with the techniques involved, which continue to improve. It is in the realm of supramolecular chemistry that ISE researchers find themselves, in a field in its relative infancy which continues to grow. The authors attempted to optimize their electrode by altering the plasticizers used. In subsequent findings, little or no response was established from the use of *ortho*-nitrophenyl octyl ether, *o*-NPOE. It would have been interesting to see whether this difficulty (loss of potentiometric response) with this plasticizer is unique to the receptor tested. If not, it could be highlighted and removed from the research or flagged as a necessary variable which must be considered by the researcher, as distorted results could be produced from employing the wrong plasticizers with receptors. This highlights again the necessity for proper and thorough analysis of the receptor before mounting into the membrane of an ISE. Importantly the authors found their calix[2]furano[2]pyrrole membrane ISE to be unaffected by changes in pH, with minor changes (<25 mV) when the pH was adjusted from 2.5 to 12 (with nitric acid and sodium hydroxide). The authors concluded that the electrode is pH independent. Again as discussed above, calix[4]pyrrole has been found to interact with the hydroxyl (OH^-) anion at high pH. Comparison was made to the commercially available *O,O''*-bis[2-(methylthio)ethyl]-*tert*-butylcalix[4]arene (Fluka silver ionophore IV) in their efforts, which produced remarkably similar results even though the receptor donor atoms are different (S and N in the case of the calixarene, O and N in the calixpyrrole).

The general procedure for membrane preparation involves producing a solution of 1% by weight of the receptor with 33% PVC and 66% plasticizer in THF in small crucibles and allowing the THF to evaporate slowly to produce a small disc which can be shaped and sized to fit the electrode.[73] This simple method allows for prototypes to be easily produced and tested once a lead-receptor compound has been produced. As such, the topic of ISEs is one which may be pursued by the physical chemist at low cost with the rapid production of results. Optimization once an electrode has been produced could prove intensive; the number of variables is small, even if one introduces an additional component to increase the magnitude of response. As such, the researcher is able to concentrate on the processes which matter most to the development of an ISE, namely the production of the receptors, their screening and the ISE testing. It is the production of receptors which is the most pressing issue to address due to the inherent problems of pilot chemical synthesis, which must of course be well planned. It has been our finding that the production of functionalized calixpyrroles requires the synthesis of a large quantity of the initial parent calix[*n*]pyrrole. As conformation of the receptor molecule must be taken into account, the separation is always at the detriment of yield; later functionalization and purification is then permitted by the quantity available.

6.5 Chemically Modified Electrodes

An alternative to PVC-based ISEs is the production of a matrix from a receptor bound to an electropolymerized moiety,[52] *e.g.* polypyrrole with polymerization at the vacant 2- and 5-positions known to form electropolymerized films *via* either potential cycling (cyclic voltammetry) or chronoamperometry. The ability for chemically modified electrodes to detect ionic species in solution was first demonstrated by Chen *et al.*[52] using a modified calixarene containing N-attached pyrrolic 'pendant arms' 27 which were electro-copolymerized with pyrrole to a poly(pyrrole-calixarene-*N*-pyrrole) coated electrode able to detect Ag^+ in an acetonitrile solution. Copolymerization was identified as required as the polymerization of 27 produced an insulating polymer; the product was compared with polypyrrole and showed numerous advantages.

6.6 Fluorescent Anion Sensors

Much research has been dedicated to the use of calixpyrroles as fluorescent or colorimetric anion detectors and some excellent papers have been published in this area.[10,68] A recent publication[74] displayed the level of complexity currently achieved in this field, where the production of a calixpyrrole strapped with azocalixarene has produced a solution-based fluoride sensor which alters the colour of the solution upon molecular detection and can be observed by eye. This is the latest in a large field of work where calixpyrrole optical sensors have been produced previously.[12] There are two methods for producing a calixpyrrole optical sensor: the first requires the covalent attachment of an optically or fluorescently active group directly onto the calixpyrrole; the second necessitates the attachment of a reporter unit as a displacement assay,[75] where displacement by the target species produces an optically measurable change. For those interested, the review by Gale *et al.*[12] serves as an introduction, although many more publications on the subject now exist since its first publication. A few examples are given here that demonstrate the relevant information which can be gained from thermodynamic data in the selection of the receptor to be incorporated in sensing devices. Thus the presence of anthracene units in the structure of calix[4]-pyrrole (**13**, **14**, **15**) facilitates the macrocycle to serve as a sensor. Fluoride was the anion which caused the greatest fluorescence quenching, followed by the hydrogen phosphate anion and then chloride ion, as shown in the stability constant data in acetonitrile.

As far as **16** and **17** are concerned, results from cyclic voltammetry showed that among the halides, the F^- ion induced the greatest cathodic shift of the ferrocene/ferricinium ion (Fc/Fc^+) couple followed by Cl^- and Br^-. Among the non-spherical anions, the CH_3COO^- ion caused the largest cathodic shift, which was more pronounced than that of $H_2PO_4^-$ and even larger than those for the halides. Additionally, the HSO_4^- anion caused a small anodic shift; see Table 6.3.

Regarding receptors **18–20**, emphasis should be made about the fact that DMSO through its basic oxygen atom (Table 6.4) is known to interact with the NH moiety of the pyrrole units *via* hydrogen bond formation. Therefore it is most likely that the stability of these complexes is higher in a non-interacting solvent. This seems to be the case, from the qualitative assessment carried out on the anion sensing ability of these receptors. Indeed, dramatic colour changes were observed (by visual examination) from the addition of fluoride, acetate, pyrophosphate and phosphate, respectively, to sensors **18–20**. However, no colour changes were observed by the addition of chloride, bromide, iodide or nitrate. Absorption spectroscopy titration experiments revealed large bathochromic shifts in spectra of sensors **18–20** upon the addition of anions.

6.7 Polymers

It is possible to polymerize pyrrole for the formation of a selective membrane which may be used as part of an ISE. Sabah *et al.*[62] employed a galvanostatic method (at constant current), maintaining pulse duration and pause duration but varying the number of electrical pulses used to produce a number of polypyrrolic films for the determination of valproate in pharmaceutical preparations. Alizadeh *et al.*[63] employed a similar technique for the production of polypyrrole films for the determination of benzoate (preservative) in soft drinks; for most of their films the current was set at 1 mA, as had been used previously[62] for most of the production. In their results they state optimum conditions for the production of their selective material as 2000 mC,

1 mA, 200 s, 10 pulses (200 s followed by 60 s pause), 0.05 mol L^{-1} benzoate, 2×10^{-3} mol L^{-1}, producing a Nernstian response of -55.4 mV dec^{-1}. In the absence of SDBS (sodium dodecyl benzenesulfonate) the authors found the membrane to be thin and the polymerization very slow.

6.8 Conclusions

Of interest in this field may be the combination of the two outlined methods for the production of an ISE. Requiring firstly the production of a calixpyrrole functionalized through extended 'arm groups' to impart the desired selectivity, followed by a detailed thermodynamic study and the subsequent production of an ISE.

Researchers searching for novel ionophores for introducing selectivity should not constrain their search to only calixpyrrole when a host of macrocycles are selective for various ionic and neutral species. It has been shown how the selectivity of calixpyrrole can be altered through structural modifications and the same is true for most other macrocycles. The major advantage of employing calixpyrrole as the sensor molecule in the membrane of an ISE is that it is an innate anion receptor, although neutral species have been recognized. Subsequently, with considered functionalization it should be possible to produce a sensor molecule based on calixpyrrole which is able to selectively detect any desired organic species containing distinct moieties. The great lengths that a researcher must go to in order to produce selective organic compounds or receptor molecules for ionic species can be seen in the publication of Bühlmann et al.,[76] where an enormous variety of receptors are detailed for detection of inorganic anions and cations, organic ionic species and neutral species. Outlined among them are a number of calixpyrrole-based receptors; here we conclude that calixpyrrole-based receptors could be produced for any soluble species, which may then serve purpose once immobilized onto the surface of an ISE. A major promoter for the use of calixpyrrole-based receptors in the membrane of an ISE is the wealth of information available in the literature which outlines its sensing abilities in solution and, importantly, the methods are well defined for determining selectivity in solution.[17–20] The behaviour of the molecule is well understood and analysis of the receptor in solution by NMR, electrochemistry and calorimetry, along with other methods, have now been well proven in the understanding of the modes of action of selective recognition. By employing NMR techniques it is possible to witness conformational changes in the receptor molecule. NMR allows for rapid screening of receptors for target species; by simply analyzing the receptor, followed by the receptor plus target, an interaction will be marked by a deviation in the spectra when overlaid. The nature of the change allows the researcher to infer an understanding of the type and strength of the integration. Once an interaction has been identified, calorimetric and other techniques are required to understand the thermodynamics, which allows for quantification of selectivity through information regarding the stability of the complex. The stability must be understood; for example, in the case of a highly stable interaction the receptor and target may remain locked together. Consequently, an electrode produced from such a receptor employed in an ISE would have a saturation point which upon being reached would render the ISE inactive. In their excellent review on the subject of ISEs, Bakker and co-workers[77] state that "selectivity is clearly one of the most important characteristics of a sensor, as it often determines whether a reliable measurement in the target sample is possible". Also the kinetics of an interaction must also be considered, and it is the nature of these experiments that kinetically slow interactions (subsequently likely of little interest to the producer of an ISE) might go unidentified unless specifically searched for. Owing to the fast and efficient NMR screening process, determination can be achieved by incubation over time. Another important issue to address is the need to assess any possible interactions before selecting the components in the membrane, particularly the additives or indeed buffer systems or any addition to be made by controlling the effect of pH of the aqueous solution on the electrode response.

Emphasis must finally be placed on the advantages of ion selective devices with respect to other analytical tools available for determining the activities of ionic species in solution. Inorganic ionic species are routinely determined in the laboratory by atomic absorption spectroscopy (AAS) and inductively coupled plasma–mass spectrometry (ICP-MS). It is these techniques which are often used to check the results of measurements of concentration (activities) from samples analyzed by an ISE. Of course, a well-equipped laboratory is required to perform these analyses; thus an ISE must produce reproducible, reliable and accurate results comparable with those possible with the aforementioned techniques, *in situ*.

References

1. J. W. Steed and J. L. Atwood, *Supramolecular Chemistry*, Wiley, Chichester, 2000.
2. P. A. Gale, J. L. Sessler, V. Kral and V. Lynch, *J. Am. Chem. Soc.*, 1996, **118**, 5140.
3. J. L. Sessler, A. Antrievsky, P. A. Gale and V. Lynch, *Angew. Chem. Int. Ed. Engl.*, 1996, **35**, 2782.
4. P. A. Gale, J. L. Sessler, W. E. Allen, N. A. Tvermoes and V. Lynch, *J. Chem. Soc. Chem. Commun.*, 1997, 665.
5. A. von Baeyer, *Ber. Dtsch. Chem. Ges.*, 1886, **19**, 2184.
6. W. H. Brown and W. N. French, *Can. J. Chem.*, 1958, **36**, 371.
7. C. M. Kretz, E. Gallo, E. Solari, C. Floriani, C. A. Villa and C. Rizzoli, *J. Am. Chem. Soc.*, 1994, **116**, 10775.
8. W. E. Allen, P. A. Gale, C. T. Brown, V. M. Lynch and J. L. Sessler, *J. Am. Chem. Soc.*, 1996, **118**, 12471.
9. P. A. Gale, J. L. Sessler and V. Král, *J. Chem. Soc. Chem. Commun.*, 1998, 1.

10. P. Anzenbacher Jr., K. Jursiková and J. L. Sessler, *J. Am. Chem. Soc.*, 2000, **122**, 9350.
11. C. Bucher, R. S. Zimmerman, V. Lynch and J. L. Sessler, *J. Am. Chem. Soc.*, 2001, **123**, 9716.
12. P. A. Gale, P. Anzenbacher Jr. and J. L. Sessler, *Coord. Chem. Rev.*, 2001, **222**, 57.
13. C.-H. Lee, H.-K. Na, D.-W. Yoon, D.-H. Won, W.-S. Cho, V. M. Lynch, S. V. Shevchuk and J. L. Sessler, *J. Am. Chem. Soc.*, 2003, **125**, 7301.
14. C.-H. Lee, J.-S. Lee, H.-K. Na, D.-W. Yoon, H. Miyaji, W.-S. Cho and J. L. Sessler, *J. Org. Chem.*, 2005, **70**, 2067.
15. J. L. Sessler, D. An, W.-S. Cho, V. Lynch, D.-W. Yoon, S.-J. Hong and C.-H. Lee, *J. Org. Chem.*, 2005, **70**, 1511.
16. J. L. Sessler, W.-S. Cho, D. E. Gross, J. A. Shriver, V. M. Lynch and M. Marquez, *J. Org. Chem.*, 2005, **70**, 5982.
17. A. F. Danil de Namor and M. Shehab, *J. Phys. Chem. B*, 2003, **107**, 6462.
18. A. F. Danil de Namor and M. Shehab, *J. Phys. Chem. A*, 2004, **108**, 7324.
19. A. F. Danil de Namor and M. Shehab, *J. Phys. Chem. B*, 2005, **109**, 17440.
20. A. F. Danil de Namor, I. Abbas and H. H. Hammud, *J. Phys. Chem. B*, 2006, **110**, 2142.
21. A. F. Danil de Namor, M. Shehab, I. Abbas, M. V. Withams and J. Zvietcovich-Guerra, *J. Phys. Chem. B*, 2006, **110**, 12653.
22. A. F. Danil de Namor, M. Shehab, R. Khalife and I. Abbas, *J. Phys. Chem. B*, 2007, **111**, 12177.
23. A. F. Danil de Namor, I. Abbas and H. H. Hammud, *J. Phys. Chem. B*, 2007, **111**, 3098.
24. A. F. Danil de Namor and I. Abbas, *J. Phys. Chem. B*, 2007, **111**, 5803.
25. A. F. Danil de Namor and R. Khalife, *J. Phys. Chem. B*, 2008, **112**, 15766.
26. A. F. Danil de Namor and I. Abbas, in *Advances in Fluorine Science*, ed. A. Tressaud, Elsevier, New York, 2006, vol. 2, p. 81.
27. A. F. Danil de Namor and I. Abbas, *Anal. Methods*, 2010, **2**, 63.
28. B. G. Cox and H. Schneider, *Coordination and Transport Properties of Macrocyclic Compounds in Solution*, Elsevier, New York, 1992.
29. A. F. Danil de Namor, R. M. Cleverley and M. L. Zapata-Ormachea, *Chem. Rev.*, 1998, **98**, 2495.
30. A. F. Danil de Namor, in *Calixarenes 2001*, ed. M. Z. Asfari, V. Böhmer, J. Harrowfield and J. Vicens, Kluwer, Dordrecht, 2001, p. 346.
31. A. F. Danil de Namor, T. T. Matsufuji-Yasuda, K. Zegarra-Fernandez, O. A. Webb and A. El Gamouz, *Croat. Chem. Acta*, 2013, **86**, 1.
32. R. I. Vessman, F. F. Stefan, K. van Staden, W. Danzer, D. T. Lindner, D. T. Burns, A. Fajgelj and H. Muller, *Pure Appl. Chem.*, 2001, **73**, 1381.
33. A. J. Parker, *Q. Rev. Chem. Soc.*, 1962, **16**, 133.
34. F. Danil de Namor, T. Hill and E. E. Sigstad, *J. Chem. Soc., Faraday Trans. 1*, 1983, **79**, 2713.
35. S. Wilke and T. Zerihun, *J. Electroanal. Chem.*, 2001, **515**, 52.
36. F. Scholz, R. Gulaboski and K. Caban, *Electrochem. Commun.*, 2003, **5**, 929.

37. L. J. Sanchez Vallejo, J. M. Ovejero, R. A. Fernandez and S. A. Dassie, *Int. J. Electrochem.*, 2012, ID 462197.
38. A. F. Danil de Namor, *J. Chem. Soc., Faraday Trans. 1*, 1988, **84**, 2241.
39. C. D. Gutsche, in *Monographs in Supramolecular Chemistry*, ed. J. F. Stoddart, Royal Society of Chemistry, Cambridge, 1998.
40. *Calixarenes, 2001*, ed. Z. Asfari, V. Böhmer, J. Harrowfield and J. Vicens, Kluwer, Dordrecht, 2001.
41. *Calixarenes in the Nanoworld*, ed. J. Vicens and J. Harrowfield, Kluwer, Dordrecht, 2007.
42. J. L. Sessler, P. Anzenbacher Jr., J. A. Shriver, K. Jursiková, V. M. Lynch and M. Marquez, *J. Am. Chem. Soc.*, 2000, **122**, 12061.
43. J. L. Sessler, D. An, W.-S. Cho and V. Lynch, *Angew. Chem. Int. Ed.*, 2003, **42**, 227.
44. Y.-S. Jang, H.-J. Kim, P.-H. Lee and C.-H. Lee, *Tetrahedron Lett.*, 2000, **41**, 291.
45. *Heterocyclic Supramolecules*, ed. K. Matsumoto, Springer, Berlin, 2008.
46. V. Kral, P. A. Gale, P. Anzenbacher Jr., K. Jursikova, V. Lynch and J. L. Sessler, *J. Chem. Soc. Chem. Commun.*, 1998, 9.
47. R. Nishiyabu and P. Anzenbacher Jr., *J. Am. Chem. Soc.*, 2005, **127**, 8270.
48. R. Nishiyabu, M. A. Palacios, W. Dehaen and P. Anzenbacher Jr., *J. Am. Chem. Soc.*, 2006, **128**, 11496.
49. P. Anzenbacher Jr., in *Anion Recognition in Supramolecular Chemistry*, ed. P. A. Gale and W. Dehaen, Springer, Berlin, 2010, p. 372.
50. L. Sessler, A. Gibauer and P. A. Gale, *Gazz. Chim. Ital.*, 1997, **127**, 723.
51. P. A. Gale, M. B. Hursthouse, M. E. Light, J. L. Sessler, C. N. Warriner and R. S. Zimmerman, *Tetrahedron Lett.*, 2001, **42**, 6759.
52. Z. Chen, P. A. Gale and P. D. Beer, *J. Electroanal. Chem.*, 1995, **393**, 113.
53. E. J. Cho, B. J. Ryu, Y. J. Lee and K. C. Nam, *Org. Lett.*, 2005, 7, 2607.
54. M. Boiocchi, L. D. Boca, D. Esteban-Gomez, L. Fabbrizzi, M. Licchelli and E. Monzani, *J. Am. Chem. Soc.*, 2004, **126**, 16507.
55. A. Das, B. Ganguly, D. K. Kumar and D. A. Jose, *Org. Lett.*, 2004, **6**, 3445.
56. H. Miyaji, W. Sato and J. L. Sessler, *Angew. Chem. Int. Ed.*, 2000, **39**, 1777.
57. F. Sancenon, R. Martinez-Manez and J. Soto, *Angew. Chem. Int. Ed.*, 2002, **41**, 1416.
58. D. H. Lee, J. H. Im, S. U. Son, Y. K. Chung and J. Hong, *J. Am. Chem. Soc.*, 2003, **125**, 7752.
59. D. Esteban-Gomez, L. Fabbrizzi and M. Licchelli, *J. Org. Chem.*, 2005, **70**, 5717.
60. P. D. Beer, P. A. Gale and G. Z. Chen, *J. Chem. Soc., Dalton Trans.*, 1999, 1897.
61. D. R. Crow, *Principles and Applications of Electrochemistry*, Blackie, London, 1994.
62. S. Sabah, M. Aghamohammadi and N. Alizadeh, *Sens. Actuators B*, 2006, **114**, 489.
63. N. Alizadeh, N. Saburi and S. E. Hosseini, *Food Control*, 2012, **28**, 315.

64. V. Král, J. L. Sessler, T. V. Shishkanova, P. A. Gale and R. Volf, *J. Am. Chem. Soc.*, 1999, **121**, 8771.
65. T. Piotrowski, H. Radecka, J. Radecki, S. Depraetere and W. Dehaen, *Electroanalysis*, 2001, **13**, 342.
66. J. Radecki, H. Radecka, T. Piotrowski, S. Depraetere, W. Dehaen and J. Plavec, *Anal. Lett.*, 2004, **16**, 2073.
67. B. Taner, P. Deveci, Z. Üstündag, S. Keskin, E. Özcan and A. O. Solak, *Surf. Interface Anal.*, 2012, **44**, 185.
68. B. Taner, E. Özcan, Z. Üstündag, S. Keskin, A. O. Solak and H. Ekşi, *Thin Solid Films*, 2010, **519**, 289.
69. A. F. Danil de Namor, Plenary lecture, XV Argentinian Congress of Toxicology, Neuquen, Argentina, September 2007.
70. I. Abbas, *Int. J. Chem.*, 2012, **4**, 23.
71. M. B. Gholivand, A. Babakhanian and M. Joshaghani, *Anal. Chim. Acta*, 2007, **584**, 302.
72. A. F. Danil de Namor, A. El Gamouz and S. Alharthi, 2014, to be submitted for publication.
73. S. M. Lim, H. J. Chung, K.-J. Paeng, C.-H. Lee, H. N. Choi and W.-Y. Lee, *Anal. Chim. Acta*, 2002, **453**, 81.
74. P. Thiampanya, N. Muangsin and B. Pulpoka, *Org. Lett.*, 2012, **14**, 4050.
75. A. Metzger and E. V. Anslyn, *Angew. Chem. Int. Ed. Engl.*, 1998, **37**, 649.
76. P. Bühlmann, E. Pretsch and E. Bakker, *Chem. Rev.*, 1998, **98**, 1593.
77. E. Bakker, P. Bühlmann and E. Pretsch, *Chem. Rev.*, 1997, **97**, 3083.
78. A. F. Danil de Namor and R. Khalife, *Phys. Chem. Chem. Phys.*, 2010, **12**, 753.

Application of Pattern Recognition Techniques in the Development of Electronic Tongues

MAIARA O. SALLES AND THIAGO R. L. C. PAIXÃO*

Instituto de Química, Universidade de São Paulo, São Paulo, SP,
Brazil 05508-900
*Email: trlcp@iq.usp.br

7.1 Biological Receptors *versus* Synthetic Receptors (Electronic Tongues)

The human brain is a powerful tool, capable of analyzing patterns and perceiving similarities and differences between shapes, objects, and colours. Our brain recognizes patterns using qualitative procedures and heuristics in a more powerful way than do any quantitative methods yet described using knowledge database discovery. From the beginning of our learning process in pre-school and kindergarten, we start training our pattern recognition ability by discriminating circles from squares, and different objects from each other, using different toys and games.

Our learned pattern recognition ability considers a large number of different and complex inputs and allows us to make a decision based on a comparison of these inputs through a mental database learned previously under different situations. These inputs are extracted from the environment

RSC Detection Science Series No. 3
Advanced Synthetic Materials in Detection Science
Edited by Subrayal Reddy

Published by the Royal Society of Chemistry, www.rsc.org

using our sight, hearing, taste, smell, and touch, *i.e.* our receptor system. For example, chemical compounds responsible for taste are detected by human taste receptors on our tongue. The taste information extracted by our taste buds is then transmitted to taste nerves as the result of a release of neurotransmitters and finally reaches the gustatory region in the brain, where the information is decoded and compared with our learned database. At this point, we can then discriminate between different objects or, for example, between different types of beverages, as shown in Figure 7.1.

Our sensorial taste properties result from the ability of our tongue to identify five basic tastes: sweetness, saltiness, bitterness, sourness, and umami or savouriness (taste from monosodium glutamate).[1–6] Contrary to popular belief, these biological receptors are uniformly distributed throughout all areas of the tongue,[1–6] as shown in Figure 7.2,[1] and can be stimulated together, creating a unique combination of these five characteristics; in addition, our tongue can distinguish the subtleness of these combinations with great accuracy, creating a fingerprint of the tasted substance. Human beings can therefore recognize complex mixtures of taste, such as in wines, but cannot recognize individual chemical substances without, as an important related point, quantifying each substance.

A large number of researchers have tried to develop prototypes for an artificial tongue resembling a mammalian tongue. An artificial tongue normally consists of an array of synthetic receptors (chemical sensors) using

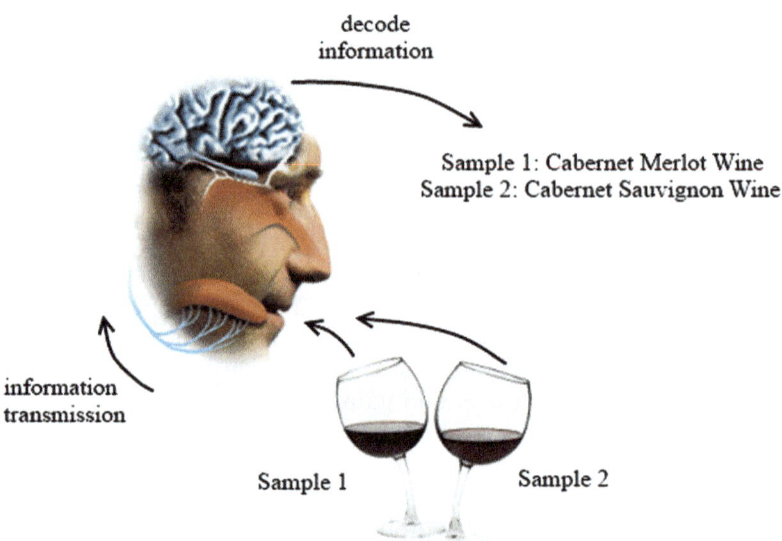

Figure 7.1 Schematic representation of the pattern recognition process performed using input information extracted by our sense of taste.
[Adapted from a graphic image by Patterson Clark, *The Washington Post*, March 10, 2008 (http://www.washingtonpost.com/wp-dyn/content/graphic/2008/03/10/GR2008031000119.html).]

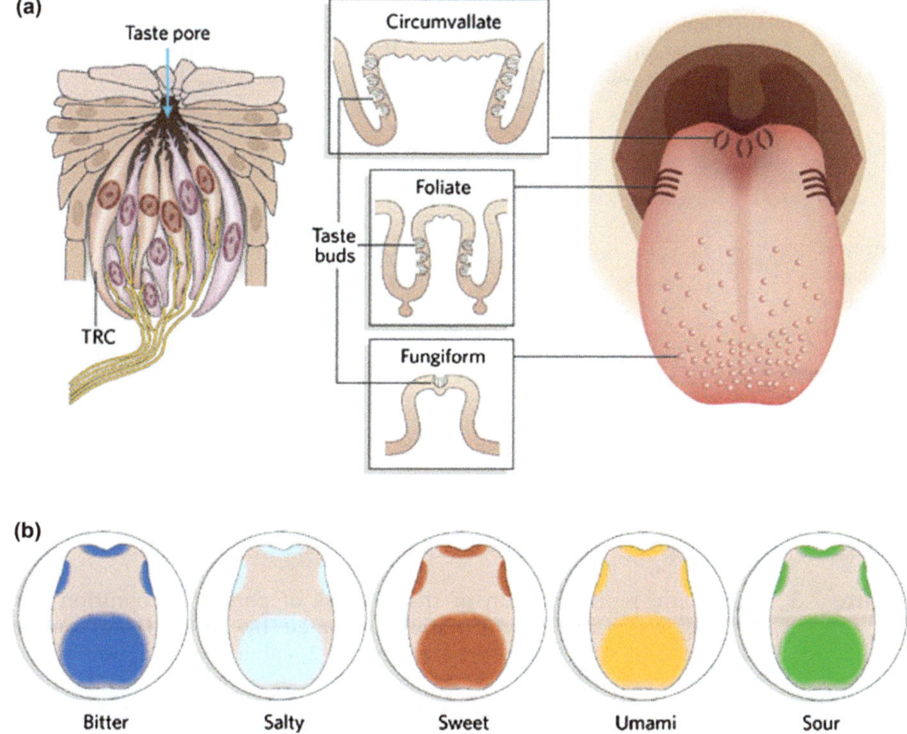

Figure 7.2 Taste-receptor cells, buds, and papillae: (a) Taste buds (*left*) are composed of 50–150 taste-receptor cells (depending on the species), distributed across different papillae. Circumvallate papillae are found at the very back of the tongue and contain hundreds (mice) to thousands (humans) of taste buds. Foliate papillae are present at the posterior lateral edge of the tongue and contain a dozen to hundreds of taste buds. Fungiform papillae contain one to a few taste buds and are found in the anterior two-thirds region of the tongue. Taste-receptor cells project microvillae to the apical surface of the taste bud, where they form a taste pore; this is the site of interaction with the tastants. (b) Responsiveness to the five basic modalities, *i.e.* bitterness, sourness, sweetness, saltiness, and umami, is present in all areas of the tongue. (Reproduced from Chandrashekar *et al.*[1] with permission from Macmillan.)

different instrumental techniques to extract information from the sample and create a dataset. To mimic the brain, we use a computer with statistical pattern recognition protocols to extract information from the dataset and create a graphical representation that clearly shows certain patterns, as well as differences and similarities between the samples, as shown in Figure 7.3. Human pattern-recognition ability allows us to differentiate between numbers in matrices, without the need for a computer. However, such recognition is easy when we have simple dataset matrices, such as four rows and four lines. In the next section, we show how mathematical algorithms

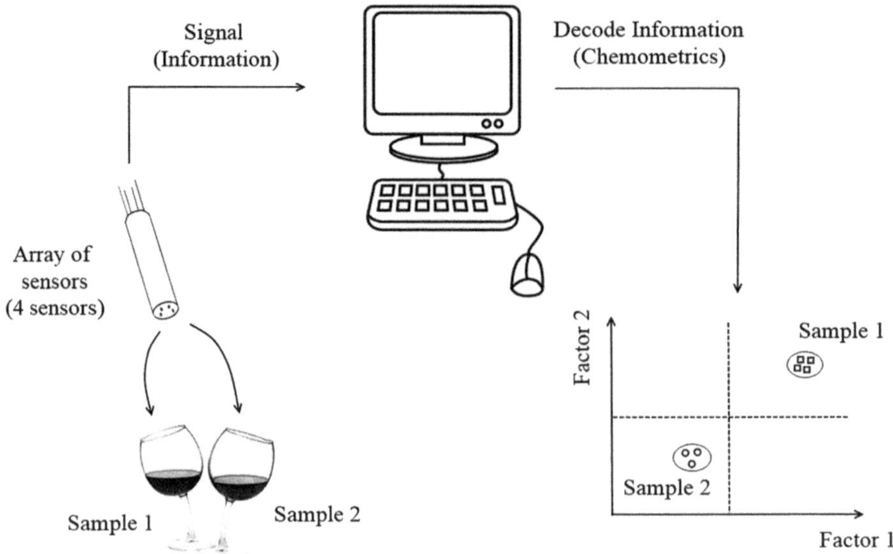

Figure 7.3   Schematic representation of the process of pattern recognition performed using input information extracted by an artificial tongue. Sample 1, cabernet merlot wine; sample 2, cabernet sauvignon wine.

are a powerful ally in the development of such a synthetic receptor, *i.e.* an artificial tongue.

7.2   Unsupervised Pattern Recognition Method Applied to the Development of Synthetic Receptors (Electronic Tongues)

Chemists often use illustrations to present their data. It is difficult to find articles showing a spectrum or voltammogram in a table form; rather, we can more easily notice the presence or absence of any peaks in absorbance (or current) resulting from a comparison between different chemical species using this type of chemical fingerprint. This is also true for simple graphs and non-complex images. However, a computer does not work with images like we do; a voltammogram consists of a large amount of data, and a computer needs to work with such a large data amount to find the differences between two voltammograms, as an example, which is not easy for the human brain. This duality regarding computer and human views is represented in Figure 7.4, which shows some cyclic voltammograms recorded for different samples. The matrix shown in Figure 7.4 represents the current values (variable measures for each sample) recorded for each potential applied to the working electrode.

Another important point that should be highlighted when working with pattern recognition is that we are presumably working with a large number

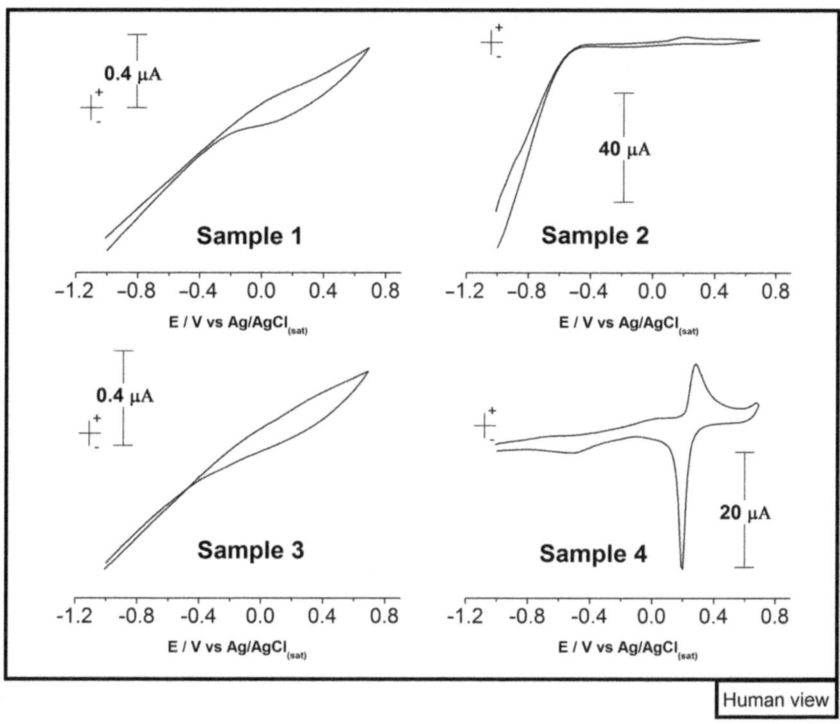

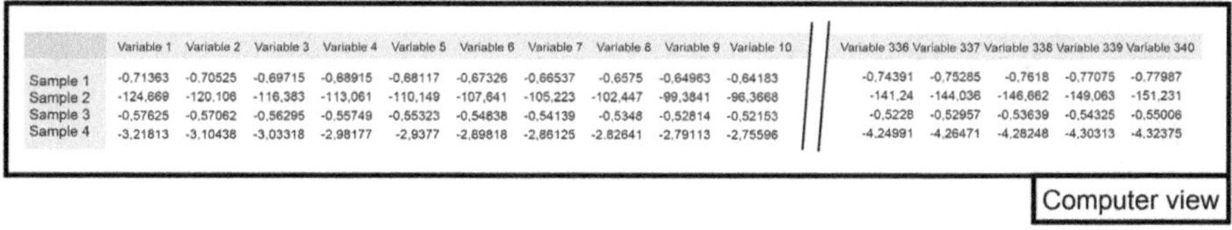

	Variable 1	Variable 2	Variable 3	Variable 4	Variable 5	Variable 6	Variable 7	Variable 8	Variable 9	Variable 10		Variable 336	Variable 337	Variable 338	Variable 339	Variable 340
Sample 1	-0,71363	-0,70525	-0,69715	-0,68915	-0,68117	-0,67326	-0,66537	-0,6575	-0,64963	-0,64183		-0,74391	-0,75285	-0,7618	-0,77075	-0,77987
Sample 2	-124,669	-120,106	-116,383	-113,061	-110,149	-107,641	-105,223	-102,447	-99,3841	-96,3668		-141,24	-144,036	-146,662	-149,063	-151,231
Sample 3	-0,57625	-0,57062	-0,56295	-0,55749	-0,55323	-0,54838	-0,54139	-0,5348	-0,52814	-0,52153		-0,5228	-0,52957	-0,53639	-0,54325	-0,55006
Sample 4	-3,21813	-3,10438	-3,03318	-2,98177	-2,9377	-2,89818	-2,86125	-2,82641	-2,79113	-2,75596		-4,24991	-4,26471	-4,28248	-4,30313	-4,32375

Computer view

Figure 7.4 Matrix (computer) and graphical (human) data views.

of variables to be measured, *i.e.* we are working with multivariate data. For example, in Figure 7.4 a single voltammetric curve was recorded having 340 variables. Each variable, a current value, was measured for a different potential applied to the system and the order of data in a matrix is, by convention, where the samples constitute the rows (the solution where the voltammograms were measured in Figure 7.4) and the variables (current values for each potential) are the columns. As we noted in Figure 7.4, it is easier to say that samples 1 and 3 are similar when using the human view instead of the computer view.

To start our discussion on pattern recognition, we will first show a simple non-chemical example to describe a pattern recognition analysis using simple mathematics. Table 7.1 shows the measurements of four features from different iris flower species (*Iris setosa*, *Iris virginica*, and *Iris versicolor*).

Table 7.1 Measurements of sepal and petal lengths and widths for three iris flower species (50 samples per species).

Iris species	Sepal length (cm)	Sepal width (cm)	Petal length (cm)	Petal width (cm)	Iris species	Sepal length (cm)	Sepal width (cm)	Petal length (cm)	Petal width (cm)
setosa	5.0	3.3	1.4	0.2	*versicolor*	5.7	2.9	4.2	1.3
virginica	6.4	2.8	5.6	2.2	*virginica*	7.2	3.0	5.8	1.6
versicolor	6.5	2.8	4.6	1.5	*setosa*	5.4	3.4	1.5	0.4
virginica	6.7	3.1	5.6	2.4	*setosa*	5.2	4.1	1.5	0.1
virginica	6.3	2.8	5.1	1.5	*virginica*	7.1	3.0	5.9	2.1
setosa	4.6	3.4	1.4	0.3	*virginica*	6.4	3.1	5.5	1.8
virginica	6.9	3.1	5.1	2.3	*virginica*	6.0	3.0	4.8	1.8
versicolor	6.2	2.2	4.5	1.5	*virginica*	6.3	2.9	5.6	1.8
versicolor	5.9	3.2	4.8	1.8	*versicolor*	4.9	2.4	3.3	1.0
setosa	4.6	3.6	1.0	0.2	*versicolor*	5.6	2.7	4.2	1.3
versicolor	6.1	3.0	4.6	1.4	*versicolor*	5.7	3.0	4.2	1.2
versicolor	6.0	2.7	5.1	1.6	*setosa*	5.5	4.2	1.4	0.2
virginica	6.5	3.0	5.2	2.0	*setosa*	4.9	3.1	1.5	0.2
versicolor	5.6	2.5	3.9	1.1	*virginica*	7.7	2.6	6.9	2.3
virginica	6.5	3.0	5.5	1.8	*virginica*	6.0	2.2	5.0	1.5
virginica	5.8	2.7	5.1	1.9	*setosa*	5.4	3.9	1.7	0.4
virginica	6.8	3.2	5.9	2.3	*versicolor*	6.6	2.9	4.6	1.3
setosa	5.1	3.3	1.7	0.5	*versicolor*	5.2	2.7	3.9	1.4
versicolor	5.7	2.8	4.5	1.3	*versicolor*	6.0	3.4	4.5	1.6
virginica	6.2	3.4	5.4	2.3	*setosa*	5.0	3.4	1.5	0.2
virginica	7.7	3.8	6.7	2.2	*setosa*	4.4	2.9	1.4	0.2
versicolor	6.3	3.3	4.7	1.6	*versicolor*	5.0	2.0	3.5	1.0
virginica	6.7	3.3	5.7	2.5	*versicolor*	5.5	2.4	3.7	1.0
virginica	7.6	3.0	6.6	2.1	*versicolor*	5.8	2.7	3.9	1.2
virginica	4.9	2.5	4.5	1.7	*setosa*	4.7	3.2	1.3	0.2
setosa	5.5	3.5	1.3	0.2	*setosa*	4.6	3.1	1.5	0.2
virginica	6.7	3.0	5.2	2.3	*virginica*	6.9	3.2	5.7	2.3
versicolor	7.0	3.2	4.7	1.4	*versicolor*	6.2	2.9	4.3	1.3
versicolor	6.4	3.2	4.5	1.5	*virginica*	7.4	2.8	6.1	1.9
versicolor	6.1	2.8	4.0	1.3	*versicolor*	5.9	3.0	4.2	1.5
setosa	4.8	3.1	1.6	0.2	*setosa*	5.1	3.4	1.5	0.2
virginica	5.9	3.0	5.1	1.8	*setosa*	5.0	3.5	1.3	0.3
versicolor	5.5	2.4	3.8	1.1	*virginica*	5.6	2.8	4.9	2.0
virginica	6.3	2.5	5.0	1.9	*versicolor*	6.0	2.2	4.0	1.0
virginica	6.4	3.2	5.3	2.3	*virginica*	7.3	2.9	6.3	1.8
setosa	5.2	3.4	1.4	0.2	*virginica*	6.7	2.5	5.8	1.8
setosa	4.9	3.6	1.4	0.1	*setosa*	4.9	3.1	1.5	0.1
versicolor	5.4	3.0	4.5	1.5	*versicolor*	6.7	3.1	4.7	1.5
virginica	7.9	3.8	6.4	2.0	*versicolor*	6.3	2.3	4.4	1.3
setosa	4.4	3.2	1.3	0.2	*setosa*	5.4	3.7	1.5	0.2
virginica	6.7	3.3	5.7	2.1	*versicolor*	5.6	3.0	4.1	1.3
setosa	5.0	3.5	1.6	0.6	*versicolor*	6.3	2.5	4.9	1.5
versicolor	5.8	2.6	4.0	1.2	*versicolor*	6.1	2.8	4.7	1.2
setosa	4.4	3.0	1.3	0.2	*versicolor*	6.4	2.9	4.3	1.3
virginica	7.7	2.8	6.7	2.0	*versicolor*	5.1	2.5	3.0	1.1
virginica	6.3	2.7	4.9	1.8	*versicolor*	5.7	2.8	4.1	1.3
setosa	4.7	3.2	1.6	0.2	*virginica*	6.5	3.0	5.8	2.2
versicolor	5.5	2.6	4.4	1.2	*virginica*	6.9	3.1	5.4	2.1
virginica	7.2	3.2	6.0	1.8	*setosa*	5.4	3.9	1.3	0.4
setosa	4.8	3.0	1.4	0.3	*setosa*	5.1	3.5	1.4	0.3

Table 7.1   (*Continued*)

Iris species	Sepal length (cm)	Sepal width (cm)	Petal length (cm)	Petal width (cm)	Iris species	Sepal length (cm)	Sepal width (cm)	Petal length (cm)	Petal width (cm)
setosa	5.1	3.8	1.6	0.2	*virginica*	7.2	3.6	6.1	2.5
virginica	6.1	3.0	4.9	1.8	*virginica*	6.5	3.2	5.1	2.0
setosa	4.8	3.4	1.9	0.2	*versicolor*	6.1	2.9	4.7	1.4
setosa	5.0	3.0	1.6	0.2	*versicolor*	5.6	2.9	3.6	1.3
setosa	5.0	3.2	1.2	0.2	*versicolor*	6.9	3.1	4.9	1.5
virginica	6.1	2.6	5.6	1.4	*virginica*	6.4	2.7	5.3	1.9
virginica	6.4	2.8	5.6	2.1	*virginica*	6.8	3.0	5.5	2.1
setosa	4.3	3.0	1.1	0.1	*versicolor*	5.5	2.5	4.0	1.3
setosa	5.8	4.0	1.2	0.2	*setosa*	4.8	3.4	1.6	0.2
setosa	5.1	3.8	1.9	0.4	*setosa*	4.8	3.0	1.4	0.1
versicolor	6.7	3.1	4.4	1.4	*setosa*	4.5	2.3	1.3	0.3
virginica	6.2	2.8	4.8	1.8	*virginica*	5.7	2.5	5.0	2.0
setosa	4.9	3.0	1.4	0.2	*setosa*	5.7	3.8	1.7	0.3
setosa	5.1	3.5	1.4	0.2	*setosa*	5.1	3.8	1.5	0.3
versicolor	5.6	3.0	4.5	1.5	*versicolor*	5.5	2.3	4.0	1.3
versicolor	5.8	2.7	4.1	1.0	*versicolor*	6.6	3.0	4.4	1.4
setosa	5.0	3.4	1.6	0.4	*versicolor*	6.8	2.8	4.8	1.4
setosa	4.6	3.2	1.4	0.2	*setosa*	5.4	3.4	1.7	0.2
versicolor	6.0	2.9	4.5	1.5	*setosa*	5.1	3.7	1.5	0.4
versicolor	5.7	2.6	3.5	1.0	*setosa*	5.2	3.5	1.5	0.2
setosa	5.7	4.4	1.5	0.4	*virginica*	5.8	2.8	5.1	2.4
setosa	5.0	3.6	1.4	0.2	*versicolor*	6.7	3.0	5.0	1.7
virginica	7.7	3.0	6.1	2.3	*virginica*	6.3	3.3	6.0	2.5
virginica	6.3	3.4	5.6	2.4	*setosa*	5.3	3.7	1.5	0.2
virginica	5.8	2.7	5.1	1.9	*versicolor*	5.0	2.3	3.3	1.0

The features measured were the length and width of the sepals and petals in cm. These data were used by Fisher[7] to introduce a discriminant analysis.

When analyzing Table 7.1, it is difficult to see the measurements of the sepal length and width and state which type of flower it indicates without knowing the species or class (columns 1 and 6). This is the approach of the unsupervised methods, in which the class information is known but is not initially used. An unsupervised method can be used to find the natural grouping of flower species, as an example. Using the values of sepal length and width, we can plot a graph in Figure 7.5 using only the species *setosa* and *virginica*.

Using Figure 7.5 and the human eye as the ultimate tool for pattern recognition, we can easily see groups and outliers within the data, and discriminate the analyzed samples (*Iris setosa* and *Iris virginica*) using only the simple flower dimension measurements. However, when we try to discriminate all types of flowers, a proper discrimination is difficult to achieve, as shown in Figure 7.6.

From Figure 7.6, is difficult to say which plots are for *Iris virginica* and which are for *Iris setosa* without knowing the class of each sample. We therefore need to include more information extracted from the sample to

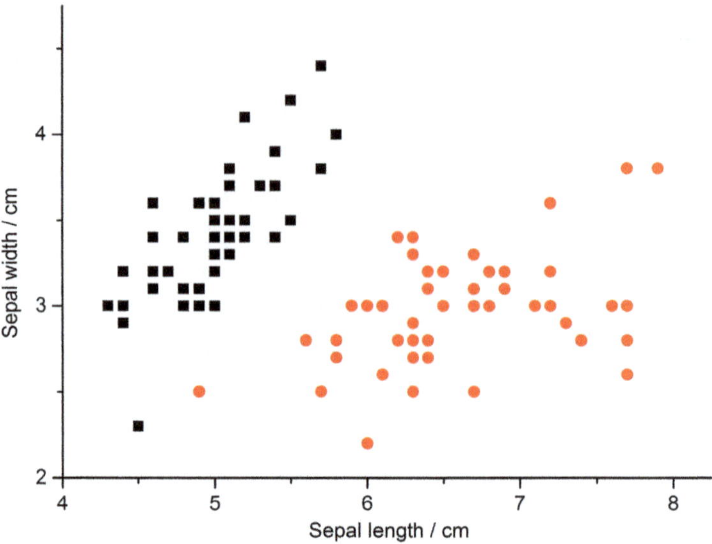

Figure 7.5   Plots of the length *versus* width of the sepal for *Iris setosa* (*black squares*) and *Iris virginica* (*red circles*).

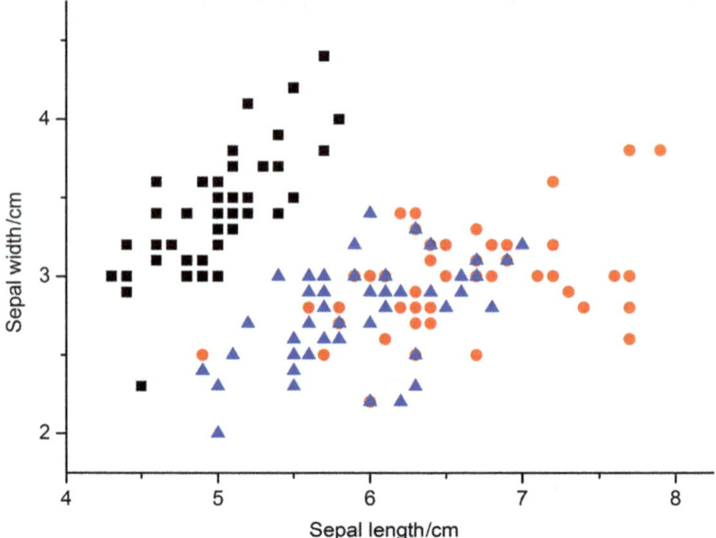

Figure 7.6   Plots of the length *versus* width of the sepal for *Iris setosa* (*black squares*), *Iris virginica* (*red circles*), and *Iris versicolor* (*blue triangles*).

discriminate the species, *e.g.* the petal length and width. However, the human eye has difficulty seeing and projecting in more than three dimensions. For this reason, multivariate techniques have been used as discriminating tools in different fields, with the principal component analysis (PCA) being the most frequently used technique.

7.2.1 Principal Component Analysis as an Analytical Tool (Brain) of an Electronic Tongue

As previously discussed, analyzing the variables extracted from a sample is an effective way to discriminate different samples if the number of measures, *i.e.* variables, is less than three per sample. Hence, the mathematical key of the PCA is manipulating the data matrix with the goal of representing the variations present in the many variables using a small number of "new factors or variables". Such new variables consist of a linear combination of the original values extracted from the samples, and a new row space is plotted instead of the original extracted values. These new axes, called principal components, allow the human eye to probe matrices with many variables and recognize patterns within the samples.

To better understand the PCA approach, we can use a simple two-variable example, as shown in Figure 7.5. It is a good idea to highlight that, for a two-variable system, it is possible to plot a graph without reducing the number of variables, *i.e.* use a PCA. This example is only used as a tutorial, and all of the PCA steps are summarized in Figure 7.7.

As we can see in Figure 7.7, the literal idea of the PCA is to look at the data from a different point of view. The first step is to calculate the first principal component that explains the greatest variation in the data in one particular direction. Furthermore, the percentage of the total variation (which is the percentage of the original dataset) can be precisely calculated. In this example, the first principal component describes 60.87% of the variation. Successive principal components that describe a portion of the remaining variation can be estimated, and the second component in the iris example contains 39.13%. Every new subsequent component is orthogonal to the previous component, and again, the direction of the component contains the greatest amount of remaining variation of the data. The total information in this case is 100% [60.87% (PC 1) + 39.13% (PC 2)] for two components, which is expected since we have two original measurements and two new variables after the PCA.

Generally, 70% of the original information (variation of the original dataset) is described for the first and second principal components. However, a two-dimensional plot will not be satisfactory if less than 70% of the variation is described by PC 1 and PC 2, since too much of the original information is missing and the discrimination process cannot be applied.

The second step is to calculate the projection of the original data to the principal component axes. This can be done by drawing a perpendicular line from the sample to the PC 1 or PC 2 axis, resulting in a new dataset. For the iris example, we have a new matrix with 50 projections to PC 1 and PC 2. The plot of the new dataset results in the graph shown in Figure 7.7C. The new coordinates of the original data relative to the principal component axes are called "scores", and result in a score plot (Figure 7.7C).

In most cases, it is interesting to know which original variable contributes the most significantly to the individual principal components once each PC

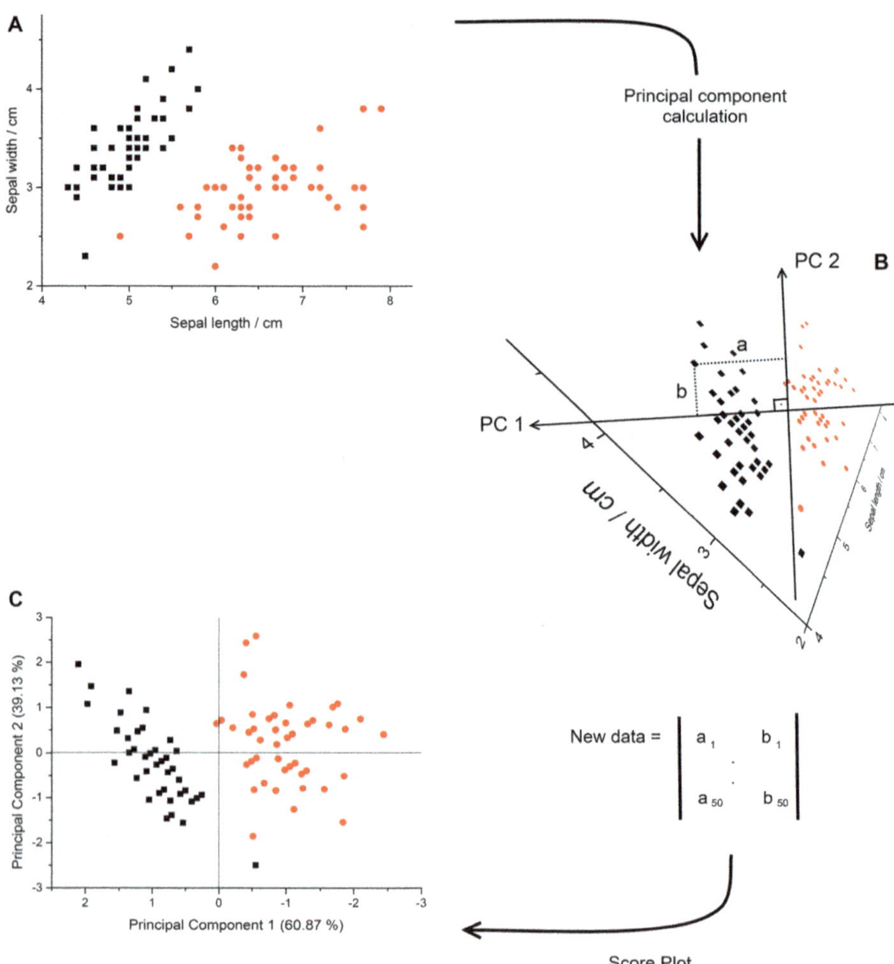

Figure 7.7   (A) A row plot of iris data for *Iris versicolor* and *Iris setosa* in a two-measurement system (length and width of the sepal). (B) The first two principal component axes calculated by the algorithm with the new coordinates to the principal component axes (*dashed lines*), and (C) a score plot obtained using the PCA.

is constructed from a combination of the original measurement variables. In addition, the extent to which the measurement variable contributes to a PC depends on the relative orientation in the PC space and the variable axes.

Mathematically, the contribution of each axis to a principal component is the cosine of the angle between the variable axis and the principal component axis. For the principal component parallel to the variable axis, the angle between the individual variable and the principal component is 0, *i.e.* a cosine of 1. A cosine of 1 indicates that the PC describes all of the variations in that variable axis. Conversely, a cosine of 0 indicates that none of the variations are contained in the PC. These cosine values are often called "loadings", and their values can range from −1 to 1.

Having briefly explained the concept of PCA, we can now return to the problem of the non-discrimination shown in Figure 7.6. In this example, we tried to discriminate the three species of iris flowers using two measurements (the length and width of the sepal) without success. However, we can introduce more variables to try to discriminate between the different iris flower species. In the example provided in Table 7.1, we have more variables than we can represent within a 3D space. In order to discriminate all species, we have to use all datasets reported in Table 7.1 as input into the PCA. The score and loading plots obtained after such analysis are shown in Figure 7.8. Different algorithms can be used to calculate the loadings and scores of the PCA, and different software can be used to calculate the PCA itself (Statistica®, Pirouette®, Mathlab®, Unscramble®, and Origin® Professional, version 8.6 or higher, from OriginLab®).

As can be seen in Figure 7.8A (score plot), we can increase the discrimination between the *setosa* and *virginica* species, as well as the distance of the cluster of the *setosa* species from the other two species, by introducing two more new variables. The analysis of the score plot and loading plot in Figure 7.8B can provide information regarding the correlation between variables: all arrows pointing in roughly the same direction are correlated. On the other hand, arrows pointing in perpendicular directions are not. In Figure 7.8B, we can see, for example, that variables in petal length and width are highly correlated with each other, while variables in sepal width are more or less uncorrelated. Furthermore, we can see from this analysis which original variables are important in the individual PCs. The variables with higher positive loadings on PC 1 are petal length and width, and sepal length; the item with the largest negative loadings is sepal width. This last affirmation indicates that *Iris setosa* types have the largest sepal width in most cases, based on the region of the *setosa* cluster shown in Figure 7.8A, with negative values found at the PC1 axis.

To conclude this section, the main idea of a PCA is to reduce the original dimensionality of the original dataset, and in the example given in Figure 7.8 we reduced four original variables to two new variables (new dataset). Such new datasets can be symbolized through a 2D or 3D plot representing the majority of variations in the original dataset, and are useful for visualizing the similarities and differences of the studied samples. Based on this property, to reduce the dimensionalities of the original data and the graphical representation, a PCA is a vital tool to simulate the brain functions in the development of an electronic tongue, *i.e.* in recognizing patterns. Additionally, our main idea here is to show how we can interpret and use PCA for the development of an electronic tongue, while not being a PCA expert.

At this stage, we can imagine that an array of chemical/electrochemical sensors can be used to extract information from a sample to produce a large amount of information, as shown in Table 7.2, that can be analyzed using an

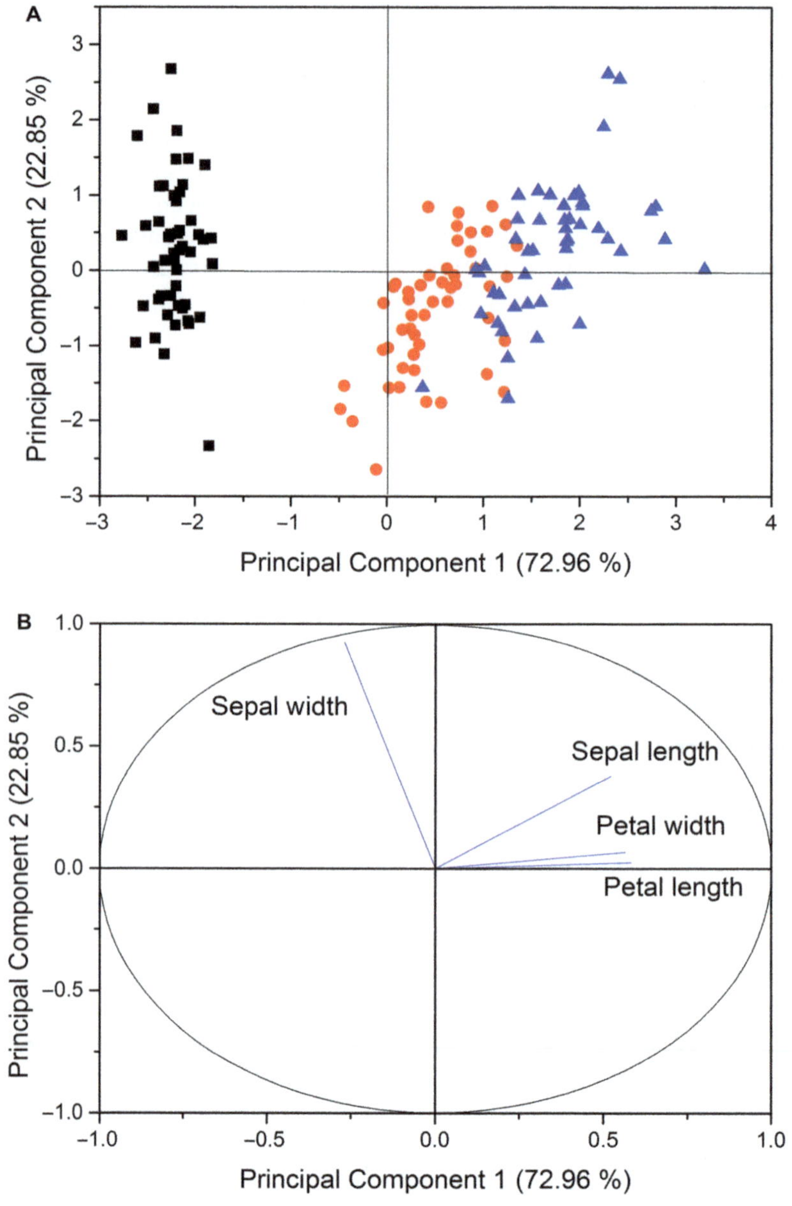

Figure 7.8   (A) Score plot of the iris data from Table 7.1: *Iris setosa* (*black squares*), *Iris virginica* (*red circles*), and *Iris setosa* (*blue triangles*), and (B) a loading plot.

unsupervised pattern recognition method, *e.g.* a PCA. It is important to mention that there are other algorithms available that can also be used. Additionally, we can use other techniques to extract more than a single information, variable or measurement per sensor, *e.g.* voltammetric

Table 7.2 An example matrix showing how information can be extracted for a sample using an array of electrochemical sensors. If transmitted from the sensor, this information can be used as input for an unsupervised pattern recognition method, as shown in Figure 7.3.

Sample	Sensor 1	Sensor 2	Sensor 3	Sensor 4	Sensor 5	...	Sensor 10
Sample 1	Value	Value	Value	Value	Value	...	Value
Sample 2	Value	Value	Value	Value	Value	...	Value
Sample 3	Value	Value	Value	Value	Value	...	Value
.	.	.	.	.	.	...	.
Sample 50	Value	Value	Value	Value	Value	...	Value

techniques. Using this technique, we can extract an electrochemical spectrum or a fingerprint analogous to a conventional spectrum obtained by UV/Vis techniques and this information could be used in a row of the matrix in Table 7.2.

In voltammetric techniques, potential–time waveforms are used to sweep the working electrode potential between two different values of potential using a constant scan rate to change the values between the limits of potential. The resulting current from the electrochemical process that occurs at the surface of the electrode is recorded as a function of the applied potential controlled by the user. The graph plotted between the applied potential (x-axis) and the recorded current (y-axis) is called a voltammogram or cyclic voltammogram, A more in-depth description of cyclic voltammetry can be found in Bard and Faulkner's textbook.[8] In the next section, we will focus on the development and application of PCA as a pattern recognition tool for use in electronic tongues and will show some examples of how we can combine cyclic voltammetric data with PCA.

7.3 Development and Application of an Electronic Tongue

According to the IUPAC technical report,[9] an *"electronic tongue is a multi-sensor system, which consists of a number of low selective sensors and uses advanced mathematical procedures for signal processing based on the pattern recognition (PARC) and/or multivariate analysis (artificial neural networks (ANNs), principal component analysis (PCA), etc.)"*. The term 'electronic tongue' was first introduced by Vlasov *et al.*[10] in 1996, six years before Toko introduced the term 'taste sensor'.[11] A taste sensor is an array of sensors that can differentiate between the five different tastes that are detectible by the human tongue (saltiness, sweetness, sourness, bitterness, and umami), and can be used to recognize the different types of food and beverage tastes, as well as for quantification proposes.[11–13] An electronic tongue is a broader term, *i.e.* a device capable of differentiating between liquid samples based on their different characteristics, including differences in taste, but not relying solely on these characteristics.

Both an electronic tongue and a taste sensor are based on global selectively, a term proposed by Toko *et al.*[12] This means that the goal of such device is not to separately analyze different species, but to bring together all information into a pattern. This concept is the same as that used by the human tongue. For instance, when a person takes a sip of coffee, while they do not comprehend that the coffee consists of more than a thousand different molecules, they nevertheless can recognize different coffee types. In other words, the principle of an electronic tongue is to evaluate, using a chemometric tool such as PCA, the variability between samples, taking into account similar characteristics, and combine them into specific groups. In this sense an electronic tongue, also called a smart device, provides a global response, a fingerprint of the sample characterizing and recognizing its substances.[14] Thus, highly selective techniques are unnecessary and, consequently, techniques that can provide a large amount of information, preferably through a simple measurement, are required. Electrochemical sensors, which are usually employed to obtain highly selective responses, can also be used through a different approach. They can be employed to obtain and extract a large amount of chemical information from a sample, and treating what might earlier have been considered as interference as important information for the discrimination of different samples.[15] The highest amount of information can be obtained in different ways, for instance the recording of a voltammogram can generate a number of variables as large as the number of current values registered by the potentiostat. Another approach is the use of an array of sensors, called a multi-sensor device. Figure 7.9 shows an example of such a device with an array of nine potentiometric sensors used to evaluate wine spoilage.[16]

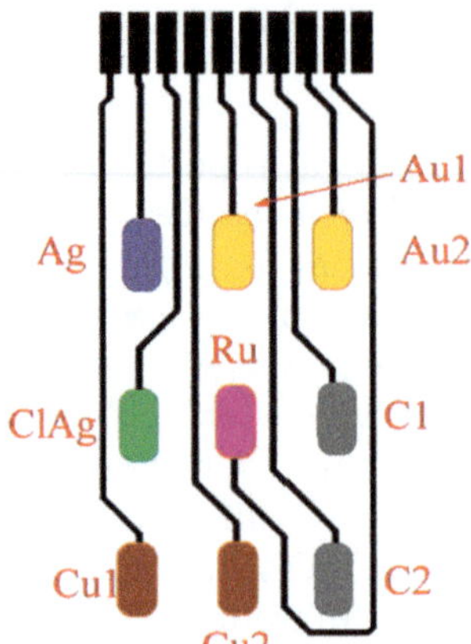

Figure 7.9 Potentiometric sensors for an electronic tongue.
(Reproduced from Gil-Sánchez *et al.*[16] with permission from Elsevier.)

Earlier electronic tongues, which are still the most widely used, are based on potentiometric devices.[17,18] A potentiometric sensor provides as an output the difference in potential (voltage) that develops through a membrane separating the inner solution of the electrode from the bulk solution containing the analyte, where the difference in the electrical charge between the solutions causes this potential difference. By changing the interface composition, *i.e.* using different polymeric membranes, proprieties such as selectivity and sensitivity can be modified.[13,19] Typically, potentiometric electrodes are highly selective for certain ions, the most famous one being a pH electrode; however, some sensors respond to a group of analytes, where the potential difference can be observed in the presence of several cations or anions, for instance. For this reason, potentiometric electronic tongues often use an array of numerous electrodes. As a result, signal overlap can occur, resulting in a cross-selective array of electrodes, which is frequently valuable for the discrimination process.

The first voltammetric electronic tongue was introduced by Winquist *et al.* and used two working electrodes (Pt and Au) and several voltammetric techniques for an analysis of fruit juices, still drinks, and milk.[20] In contrast to potentiometric sensors that usually respond to only one analyte, a voltammetric electronic tongue has the advantage of providing a much broader response, since a voltammogram obtained in a solution provides information regarding all electro-active species present in the solution, as well as the adsorption of certain species and diffusion coefficients.[20] Electronic tongues based on impedance or amperometric measurements can also be found in the literature, but to a lesser degree.[17,21–23]

In the same way in which the human taste receptors from our tongue recognize the chemical responses from different tastes, synthetic receptors, based on different membranes for potentiometric sensors or different compounds used in the modification of electrode surfaces, are applied for the recognition and discrimination of a wide range of samples such as food,[24] beverages,[25] pharmaceuticals,[26] and environmental analyses.[27] In almost all cases, these potentiometric sensors for fabrication of an electronic tongue are modified with ion-selective poly(vinyl chloride) (PVC) membranes using different plasticizers [bis(1-butylpentyl) adipate, dioctyl sebacate, *o*-nitrophenyl octyl ether, dioctyl phenyl phosphate, tris(ethylhexyl) phosphate, and dibutyl sebacate] and ionophores, or recognition elements {nonactin, valinomycin, 2,3:11,12-didecalino(16-crown-5), lasalocid, dibenzo(18-crown-6), tridodecylamine, 2,9-di-*n*-butyl-1,10-phenanthroline, bis[bis[4-(1,1,3,3-tetramethylbutyl)phenyl]phosphato]calcium(II), bis[(12-crown-4)methyl]-2-dodecyl-2-methyl malonate, tridodecylamine, 4-*tert*-butylcalix[8]arene-octaacetic acid octaethyl ester, monensin sodium salt, tetraoctylammonium nitrate, and the sodium salt of the antibiotic tetronasin} and they have been used extensively by del Valle's group in Spain.[24,25]

One of the major applications of an electronic tongue is the analysis of foods, beverages, and tastes.[18,22,28–33] The range of samples possible within these classes is enormous, including analyses of water, wine, tea, coffee, milk, beer, honey, oils, fish, fruits, rice, tomato, and onions; the recognition of different tastes (sweetness, saltiness, bitterness, sourness, and umami or savouriness); and the evaluation of drug bitterness.[18,22,28–33] Since the number of publications focusing on the topics mentioned above is huge, to avoid an exhaustive overview of this topic, and with the aim of focusing on the parallel aspects between a human and an electronic tongue, only analysis of beverages using PCA is discussed herein. Other applications of an electronic tongue can be found in the review articles cited.[18,22,28–33]

The goal is to use an electronic tongue to find, within a set of samples, different groups that differ from each other based on their characteristics, such as brand, quality, year of production, and adulteration. Numerous papers can be found focusing on the discrimination of wine, milk, tea, beer, juice, coffee, or other alcoholic beverages and such discriminations reported in the literature, including the majority of manuscripts reported since the development and use of the term electronic tongue, are based on the differences in brand, year of manufacturing, type of beverage, storage conditions, and origin.[20,23,24,34–88] Some of the works will be described later in the text. Electronic tongues are also being used for different proposes, for instance to evaluate the aging process and its effect on products, such as milk, beer, and wine,[89–95] to determine beverage quality,[16,96–103] and to detect adulteration processes.[104–106]

The use of an electronic tongue in the discrimination of different beverages according to their taste is a field of high research interest worldwide. Well-trained and certified persons called panellists are usually employed in the evaluation of beverages, assessing their flavours, general characteristics, aromas, and potential faults. Ultimately, such panellists can evaluate the quality of a beverage. However, despite being extremely well trained, panellists are more prone to errors than instrumental techniques, which also provide faster results. As a result, electronic tongues are being applied to discriminate different beverages based on their brands, taste, and other characteristics. Ciosek *et al.*[34] performed the analysis of mineral waters and apple juices using an array of 17 potentiometric electrodes. Each electrode had a synthetic receptor sensitive to a set of ions. The authors used one standard pH electrode, and two of each of the following selective electrodes: Cl^-, F^-, NO_3^-, HCO_3^-, K^+, Na^+, NH_4^+, and Ca^{2+}. The proposed array can successfully discriminate between 10 brands of mineral water, as shown in Figure 7.10, where each sample of water is placed on a different region of the graph.

With the aid of a mathematical equation that relates the variances between the samples clustered together and the sum of the internal variances in each cluster, the authors were able to reduce the number of sensors utilized to nine. For the water analysis, sensors for the pH and for F^-, HCO_3^-, K^+, and Na^+ were used, and for the juice analysis, sensors for the

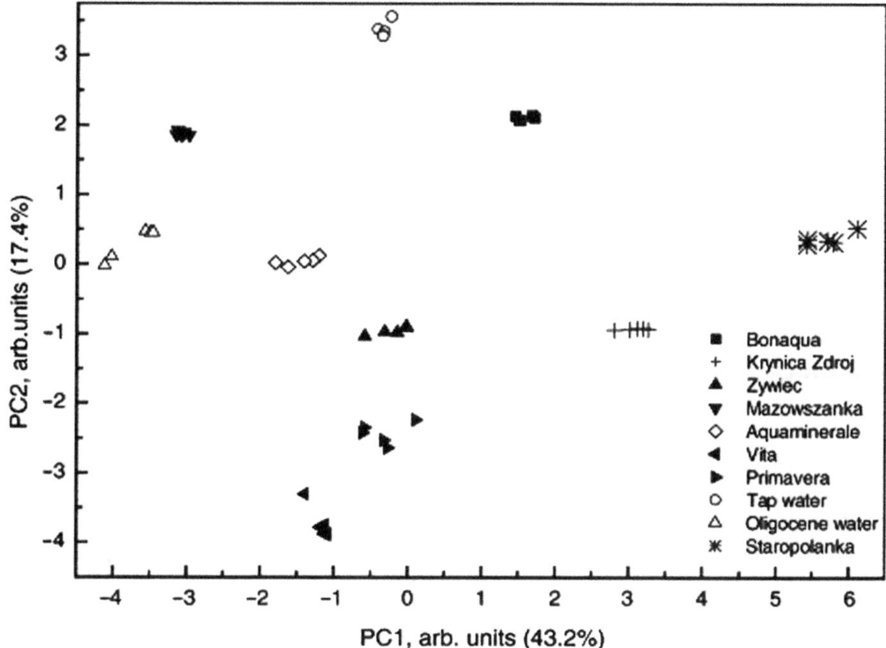

Figure 7.10   PCA plot of different brands of mineral water before reducing the number of sensors in the sensor array.
(Reproduced from Ciosek *et al.*[34] with permission from Elsevier.)

pH and for Cl^-, HCO_3^-, NH_4^+, and Ca^{2+} were chosen. Figure 7.11 shows that different mineral water samples can still be discriminated from each other even when using a smaller number of electrodes, suggesting that an electronic tongue can be more simply designed.

The same research group applied an electronic tongue to discriminate between five brands of milk, five different types of beer from the same brewery, and three brands of orange juice.[35] The electronic tongue consisted of six potentiometric sensors, three of them selective to NH_4^+, Cl^-, and H^+ ions, and the other three partially selective to Na^+/K^+, cation selective, and $F^-/H_2PO_4^-$. The selectivity or partial selectivity was gained using different membranes, *i.e.* synthetic receptors on the potentiometric sensor. The authors also evaluated the results of performing the experiments under static and dynamic conditions. Despite being able to distinguish between juices in flow mode using only the selective electrodes, and discriminate between beer samples in flow mode using only the partially selective electrodes, the authors concluded that the best results for all samples were obtained when simultaneously using both types of sensors.

Arrieta *et al.*[36] applied a voltammetric electronic tongue based on six platinum surfaces modified with six different polypyrroles as synthetic

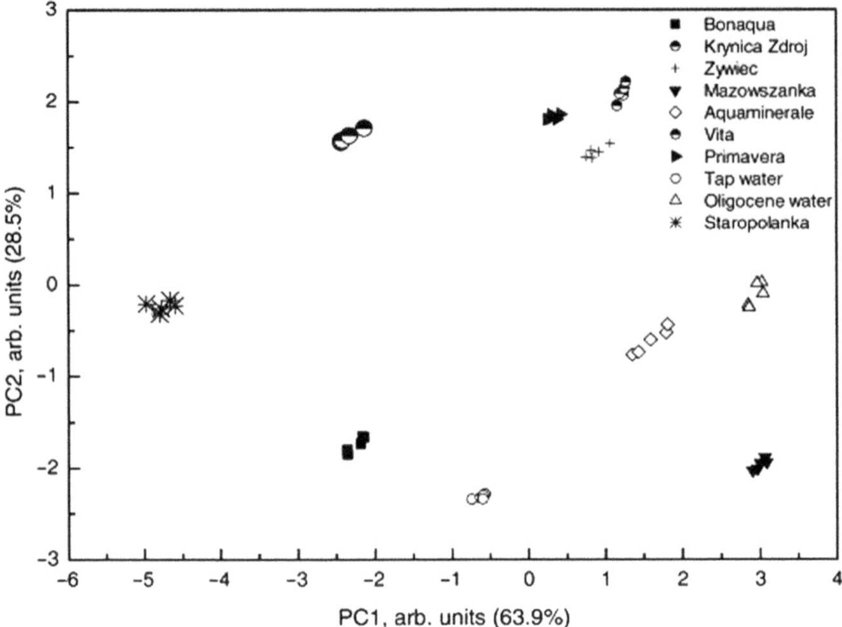

Figure 7.11   PCA plot of different brands of mineral water after reducing the number of sensors in the sensor array.
(Reproduced from Ciosek *et al.*[34] with permission from Elsevier.)

receptors (disulfonic acid disodium salt, 1-decanesulfonic acid sodium salt, potassium hexacyanoferrate, sulfuric acid, phosphotungstic acid, *p*-toluenesulfonic acid, and anthraquinone-2,6-disulfonic acid disodium salt) for the ions, and iso-α-acid present in the solution, and successfully discriminated between 21 different commercial beers. Bhondekar *et al.*[45] used an electronic tongue based on impedance measurements for the discrimination of different tea samples. Six electrodes were used in the impedance array (glassy carbon, gold, platinum, silver, and ITO-coated glass covered with polyaniline or polypyrrole), and the authors were able to successfully distinguish between the evaluated teas and determine which electrode was responsible for the discrimination of each tea sample.

Qualitative and quantitative analyses of cava wines were performed by Ceto *et al.*[37] using a voltammetric electronic tongue consisting of an array of five graphite-epoxy electrodes unmodified and modified using copper, platinum, polyaniline, or polypyrrole nanoparticles. A good discrimination between the samples was reported, which the authors related to the different amounts of sucrose added during the cava wine production. A linear relationship was obtained through PC 1 *versus* the amount of sucrose, through which the authors also performed a quantitative analysis. Qualitative and quantitative analyses were also performed by Evtugyn *et al.*[38] using apple juices and herbal liquors; here, a different approach was used, in which known amounts of Fe^{3+} were added to all samples such that the organic ligands and antioxidants present in the samples could react with the ions, resulting in different responses of the ion-selective electrodes (ISEs) covered only with synthetic receptors, namely electropolymerized polyaniline, or with electropolymerized polyaniline and two kinds of thiacalix[4]arenes. Quantitative responses were obtained for ascorbic acid, malic acid, oxalic acid, hydroquinone, and quercetin, while a PCA analysis showed a good discrimination between the different apple juice and herbal liquor samples. The authors proposed that the different signals were from three factors: a partial oxidation of the components by the Fe^{3+}; changes in the pH and redox potentials owing to the hydrolytic instability of the Fe^{3+}; and the complexation of Fe^{3+} ions with organic species.

Some of the parameters influencing the quality of wine are its age, and where and in what manner the aging process takes place. Studies using different synthetic receptors have been conducted to compare the different aging processes, *i.e.* aged in oak barrels or in steel tanks in contact with oak wood chips or staves, and through micro-oxygenation.[89,90] For instance, Apetrei *et al.*[89] proposed the use of a voltammetric electronic tongue consisting of an array of six electrodes, *i.e.* one platinum electrode and five carbon paste electrodes, one of them unmodified and the other four modified using different synthetic receptors; two of the four were modified using two rare-earth bisphthalocyanine molecules [one with lutetium(III) and the other with gadolinium(III) as a centre], and the other two used carbon paste electrodes modified with cobalt(II) monophthalocyanine and ferrocene. Square wave voltammograms were registered in the differently aged wines, and distinct responses were obtained with each voltammogram, which were related to the different electro-activities of the phenolic compounds present in red wines, and to the oxidation and reduction processes of other electro-active substances. The proposed method is able to discriminate between the different aging processes. Analogous to an analysis of the wine aging process, Ghasemi-Varnamkhasti *et al.*[91] evaluated beer aging; however, in contrast to wine, which increases in quality over time, the concern of brewers is not having the characteristic flavour of each beer change as time passes; they are therefore more concerned with flavour stability. Three tyrosinase biosensors based on one unmodified carbon paste electrode (CPE) and two CPEs based on cobalt phthalocyanine and iron phthalocyanine were used for this purpose. Voltammograms were registered in the beer samples, and a good discrimination was obtained based on a PCA. The differences were attributed to the phenolic composition, in particular, to changes in the flavonoid levels that occur during aging. These changes were observed in the voltammograms through variations in the peak position and current intensity values.

An interesting approach to the use of an electronic tongue is the detection of adulterated samples. For instance, Dias *et al.*[104] used 36 cross-sensitivity potentiometric sensors for the detection of goat milk within bovine milk. The sensors were built using synthetic receptors to recognize the five basic tastes, *i.e.* saltiness (sensors for sodium chloride, potassium chloride, and ammonium chloride), sweetness (sensors for glucose, fructose, and sucrose), bitterness (sensors for caffeine, urea, and magnesium sulfate), sourness (sensors for citric acid, ascorbic acid, and hydrochloric acid), and umami (sensor for monosodium glutamate). The sensors proved to be more effective toward the recognition of acidity, saltiness, and umami; nevertheless, the proposed electronic tongue did discriminate between goat, cow, and goat/cow raw skimmed milk samples. Milk adulteration was also evaluated by Paixão *et al.*,[105] who proposed the use of a disposable voltammetric electronic tongue made of copper, gold, and gold modified with a Prussian blue film for the detection of milk adulterated with hydrogen peroxide. Experiments also showed the electronic tongue's ability to discriminate between different pasteurization processes [homogenized/pasteurized, ultra high temperature (UHT) pasteurized, and UHT pasteurized with low fat content].

Few studies can be found in the literature regarding the use of an electronic tongue for the analysis of food/beverages containing diseases from their source materials (plants or animals). For example, Mottram *et al.*[96] used 15 sensors with cross-sensitivity to inorganic and organic cations and anions; chalcogenide glass sensors; chloride-, potassium- and sodium-selective electrodes; and a glass pH electrode to discriminate between milk samples obtained from healthy cows and from cows diagnosed with mastitis, an inflammation of the bovine mammary gland caused by a pathogen infection. The discrimination obtained from the proposed synthetic receptors reached a specificity of 96% and a sensitivity of 93%. In a similar way, Wei *et al.*[97] developed an electronic tongue composed of five working electrodes (gold, silver, platinum, palladium, and titanium) to detect antibiotic residue in bovine milk. Despite the fact that the PCA was not able to distinguish between all antibiotics, the authors could discriminate between samples using other chemometric techniques.

A hybrid electronic tongue has also been used, meaning that more than one type of technique is utilized. The range of possible techniques to be used in combination is enormous; for instance, such a fusion can be between two electrochemical techniques,[39] between electrochemical and spectroscopic techniques,[40,41] or between an electronic nose, electronic eye, and electronic tongue.[16,41,42,92,98] An electronic nose is a multi-sensor applied to gas analyses, while an electronic eye can be defined as a device used in colorimetric analyses. Pietro and co-workers[92] developed a device, called an e-panel, to analyze the influence of oxygen pick-up before and after wine bottling. The term 'e-panel' is a reference to the panellists (certified persons employed in an evaluation of beverages) mentioned previously, and is composed of an electronic nose, an electronic eye, and an electronic tongue. The voltammetric electronic tongue consists of an array of eight electrodes: a platinum electrode; an unmodified carbon paste electrode, and six modified carbon paste electrodes with two bisphthalocyanines (lutetium bisphthalocyanine and the octatert-butyl substituted analogue), cobalt metallophthalocyanine, *N,N*-bis(methylpiperidine)-3,4,9,10-perylenebis(dicarboximide), the *N*-octyl-3,4,9,10-perylenebis-(dicarboximide), and the *N*-butyl-3,4,9,10-perylenebis(dicarboximide); the electronic eye, a spectrophotometer; and an electronic nose array of 15 gas sensors. Using only the electronic tongue, some degree of discrimination was observed between wines with different amounts of polyphenol. The amount of phenolic compounds can be related to the oxygenation of the wine. Since the information obtained from the electronic tongue is not sufficient to discriminate all samples, different kinds of information are necessary, which in this work were obtained using an electronic nose and electronic eye. Thus, the discrimination obtained using the e-panel is better than that obtained using only the electronic tongue.

The main idea of using a hybrid electronic tongue is to obtain the most from each technique, *i.e.* obtaining information that could not be possible with just one technique alone, making the analysis more reliable. For instance, with the fusion of an electronic tongue and an electronic nose, information regarding volatile and non-volatile compounds can be obtained simultaneously. However, this concept can be employed in simpler ways, without having to resort to other types of techniques and avoiding the use of more equipment. For example, Novakowski *et al.*[43] used a copper electrode together with a gold electrode to discriminate between different wine samples. Both copper and gold electrodes alone can partially discriminate different samples; however, by gathering information from both electrodes, all samples tested could be discriminated. This fusion was possible since the authors understood that different information could be obtained from different electrodes, *i.e.* polyphenol compounds are electroactive on gold surfaces, while sugars, sulfite, and alcohols are electrocatalytically oxidized on copper electrodes.

More extended information regarding the use of voltammetric, potentiometric, amperometric, and impedance-based electronic tongues focusing on PCA of different beverages can be found in Table 7.3.

Based on the above discussion, we can note that the field of research focusing on the manufacture of artificial receptors, such as an electronic tongue, is very promising, and that advances in this area may help in understanding and replacing the human sensory panel used in repetitive activities, and in activities that may be harmful to human health. It is important to highlight also that these artificial receptors can help in human palate reconstruction or in the future development of an artificial palate system for intelligent machines, such as robots.

Table 7.3 PCA of voltammetric, potentiometric, amperometric, and impedance-based electronic tongues, focusing on beverage analyses.

Sensor electrodes or arrays (*i.e.* synthetic receptors)	Electronic tongue	Sample	Reference
Two cation sensitive membranes, two anion-sensitive membranes, and one hydrogen ion-sensitive membrane	Potentiometric	Wine	44
Six porphyrin-based electrodes and a conventional glass pH electrode	Potentiometric	Wine	46
23 sensors sensitive to inorganic cations and anions, pH, organic substances, and chalcogenide glass membrane	Potentiometric	Wine	47
Three platinum electrodes modified using a conducting polymer, (poly(3,4-ethylenedioxythiophene)), two of them with composite materials of Au and Pt nanoparticles	Potentiometric	Wine	48
Seven sensors with the following membranes: H_2TPP-based membrane, N1; $Co(III)TPPBr$-based membrane, N2; $Co(II)Por$-based membrane, N4; dummy membrane, N5; $Pt(II)TPP$-based membrane, N8; and two $Pt(IV)TPPCl_2$-based membranes, NN11 and 12	Potentiometric	Wine	49
Eighteen sensors with plasticized PVC and chalcogenide glass membranes and a pH glass electrode	Potentiometric	Beer	50
29 sensors, *i.e.* plasticized PVC membranes displaying sensitivity to organic anions and with chalcogenide glass membranes displaying a redox response, and a glass pH electrode	Potentiometric	Beer	51
36 polymeric membranes applied to two-sensor arrays	Potentiometric	Juice	52
Seven electrodes with membranes made of dioctyl phosphate with different precondition periods	Potentiometric	Milk	53
Six potentiometric sensors, *i.e.* one standard pH electrode, two selective electrodes for NH_4^+ and Cl^-, and three partially selective electrodes: Na^+/K^+, cation selective, and $F^-/H_2PO_4^-$	Potentiometric	Milk/Orange Juice/Beer	34
Six potentiometric sensors, *i.e.* one standard pH electrode, two selective electrodes for NH_4^+ and Cl^- and three partially selective electrodes: Na^+/K^+, cation selective, and $F^-/H_2PO_4^-$	Potentiometric	Beer/Juice/ Milk	54
Seventeen potentiometric sensors, *i.e.* a standard pH electrode plus two of each the following selective electrodes: Cl^-, F^-, NO_3^-, HCO_3^-, K^+, Na^+, NH_4^+, and Ca^{2+}	Potentiometric	Mineral water/ apple juice	34
Seventeen potentiometric sensors, *i.e.* standard pH electrode plus two of each of the following selective electrodes: Cl^-, $F^-/H_2PO_4^-$, HCO_3^-, anion selective, K^+/Na^+, NH_4^+, Ca^{2+}, and cation selective	Potentiometric	Orange juice/ tonic/milk	55

Sensor description	Method	Sample	Ref.
Sixteen potentiometric sensors, *i.e.* two of each of the following selective electrodes: Cl^-, $F^-/H_2PO_4^-$, HCO_3^-, anion selective, K^+/Na^+, NH_4^+, Ca^{2+}, and cation selective	Potentiometric	Milk/fruit juice/tonic	56
Eight potentiometric sensors selective to Cl^-, $F^-/H_2PO_4^-$, HCO_3^-, K^+/Na^+, NH_4^+, Ca^{2+}, and cation selective, and one with an ionic liquid membrane	Potentiometric	Tea/herbals	57
Two sets of multisensors, one with sensors sensitive to Na^+, K^+, Ca^{2+}, NH_4^+, and Cl^-, and another sensitive to heavy metal ions (Pb^{2+}, Cd^{2+}, Cu^{2+}, Tl^+, and Ag^+). Both multi-sensor chips also contain a pH sensor	Potentiometric	Grape juice/wine	58
Three ion-selective electrodes covered with only electropolymerized polyaniline or electropolymerized polyaniline and two kinds of thiacalix[4]arenes	Potentiometric	Apple juice/herbal liquors	38
Three ion-selective electrodes covered with only electropolymerized polyaniline or electropolymerized polyaniline and two kinds of thiacalix[4]arenes	Potentiometric	Tea/ions	59
Seven potentiometric sensors selective to H^+ (x2), Li^+, NH_4^+, K^+, Na^+, Ca^+, and two generic sensors	Potentiometric	Commercial waters/orange based drinks/tea	24
18 to 21 selective sensors and non-specific sensors	Potentiometric	Tea/coffee/soft drink/mineral water /orange and grape juice/beer	107
Twenty sensors sensitive to cations, anions, pH, and with a chalcogenide glass membrane	Potentiometric	Ethanol/vodka/eau-de-vie	60
Twelve carbon paste electrodes sensitive to inorganic cations and anions, pH, and organic substances	Potentiometric	Potable waters/soft drinks/beers	61
Twelve carbon paste electrodes sensitive to inorganic cations and anions, pH, and organic substances	Potentiometric	Coffee/tea	62
Eight potentiometric solvent polymeric membrane sensors	Potentiometric	Wine/grappa/beer	63

Table 7.3 (*Continued*)

Sensor electrodes or arrays (*i.e.* synthetic receptors)	Electronic tongue	Sample	Reference
Two sensor arrays, each containing thirty sensors, containing sensors with chalcogenide glass and polyvinyl chloride (PVC)-plasticized membranes, a pH glass electrode, and crystalline sensors. The first array has twelve solid-state sensors and eighteen PVC plasticized sensors. The second array has seven solid-state sensors and twenty-three PVC sensors	Potentiometric	Mineral waters/ coffee/soft drinks/fish	64
Paste electrodes with active elements, *i.e.* Ag, Au, Cu, Ru, AgCl, C	Potentiometric[a]	Wine	16
Fifteen sensors with cross-sensitivity to inorganic and organic cations and anions; chalcogenide glass sensors; chloride-, potassium- and sodium-selective electrodes;and a glass pH electrode	Potentiometric	Milk (mastitic and healthy)	96
36 cross sensitivity sensors	Potentiometric	Milk	97
α-ASTREE, *i.e.* seven potentiometric sensors with different membranes (ZZ36, BB06, CA07, BA07, AB07, HA06, CB07)	Potentiometric[a] (commercial from Alpha MOS)	Juice	98
α-ASTREE, *i.e.* seven potentiometric sensors with different membranes (ZZ36, BB06, CA07, BA07, AB07, HA06, and CB07)	Potentiometric (commercial from Alpha MOS)	Tea	65
α-ASTREE, *i.e.* seven potentiometric sensors with different membranes (ZZ36, BB06, CA07, BA07, AB07, HA06, and CB07)	Potentiometric[a] (commercial from Alpha MOS)	Tea	42
α-ASTREE, *i.e.* seven potentiometric sensors with different membranes (ZZ36, BB06, CA07, BA07, AB07, HA06, and CB07)	Potentiometric (commercial from Alpha MOS)	Tea	66
α-ASTREE, *i.e.* seven potentiometric sensors with different membranes (ZZ36, BB06, CA07, BA07, AB07, HA06, and CB07)	Potentiometric (commercial from Alpha MOS)	Coffee	67
α-ASTREE, *i.e.* seven potentiometric sensors with different membranes (ZZ36, BB06, CA07, BA07, AB07, HA06, and CB07)	Potentiometric (commercial from Alpha MOS)	Orange beverages	68

Sensors	Technique	Sample	Ref.
α-ASTREE, *i.e.* seven potentiometric sensors with different membranes (ZZ36, BB06, CA07, BA07, AB07, HA06, and CB07)	Potentiometric (commercial from Alpha MOS)	Orange beverages/ Chinese vinegar	69
α-ASTREE, *i.e.* seven potentiometric sensors with different membranes (ZZ36, BB06, CA07, BA07, AB07, HA06, and CB07)	Potentiometric (commercial from Alpha MOS)	Soy sauce	70
α-ASTREE, *i.e.* seven potentiometric sensors with different membranes (ZZ36, BB06, CA07, BA07, AB07, HA06, and CB07)	Potentiometric (commercial from Alpha MOS)	Milk/yogurt	71
α-ASTREE, *i.e.* seven potentiometric sensors with different membranes (ZZ36, BB06, CA07, BA07, AB07, HA06, and CB07)	Potentiometric (commercial from Alpha MOS)	Apple juice	72
α-ASTREE, *i.e.* seven potentiometric sensors with different membranes (ZZ36, BB06, CA07, BA07, AB07, HA06, and CB07)	Potentiometric (commercial from Alpha MOS)	Tea	73
Six potentiometric sensors, *i.e.* Na^+, K^+, Ca^{2+}, Cl^-, NO_3^-, and pH; and two amperometric sensors, *i.e.* a composite planar electrode and a Au microelectrode	Potentiometric and amperometric[a]	Wine	74
Six potentiometric sensors, *i.e.* Na^+, K^+, Ca^{2+}, Cl^-, NO_3^-, and pH; and two amperometric sensors, *i.e.* a Au microelectrode unmodified and modified with a conductive composite based on oxide catalysts	Potentiometric and amperometric[a]	Wine	75
Eight different lipid/polymer potentiometric sensors and four voltammetric electrodes: gold, iridium, platinum, and rhodium	Potentiometric and voltammetric	Tea/detergent	76
Six voltammetric sensors (graphite-epoxy composites not modified plus five modified with copper nanoparticles, platinum nanoparticles, phthalocyanine, glucose oxidase (biosensor), or polypyrrole, and fifteen potentiometric sensors (Na^+, K^+, NH_4^+, Ca^{2+} (x3), Mg^{2+}, Ba^{2+}, H^+, generic cations (x 2), NO_3^-, Cl^-, SO_4^2, and generic anions)	Potentiometric and voltammetric	Beer	39

Table 7.3 (*Continued*)

Sensor electrodes or arrays (*i.e.* synthetic receptors)	Electronic tongue	Sample	Reference
Platinum and carbon paste electrodes unmodified and modified (two rare-earth bisphthalocyanine molecules (one with lutetium(III) and the other with gadolinium(III) as a center); cobalt(II) monophthalocyanine; and ferrocene)	Voltammetric	Wine	89
Platinum and carbon paste electrodes unmodified and modified (lutetium phthalocyanine, gadolinium bisphthalocyanine, and cobalt phthalocyanine)	Voltammetric	Wine	90
Seven polypyrrole-based sensors (doping agents = 1-decanesulfonic acid sodium salt, potassium hexacyanoferrate, sulphate anion, phosphotungstic acid, p-toluenesulfonic, acid, anthraquinone-2,6-disulfonic acid, disodium salt, and Prussian blue); four carbon paste electrodes modified using phthalocyanines (central lanthanide = lutetium(III), gadolinium(III), and their corresponding octa-tert-butyl substituted); and two carbon paste electrodes based on perylenes (N,N'-bis(piperidino)-3,4,9,10-perylenebis(dicarboximide) and N-octyl-3,4,9,10-perylene(dicarboximide))	Voltammetric	Wine	93
Six electrodes, *i.e.* a platinum electrode; an unmodified carbon paste electrode; a modified carbon paste electrode using lutetium bisphthalocyanine, or octatert-butyl, or a substituted analogue of lutetium, or cobalt metallophthalocyanine, or N,N-bis(methylpiperidine)-3,4,9,10-perylenebis(dicarboximide), or N-octyl-3,4,9,10-perylenebis(dicarboximide) or N-butyl-3,4,9,10-perylenebis(dicarboximide)	Voltammetric[a]	Wine	92
Six electrodes, *i.e.* a platinum electrode; an unmodified carbon paste electrode; and modified carbon paste electrodes with lutetium (III) and gadolinium(III) bisphthalocyaninates, cobalt (II) monophthalocyanine, and ferrocene	Voltammetric[a]	Wine	94
Six electrodes, *i.e.* gold, silver, platinum, palladium, tungsten, and titanium	Voltammetric	Rice wine	95
Three tyrosinase biosensors based on one unmodified carbon paste electrode (CPE), and two CPEs based on cobalt phthalocyanine and iron phthalocyanine	Voltammetric	Beer	91
Three graphite-epoxy electrodes incorporated with tyrosinase, laccase, or copper nanoparticles	Voltammetric	Wine	77
Five graphite-epoxy composites, one not modified, and the other four modified with copper nanoparticles, platinum nanoparticles, polyaniline or polypyrrole	Voltammetric	Wine	78

Carbon paste electrodes modified with three rare-earth bisphthalocyaninate compounds: lutetium(III), gadolinium(III), and praseodymium	Voltammetric	Wine	79
Seven chemically modified electrodes based on polypyrrole (x1), phthalocyanines (cobalt (II) monophthalocyanine, central lanthanide, *i.e.* lutetium(III), gadolinium(III), and gadolinium octa-tert-butyl), and perylene (*N,N'*-bis(piperidino)-3,4,9,10-perylenebis(dicarboximide) and *N*-octyl-3,4,9,10-perylene(dicarboximide))	Voltammetric	Wine	80
Five graphite-epoxy electrodes unmodified and modified with copper, platinum, polyaniline, or polypyrrole nanoparticles	Voltammetric	Cava wine	37
Six platinum electrodes modified with doped polypyrrole with different agents	Voltammetric	Beer	36
Four graphite-epoxy electrodes, one unmodified and the other three incorporated with tyrosinase, laccase, or copper nanoparticles	Voltammetric	Beer	81
Five electrodes, *i.e.* gold, iridium, palladium, platinum, and rhodium	Voltammetric[a]	Tea	82
Two electrodes, *i.e.* platinum and glassy carbons	Voltammetric	Tea	83
Three electrodes, *i.e.* iridium, platinum, and rhodium	Voltammetric	Tea	84
Five electrodes, *i.e.* gold, iridium, palladium, platinum, and rhodium	Voltammetric	Tea	85
Five electrodes, *i.e.* gold, iridium, palladium, platinum, and rhodium	Voltammetric	Tea	108
Six electrodes, *i.e.* gold, iridium, palladium, platinum, rhenium and rhodium	Voltammetric[a]	Milk	86
Four electrodes, *i.e.* gold, silver, platinum, and palladium	Voltammetric	Yogurt	87
Three electrodes, *i.e.* gold, platinum and rhodium	Voltammetric	Apple juice	88
Gold and copper electrodes	Voltammetric	Wine/whiskey	43
Gold and platinum electrodes	Voltammetric	Fruit juices/ still drinks/ milk	20
Five electrodes, *i.e.* gold, iridium, palladium, platinum, and rhodium	Voltammetric	Tea	99
Five electrodes, *i.e.* gold, iridium, palladium, platinum, and rhodium	Voltammetric	Tea	100
Three electrodes, *i.e.* glassy carbon, gold, and platinum	Voltammetric	Rice wine	101
Five electrodes, *i.e.* gold, silver, platinum, palladium, and titanium	Voltammetric	Milk	97
Four electrodes, *i.e.* gold, silver, platinum, and palladium	Voltammetric	Milk	102
Five electrodes, *i.e.* gold, platinum, iridium, rhodium, and palladium	Voltammetric	Milk	103
Three electrodes, *i.e.* gold, copper, and gold modified with Prussian blue	Voltammetric	Milk	105

Table 7.3   (*Continued*)

Sensor electrodes or arrays (*i.e.* synthetic receptors)	Electronic tongue	Sample	Reference
Eleven electrodes, *i.e.* three carbon paste electrodes modified with cobalt phthalocyanine, two modified with lanthanide bis-phthalocyanines (lutetium and its octatert-butyl derivative); six polypyrrole-based electrodes doped with different doping anions (sodium sulphate, sodium 1-decanesulfonate, potassium ferrocyanide, anthraquinone-2,6-disulfonic acid, disodium salt, phosphotungstic acid, and p-toluenesulfonic acid); two bare electrodes: carbon paste; and gold	Voltammetric	Wine	106
Two screen printed sensors (DRP-110 for the "sweetness" sensor and DRP-410 for the "bitterness" sensor)	Voltammetric (commercial from Dropsens)[a]	Soft drinks fortified with plant extracts of green tea	40
A gold electrode and a dual glassy carbon electrode	Amperometric[a]	Wine	41
Eight metal electrodes (platinum and gold) based on astringency	Amperometric	Tea	23
Six electrodes, *i.e.* glassy carbon, gold, platinum, silver, and an ITO-coated glass covered with polyaniline or polypyrrole	Impedance	Tea	45

[a]Hybrid electronic tongue.

References

1. J. Chandrashekar, M. A. Hoon, N. J. P. Ryba and C. S. Zuker, *Nature*, 2006, **444**, 288–294.
2. E. Adler, M. A. Hoon, K. L. Mueller, J. Chandrashekar, N. J. P. Ryba and C. S. Zuker, *Cell*, 2000, **100**, 693–702.
3. M. A. Hoon, E. Adler, J. Lindemeier, J. F. Battey, N. J. P. Ryba and C. S. Zuker, *Cell*, 1999, **96**, 541–551.
4. A. L. Huang, X. K. Chen, M. A. Hoon, J. Chandrashekar, W. Guo, D. Trankner, N. J. P. Ryba and C. S. Zuker, *Nature*, 2006, **442**, 934–938.
5. G. Nelson, J. Chandrashekar, M. A. Hoon, L. X. Feng, G. Zhao, N. J. P. Ryba and C. S. Zuker, *Nature*, 2002, **416**, 199–202.
6. G. Nelson, M. A. Hoon, J. Chandrashekar, Y. F. Zhang, N. J. P. Ryba and C. S. Zuker, *Cell*, 2001, **106**, 381–390.
7. R. A. Fisher, *Ann. Eugenics*, 1936, 7, 179–188.
8. A. J. Bard and L. R. Faulkner, *Electrochemical Methods: Fundamentals and Applications*, Wiley, New York, 2nd edn, 2001.
9. Y. Vlasov, A. Legin, A. Rudnitskaya, C. Di Natale and A. D'Amico, *Pure Appl. Chem.*, 2005, 77, 1965–1983.
10. Y. G. Vlasov, A. V. Legin, A. M. Rudnitskaya, C. DiNatale and A. Damico, *Russ. J. Appl. Chem.*, 1996, **69**, 848–853.
11. K. Hayashi, M. Yamanaka, K. Toko and K. Yamafuji, *Sens. Actuators B*, 1990, **2**, 205–213.
12. K. Toko, *Mater. Sci. Eng. C*, 1996, **4**, 69–82.
13. Y. Tahara and K. Toko, *IEEE Sens. J.*, 2013, **13**, 3001–3011.
14. A. Riul, R. R. Malmegrim, F. J. Fonseca and L. H. C. Mattoso, *Artif. Organs*, 2003, **27**, 469–472.
15. L. Lvova, R. Paolesse, C. Di Natale, E. Martinelli, E. Mazzone, A. Orsini and A. D. Amico, *Chemical Images of Liquids*, in *Imaging for Detection and Identification*, ed. J. Byrnes, Springer, Dordrecht, 2007, pp. 63–95.
16. L. Gil-Sanchez, J. Soto, R. Martinez-Manez, E. Garcia-Breijo, J. Ibanez and E. Llobet, *Sens. Actuators A*, 2011, **171**, 152–158.
17. M. del Valle, *Electroanalysis*, 2010, **22**, 1539–1555.
18. P. Ciosek and W. Wroblewski, *Sensors*, 2011, **11**, 4688–4701.
19. J. Zeravik, A. Hlavacek, K. Lacina and P. Skladal, *Electroanalysis*, 2009, **21**, 2509–2520.
20. F. Winquist, P. Wide and I. Lundstrom, *Anal. Chim. Acta*, 1997, **357**, 21–31.
21. M. Cortina-Puig, X. Munoz-Berbel, M. A. Alonso-Lomillo, F. J. Munoz-Pascual and M. del Valle, *Talanta*, 2007, **72**, 774–779.
22. M. Scampicchio, D. Ballabio, A. Arecchi, S. M. Cosio and S. Mannino, *Microchim. Acta*, 2008, **163**, 11–21.
23. M. Scampicchio, S. Benedetti, B. Brunetti and S. Mannino, *Electroanalysis*, 2006, **18**, 1643–1648.
24. J. Gallardo, S. Alegret and M. del Valle, *Talanta*, 2005, **66**, 1303–1309.
25. X. Ceto, M. Gutierrez-Capitan, D. Calvo and M. del Valle, *Food Chem.*, 2013, **141**, 2533–2540.

26. A. Rudnitskaya, D. Kirsanov, Y. Blinova, E. Legin, B. Seleznev, D. Clapham, R. S. Ives, K. A. Saunders and A. Legin, *Anal. Chim. Acta*, 2013, **770**, 45–52.

27. L. Nunez, X. Ceto, M. I. Pividori, M. V. B. Zanoni and M. del Valle, *Microchem. J.*, 2013, **110**, 273–279.

28. A. K. Deisingh, D. C. Stone and M. Thompson, *Int. J. Food Sci. Technol.*, 2004, **39**, 587–604.

29. F. Winquist, *Microchim. Acta*, 2008, **163**, 3–10.

30. A. Bratov, N. Abramova and A. Ipatov, *Anal. Chim. Acta*, 2010, **678**, 149–159.

31. L. Escuder-Gilabert and M. Peris, *Anal. Chim. Acta*, 2010, **665**, 15–25.

32. A. Riul, C. A. R. Dantas, C. M. Miyazaki and O. N. Oliveira, *Analyst*, 2010, **135**, 2481–2495.

33. E. A. Baldwin, J. H. Bai, A. Plotto and S. Dea, *Sensors*, 2011, **11**, 4744–4766.

34. P. Ciosek, Z. Brzozka and W. Wroblewski, *Sens. Actuators B*, 2004, **103**, 76–83.

35. P. Ciosek and W. Wroblewski, *Talanta*, 2007, **71**, 738–746.

36. A. A. Arrieta, M. L. Rodriguez-Mendez, J. A. de Saja, C. A. Blanco and D. Nimubona, *Food Chem.*, 2010, **123**, 642–646.

37. X. Ceto, J. M. Gutierrez, L. Moreno-Baron, S. Alegret and M. del Valle, *Electroanalysis*, 2011, **23**, 72–78.

38. G. A. Evtugyn, S. V. Belyakova, R. V. Shamagsumova, A. A. Saveliev, A. N. Ivanov, E. E. Stoikova, N. N. Dolgova, I. I. Stoikov, I. S. Antipin and H. C. Budnikov, *Talanta*, 2010, **82**, 613–619.

39. J. M. Gutierrez, Z. Haddi, A. Amari, B. Bouchikhi, A. Mimendia, X. Ceto and M. del Valle, *Sens. Actuators B*, 2013, **177**, 989–996.

40. A. Bulbarello, M. Cuenca, L. Schweikert, S. Mannino and M. Scampicchio, *Electroanalysis*, 2012, **24**, 1989–1994.

41. S. Buratti, S. Benedetti, M. Scampicchio and E. C. Pangerod, *Anal. Chim. Acta*, 2004, **525**, 133–139.

42. Q. S. Chen, J. W. Zhao and S. Vittayapadung, *Food Res. Int.*, 2008, **41**, 500–504.

43. W. Novakowski, M. Bertotti and T. R. L. C. Paixão, *Microchem. J.*, 2011, **99**, 145–151.

44. Y. M. Bae and S. I. Cho, *Trans. ASAE*, 2005, **48**, 251–256.

45. A. P. Bhondekar, M. Dhiman, A. Sharma, A. Bhakta, A. Ganguli, S. S. Bari, R. Vig, P. Kapur and M. L. Singla, *Sens. Actuators B*, 2010, **148**, 601–609.

46. C. Di Natale, R. Paolesse, A. Macagnano, A. Mantini, A. D'Amico, M. Ubigli, A. Legin, L. Lvova, A. Rudnitskaya and Y. Vlasov, *Sens. Actuators B*, 2000, **69**, 342–347.

47. A. Legin, A. Rudnitskaya, L. Lvova, Y. Vlasov, C. Di Natale and A. D'Amico, *Anal. Chim. Acta*, 2003, **484**, 33–44.

48. L. Pigani, G. Foca, K. Ionescu, V. Martina, A. Ulrici, F. Terzi, M. Vignali, C. Zanardi and R. Seeber, *Anal. Chim. Acta*, 2008, **614**, 213–222.

49. G. Verrelli, L. Lvova, R. Paolesse, C. Di Natale and A. D'Amico, *Sensors*, 2007, 7, 2750–2762.

50. E. Polshin, A. Rudnitskaya, D. Kirsanov, A. Legin, D. Saison, F. Delvaux, F. R. Delvaux, B. M. Nicolai and J. Lammertyn, *Talanta*, 2010, **81**, 88–94.

51. A. Rudnitskaya, E. Polshin, D. Kirsanov, J. Lammertyn, B. Nicolai, D. Saison, F. R. Delvaux, F. Delvaux and A. Legin, *Anal. Chim. Acta*, 2009, **646**, 111–118.

52. A. M. Peres, L. Dias, T. P. Barcelos, J. Sá Morais and A. Machado, *Procedia Chem.*, 2009, **1**, 1023–1026. DOI: 10.1016/j.proche.2009.07.255.

53. H. Yamada, Y. Mizota, K. Toko and T. Doi, *Mater. Sci. Eng. C*, 1997, **5**, 41–45.

54. P. Ciosek, Z. Brzozka and W. Wroblewski, *Sens. Actuators B*, 2006, **118**, 454–460.

55. P. Ciosek, E. Augustyniak and W. Wroblewski, *Analyst*, 2004, **129**, 639–644.

56. P. Ciosek, T. Sobanski, E. Augustyniak and W. Wroblewski, *Meas. Sci. Technol.*, 2006, **17**, 6–11.

57. P. Ciosek, Z. Kraszewska and W. Wroblenski, *Electroanalysis*, 2009, **21**, 2036–2043.

58. L. M. I. Codinachs, J. P. Kloock, M. J. Schoning, A. Baldi, A. Ipatov, A. Bratov and C. Jimenez-Jorquera, *Analyst*, 2008, **133**, 1440–1448.

59. G. A. Evtugyn, R. V. Shamagsumova, E. E. Stoikova, R. R. Sitdikov, I. I. Stoikov, H. C. Budnikov, A. N. Ivanov and I. S. Antipin, *Electroanalysis*, 2011, **23**, 1081–1088.

60. A. Legin, A. Rudnitskaya, B. Seleznev and Y. Vlasov, *Anal. Chim. Acta*, 2005, **534**, 129–135.

61. L. Lvova, S. S. Kim, A. Legin, Y. Vlasov, J. S. Yang, G. S. Cha and H. Nam, *Anal. Chim. Acta*, 2002, **468**, 303–314.

62. L. Lvova, A. Legin, Y. Vlasov, G. S. Cha and H. Nam, *Sens. Actuators B*, 2003, **95**, 391–399.

63. L. Lvova, R. Paolesse, C. Di Natale and A. D'Amico, *Sens. Actuators B*, 2006, **118**, 439–447.

64. A. L. A. Rudnitskaya, B. Seleznev and Y. Vlasov, *Int. J. Food Sci. Technol.*, 2002, **37**, 375–385.

65. W. He, X. S. Hu, L. Zhao, X. J. Liao, Y. Zhang, M. W. Zhang and J. H. Wu, *Food Res. Int.*, 2009, **42**, 1462–1467.

66. Z. Kovacs, I. Dalmadi, L. Lukacs, L. Sipos, K. Szantai-Kohegyi, Z. Kokai and A. Fekete, *J. Chemometrics*, 2010, **24**, 121–130.

67. D. Szollosi, Z. Kovacs, A. Gere, L. Sipos, Z. Kokai and A. Fekete, *IEEE Sens. J.*, 2012, **12**, 3119–3123.

68. M. Liu, J. Wang, D. Li and M. J. Wang, *J. Food Quality*, 2012, **35**, 429–441.

69. M. Liu, M. J. Wang, J. Wang and D. Li, *Sens. Actuators B*, 2013, **177**, 970–980.

70. O. Y. Qin, J. W. Zhao, Q. S. Chen, H. Lin and X. Y. Huang, *J. Food Sci.*, 2011, **76**, S523–S527.

71. M. Hruskar, N. Major, M. Krpan, I. P. Krbavcic, G. Saric, K. Markovic and N. Vahcic, *Mljekarstvo*, 2009, **59**, 193–200.
72. M. Simunek, A. R. Jambrak, M. Petrovic, H. Juretic, N. Major, Z. Herceg, M. Hruskar and T. Vukusic, *Food Technol. Biotechnol.*, 2013, **51**, 101–111.
73. H. Xiao and J. Wang, *Afr. J. Biotechnol.*, 2009, **8**, 6985–6992.
74. M. Gutierrez, A. Llobera, A. Ipatov, J. Vila-Planas, S. Minguez, S. Demming, S. Buttgenbach, F. Capdevila, C. Domingo and C. Jimenez-Jorquera, *Sensors*, 2011, **11**, 4840–4857.
75. M. Gutierrez, C. Domingo, J. Vila-Planas, A. Ipatov, F. Capdevila, S. Demming, S. Buttgenbach, A. Llobera and C. Jimenez-Jorquera, *Sens. Actuators B*, 2011, **156**, 695–702.
76. P. Ivarsson, Y. Kikkawa, F. Winquist, C. Krantz-Rulcker, N. E. Hojer, K. Hayashi, K. Toko and I. Lundstrom, *Anal. Chim. Acta*, 2001, **449**, 59–68.
77. X. Ceto, F. Cespedes and M. del Valle, *Talanta*, 2012, **99**, 544–551.
78. J. M. Gutierrez, L. Moreno-Baron, M. I. Pividori, S. Alegret and M. del Valle, *Microchim. Acta*, 2010, **169**, 261–268.
79. V. Parra, T. Hernando, M. L. Rodriguez-Mendez and J. A. de Saja, *Electrochim. Acta*, 2004, **49**, 5177–5185.
80. V. Parra, A. A. Arrieta, J. A. Fernandez-Escudero, H. Garcia, C. Apetrei, M. L. Rodriguez-Mendez and J. A. de Saja, *Sens. Actuators B*, 2006, **115**, 54–61.
81. X. Ceto, F. Cespedes and M. del Valle, *Electroanalysis*, 2013, **25**, 68–76.
82. R. Banerjee, B. Tudu, L. Shaw, A. Jana, N. Bhattacharyya and R. Bandyopadhyay, *J. Food Eng.*, 2012, **110**, 356–363.
83. R. Bhattacharyya, B. Tudu, S. C. Das, N. Bhattacharyya, R. Bandyopadhyay and P. Pramanik, *J. Food Eng.*, 2012, **109**, 120–126.
84. P. Ivarsson, S. Holmin, N. E. Hojer, C. Krantz-Rulcker and F. Winquist, *Sens. Actuators B*, 2001, **76**, 449–454.
85. M. Palit, N. Bhattacharyya, S. Sarkar, A. Dutta, P. K. Dutta, B. Tudu and R. Bandyopadhyay, *Virtual Instrumentation Based Voltammetric Electronic Tongue for Classification of Black Tea*, Industrial and Information Systems, 2008. ICIIS 2008. IEEE Region 10 and the Third International Conference on. DOI: 10.1109/ICIINFS.2008.4798485.
86. F. Winquist, S. Holmin, C. Krantz-Rulcker, P. Wide and I. Lundstrom, *Anal. Chim. Acta*, 2000, **406**, 147–157.
87. Z. B. Wei, J. Wang and W. F. Jin, *Sens. Actuators B*, 2013, **177**, 684–694.
88. F. Winquist, E. Rydberg, S. Holmin, C. Krantz-Rulcker and I. Lundstrom, *Anal. Chim. Acta*, 2002, **471**, 159–172.
89. C. Apetrei, I. M. Apetrei, I. Nevares, M. del Alamo, V. Parra, M. L. Rodriguez-Mendez and J. A. De Saja, *Electrochim. Acta*, 2007, **52**, 2588–2594.
90. M. Gay, C. Apetrei, I. Nevares, M. del Alamo, J. Zurro, N. Prieto, J. A. De Saja and M. L. Rodriguez-Mendez, *Electrochim. Acta*, 2010, **55**, 6782–6788.

91. M. Ghasemi-Varnamkhasti, M. L. Rodriguez-Mendez, S. S. Mohtasebi, C. Apetrei, J. Lozano, H. Ahmadi, S. H. Razavi and J. A. de Saja, *Food Control*, 2012, **25**, 216–224.

92. N. Prieto, M. Gay, S. Vidal, O. Aagaard, J. A. de Saja and M. L. Rodriguez-Mendez, *Food Chem.*, 2011, **129**, 589–594.

93. V. Parra, A. A. Arrieta, J. A. Fernandez-Escudero, M. Iniguez, J. A. de Saja and M. L. Rodriguez-Mendez, *Anal. Chim. Acta*, 2006, **563**, 229–237.

94. M. L. Rodriguez-Mendez, C. Apetrei, I. Apetrei, S. Villanueva, I. J. A. de Saja, I. Nevares and M. del Alamo, *Combination of an Electronic Nose, an Electronic Tongue and an Electronic Eye for the Analysis of Red Wines Aged with Alternative Methods*, Industrial Electronics, 2007. ISIE 2007. IEEE International Symposium on. DOI: 10.1109/ISIE.2007.4375050.

95. Z. B. Wei, J. Wang and L. S. Ye, *Biosens. Bioelectron.*, 2011, **26**, 4767–4773.

96. T. Mottram, A. Rudnitskaya, A. Legin, J. L. Fitzpatrick and P. D. Eckersall, *Biosens. Bioelectron.*, 2007, **22**, 2689–2693.

97. Z. B. Wei and J. Wang, *Anal. Chim. Acta*, 2011, **694**, 46–56.

98. R. N. Bleibaum, H. Stone, T. Tan, S. Labreche, E. Saint-Martin and S. Isz, *Food Qual. Preference*, 2002, **13**, 409–422.

99. A. Ghosh, P. Tamuly, N. Bhattacharyya, B. Tudu, N. Gogoi and R. Bandyopadhyay, *J. Food Eng.*, 2012, **110**, 71–79.

100. A. Ghosh, B. Tudu, P. Tamuly, N. Bhattacharyya and R. Bandyopadhyay, *Chemom. Intell. Lab. Syst.*, 2012, **116**, 57–66.

101. Q. Ouyang, J. W. Zhao and Q. S. Chen, *Food Res. Int.*, 2013, **51**, 633–640.

102. Z. B. Wei, J. Wang and X. Zhang, *Electrochim. Acta*, 2013, **88**, 231–239.

103. F. Winquist, C. Krantz-Rulcker, P. Wide and I. Lundstrom, *Meas. Sci. Technol.*, 1998, **9**, 1937–1946.

104. L. A. Dias, A. M. Peres, A. C. A. Veloso, F. S. Reis, M. Vilas-Boas and A. Machado, *Sens. Actuators B*, 2009, **136**, 209–217.

105. T. R. L. C. Paixão and M. Bertotti, *Sens. Actuators B*, 2009, **137**, 266–273.

106. V. Parra, A. A. Arrieta, J. A. Fernandez-Escudero, M. L. Rodriguez-Mendez and J. A. De Saja, *Sens. Actuators B*, 2006, **118**, 448–453.

107. A. Legin, A. Rudnitskaya, Y. Vlasov, C. Di Natale, F. Davide and A. D'Amico, *Sensor. Actuat. B-Chem.*, 1997, **44**, 291–296.

108. M. Palit, B. Tudu, N. Bhattacharyya, A. Dutta, P. K. Dutta, A. Jana, R. Bandyopadhyay and A. Chatterjee, *Anal. Chim. Acta*, 2010, **675**, 8–15.

Subject Index

References to tables and charts are in **bold** type